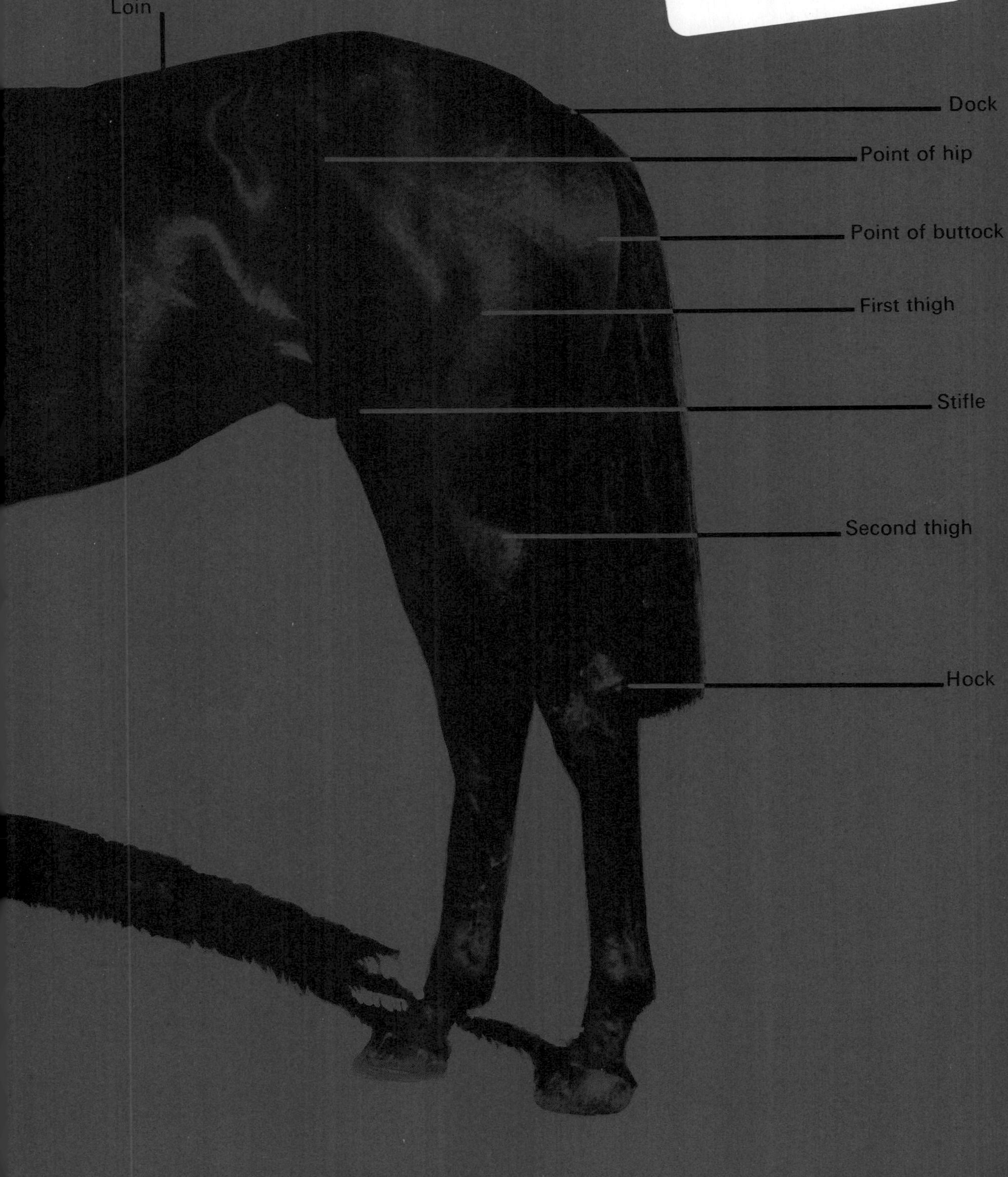
Loin
Dock
Point of hip
Point of buttock
First thigh
Stifle
Second thigh
Hock

The complete book of the horse

The complete book of the HORSE

Edited by Elwyn Hartley Edwards and Candida Geddes

Foreword by Richard Meade

ARCO PUBLISHING COMPANY, INC.
New York

Design and art direction by
Denis Wrigley MSIA
All line illustrations in text by
Christine Bousfield
Part title illustrations by
Denis and Paul Wrigley

Published by Arco Publishing Company, Inc.
219 Park Avenue South, N.Y. 10003

Library of Congress Catalog Card Number
73-80764
ISBN 0-668-03335-5

Printed in Great Britain

Contents

Part four Care and management

Part five Breeding and stud management

Part six The history of riding

Part seven Methods and techniques of training

Part eight Equestrian sports

Foreword

For centuries people in different parts of the world have ridden horses and evolved their own particular styles of riding. Seldom did one style influence another. Even in recent times, the Red Indian was riding bareback, Fillis was cantering backwards, Caprilli was teaching the forward seat while the Englishman continued to stick his feet forward and lean so far back over the fences that he was almost on the horse behind. In Britain, we hunted, we raced, we played polo. Show-jumping was artificial, three-day-eventing was too complicated, and as for dressage, that was something dreamed up by the foreigners and, worse, had a foreign name!

In the last forty years, however, there has been a profound change in man's attitude to riding. As the horse's usefulness as a beast of burden has waned the sport of riding has gained popularity. Accompanying this growth the horseman has developed such an insatiable thirst for knowledge that this book, which covers the whole spectrum of the horse world, will undoubtedly be enjoyed the world over.

As a child, I was fortunate enough to be schooled in the hunting field and the Pony Club. With that background and a taste for riding across country I very soon decided that I wanted to take part in three-day events. It was in the early sixties that this desire became reality. It was difficult at that time to learn about the sport. There were no books on the subject and very few professional trainers who had any first-hand experience. There was a great divergence of opinion on what constituted a good event horse and how he should be trained. There were those who believed that a courageous rider on a good Thoroughbred horse could win the competition on the second day's performance alone, dressage being an unnecessary evil which need not be bothered about.

There were others who maintained that since the jump was merely an elongation of the canter stride, a horse trained to a high degree of dressage, and with the obedience that goes with it, would have no problem with the second and third days–dressage was all that mattered.

Now, of course, we know better. Competition is such that the rider must aim to raise himself and his horse to the highest possible standard in every phase of the event.

To my mind there are five distinct phases in the three-day event: dressage, roads and tracks, steeplechasing, cross-country and show-jumping. Each phase involves a different balance of horse and rider, different paces and different priorities. Each must be trained for, and the horse must be equally at home in all phases.

In the early stages I set out to discover how an expert in each phase trained himself and his horse. For dressage, that meant gleaning as much as possible from a pure dressage expert, for steeplechasing from a trainer, for cross-country from hunting, for show-jumping from a top show-jumper. I was surprised when I talked to these experts how few of them seemed to be interested in other branches of horsemanship.

In the last ten years I believe that a great change has taken place in this respect. In Britain, steeplechase jockeys are beginning to ride in events and show-jumping, event riders are doing much more show-jumping and racing. Many more people play polo, and the Pony Club is spreading its wings wider and wider. May this trend long continue. The broader the knowledge and experience of a horseman the better for the horse world. It is, therefore, gratifying that between the covers of this book there is such a wealth of information. Inevitably a book of this type cannot be completely comprehensive on such a wide range of subject matter, but there is plenty of value for the cognoscenti to read and nothing too technical for the layman.

I wish this book well because I know that it will give great pleasure to its readers and above all will do much to broaden the knowledge of all of us who have the interest and welfare of the horse at heart.

RICHARD MEADE

Introduction

'After God', said the Spanish *conquistadores*, 'we owed the victory to the horses.' With equal truth it can be asserted that much of our human progress has been dependent upon the use man has made of the horse.

In the days of prehistory we know that man looked upon the horse as a source of food and that in very early times herds of wild horses were employed as instruments of war and driven against an enemy. From that point the horse plays an ever-increasing part in the human saga, right up to the time when the internal combustion engine replaced him on the roads and in the fields, and when the ever-more deadly weapons of war mercifully made his presence on the battlefield no longer a practical possibility.

Having, seemingly, fulfilled his purpose it might have been thought that to all intents the anachronism that was the horse would bow himself off the moving stage of man's progress. But such is the strength of the relationship that has grown up over the centuries between horse and man that we have witnessed, in a bare quarter of a century, an astounding revival of interest in the horse as a means of recreation and sport. This extension, almost an explosion, of interest has been one that has taken no account of wealth or social position, and riding has become one of the greatest of growth sports, with participants being counted not in hundreds or thousands but in millions.

In effect the role of the horse has been reversed. Today he is no longer the servant of man; instead it is men and, more particularly, women, who in ever-increasing numbers willingly become servants to the horse. By doing so they are proving, if proof were needed, that the horse is as well able today to captivate the minds of men as he did in the far-off days when he was worshipped as a deity.

In part it is because of the large numbers of horse-orientated people that this book has been compiled. By bringing together a great many authorities to contribute articles on particular aspects of the horse we have sought to produce a full, rounded work which would have been beyond the capacity of any single person.

We do not claim that our book contains everything there is to know about horses–that would be impossible. What we have tried to produce is a book that covers in quite considerable detail the whole vast spectrum of the world of horses, and to produce it in a way that makes it both highly instructive and also of general interest and entertainment. When you have read the book you may not know everything there is to know about horses, but we are prepared to bet that you will have learned something new. If you have added to your knowledge of, and your respect for, the horse to the extent that we both have, this book will have fulfilled our hopes.

CANDIDA GEDDES
London 1973
ELWYN HARTLEY EDWARDS
Dedham 1973

PART ONE

The history of the horse

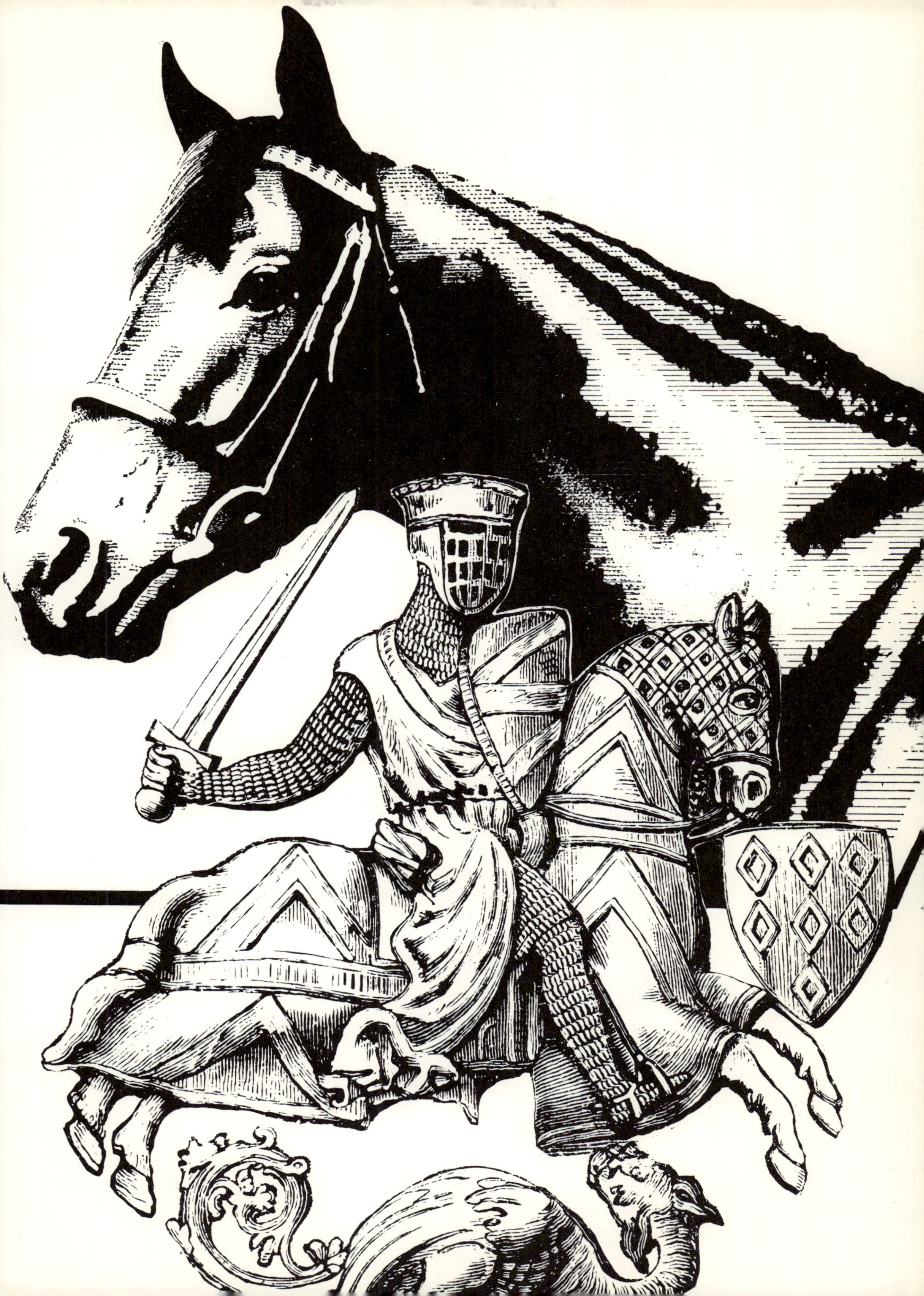

Chapter one

The evolution of the horse

The modern horse, known as *equus caballus*, has descended in gradual stages from a creature which existed some sixty million years ago called the Eohippus. Within the last hundred years it has been proved that the development of *equus caballus* took place in what is now North America. The land masses of the world were differently disposed at that time, and *equus caballus* was able to move freely, travelling over land from North America across Asia and into Africa and Europe, before the land bridges were submerged beneath the rising sea level caused by melting ice in the post-Ice Age period.

There is still no unanimity of opinion even among experts concerning the origins of the domestic horse. Debate about it has so far lasted for about 150 years. Even before Darwin's new theories changed so radically the traditional concept of evolution, the immense diversity of size, form and colour of horses had led enquiring minds to explore the possibility of various primitive races of horse being extant before the domestication of *equus caballus*. However, most of these 'primeval races' were supposed to differ from each other chiefly in colour, and the theorists have therefore been able to draw little support from archaeological remains.

Today, prehistorians are still broadly divided into two camps: the monophyletic and polyphyletic. The first of these maintains that all modern breeds are descended from one wild type of horse which was at some stage taken into domestication. This type closely resembled Przewalski's horse, if it was not actually identical to it. The second camp assumes the existence of a number of wild horse varieties postulated as having existed in times which are recent by geological standards but early in terms of human culture – i.e. coming after the Ice Age, during the Old Stone Age of European man. The number of these species varies between six and two.

Perhaps I should declare my own belief here: I will deal with the subject on the assumption that the modern horse has derived from a number of origins. A recent, and widely accepted, scheme in accordance with this is that synthesised from the work of Speed, Skorkowski, Ebhardt and d'Andrade in Scotland, Poland, Germany and Portugal respectively. It suggests that four primeval types of horse existed, all of which were domesticated, though at different times and in different parts of the world. For the sake of convenience two of these are known as ponies and two as horses, though one of the 'horses' is believed to have been quite as small originally as the smaller of the pony types.

The first of these four types, which let us call Pony I, was to be found in north-west Europe and was more or less identical to the 'Celtic Pony'; it would have resembled closely the modern English native pony, the Exmoor. As near as we can determine its height would have averaged 12·2 hands; a 'waterproof' animal, brown or bay in colour.

Pony II, native to northern Eurasia, was heavier in build and in bone than the first type, with a coarser head, and was better at trotting than at galloping. It would frequently (if not invariably) have had an eel-stripe running down its back, and was a dun colour which might be so pale as to be almost a cream, like some Przewalski foals, or might almost verge on chestnut. This pony was 'frost-proof', and has a typical modern descendant in Przewalski's horse with its stiff mane hairs, though the abundant and wavy tail it carried is not a feature of the Przewalski.

The third strain, Horse III, inhabited the central Asian areas, extending westward in pockets north of the Alps as far as Spain. One can still see examples of it in the kind of nightmare that is sometimes thrown up in Thoroughbred families: a long, narrowish head with a straight or 'Roman' profile, long straight neck, sloping croup, long ears with a tendency to lop, slab-sided, rather shallow hooves which are broader than those of other races, and a sparse, lank mane and tail. The nearest living equivalent of this type is the marsh-dwelling, clay-coloured horse discovered by d'Andrade as an unfashionable domestic breed in the Douro valley where the river marks the frontier with Spain. Evidently this was the ancestor, or *an* ancestor, of the Andalusian. It was much the largest of these early types, averaging some 15 hands – and is the 'drought-proof' horse.

Horse IV, the last of these, was native to western Asia. It was small – only about 12 hands – and fine-boned, with a straight or concave facial profile, silky, abundant mane and tail – in other words, with most of the attributes now regarded as indications of 'quality'. It had a flat-topped croup at the same height as its withers, with a high-set tail. The nearest modern representative of this 'heat-proof' horse is the recently-identified Caspian pony.

The potential increase in size of each of these early types is known to have varied. The Shetland pony, for example, is a dwarf variant of type I which, if of unmixed blood and living in the relatively mild, wet environment of western England or Ireland would not grow any larger. Type II was capable of the greatest variation in size: under favourable feeding conditions of open forest or natural water-meadows it would achieve massive proportions. Types III and IV would both grow bigger in a favourable environment, as would a cross-bred from any two of the four types. Such crosses are thought to have existed in a wild state and to have become much more numerous in even the early stages of domestication. Thus the heavy draught breeds of today are crosses of types II and III; light horses are crosses of types III and IV; while most of the ponies native to north-west Europe have the blood of types I and II in varying proportions.

STAGES IN THE EVOLUTION OF THE HORSE

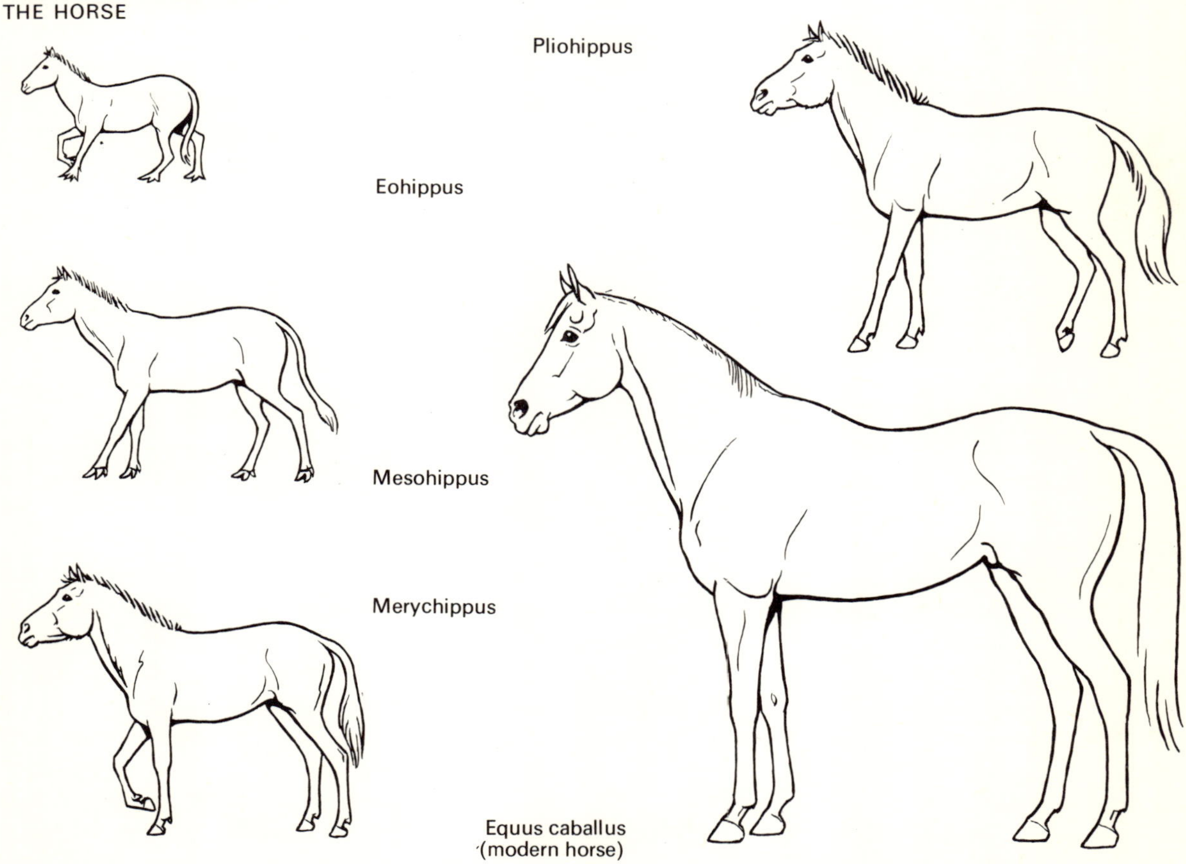

The Tarpan, if it was indeed a wild race and not the feral descendant of once-tame horses, was a cross of types I and IV. Przewalski's horse probably carries no blood other than that of type II, but developed certain characteristics which have modified it in response to the unusually harsh environment of saline or alkaline desert with extremes of heat and cold, which its increasingly restricted habitat in western Mongolia offers.

There is one other possible descendant of these primeval types, probably of unmixed descent but also modified, though in circumstances different from those of the Przewalski. From time to time there are reports of yet another deep-frozen mammoth being dug out of the 'permafrost' in Siberia, and 'white horses' are sometimes mentioned in this connection, their well-preserved remains being salvaged from this frozen swamp. There is as yet no detailed anatomical description of these prehistoric horses in any western language, but wild white horses have been known to the Yakut tribesmen of the Lena valley from time immemorial, and they are still to be found in this area. Before the Cossacks conquered Siberia, the Yakuts used to catch and break these horses, working them during the summer and using reindeer in winter. Nowadays it is easier for the Yakuts to buy horses from the Russians, and their only contact with the wild white horses is to shoot one now and again for meat. When it is possible to compare anatomical reports of these horses with one detailing the mammoth horses from the same region, we shall be able to draw more positive conclusions, but this Siberian wild horse will probably prove to be essentially of type II, with special characteristics enabling it to survive in an environment of adequate grazing and water in a short, very hot summer, an autumn in which there is still plenty of grass, though it is by now bone dry – natural hay, in fact – a long, hard, intensely cold winter during which there is a change in the horse's coat from dun to near-white, and the only means of finding food is to dig away the snow to find the 'hay' underneath, and a much more trying spring during which life is maintained chiefly by browsing on such shrubs as birch and willow and digging up roots with the fore-hooves. Only horses with special adaptations of the teeth, which would in turn affect the shape of the jaw and of the head generally, could survive conditions such as these. It should therefore be possible to substantiate or disprove this theory by studying a well-preserved skull specimen.

A Przewalski mare and foal, descendants of one of the primeval types of horse.

Chapter two

The emergence of breeds

In spite of the discussion about breeds in the last chapter, strictly speaking a breed is the result of conscious effort by a stockbreeder to produce in a domestic animal what he wants for a particular purpose. Let us take the simplest example from the oldest of domestic animals, the dog. Using the same basic raw material – the wolf, which had become dependent on early man as a parasite – early hunters and husbandmen produced, by selection and by mating like with like, three variants for use as watchdogs, hunting hounds and cattle-herding dogs respectively.

Thus the description of the 'primitive breeds' is really a misnomer, but also a convenient shorthand for 'local sub-races' or 'geographical breeding-groups' formed within the general body of wild horse stock. The characteristics which these developed were the response of the wild horse to the challenge of various environments, physical modifications which best enabled it to live in differing climates, on different vegetation, at different altitudes, on stony or marshy ground, in forest, steppe or tundra, and so on.

This is the raw material which the first horsemen inherited, a product of the celebrated test of the 'survival of the fittest'. This phrase has become so well worn since Darwin's day that we have ceased to consider quite what it means. Fittest for what? *For survival alone*, of either the individual, the race or the strain. Thus will stallions survive under wild conditions, and beget foals in their image who are the most aggressive and the dirtiest, deadliest fighters; the only foals which grow up to marehood and stallionhood belong to the band which is led by the bravest and most cunning old horse, willing and able to protect his family from predators such as wolves and men, to lead them always where the grass is greenest and the water most abundant from one season to another. But these qualities do not necessarily match those desired by the human selector of stallions, especially as regards their size, tractability, speed, or even what human standards define as elegant conformation. There was a sad example of the results of this difference recently: a New Forest breed stallion called Winston Churchill, a horse of dazzling beauty foaled in the 1950s, won every prize and premium available and was subsequently turned out on the Forest to run with the wild ponies and 'improve the breed'. He had to be taken off the Forest again without begetting a single foal, because he was such a poor fighter that the most scruffy two-year-old was able to send him packing without the slightest difficulty.

The earliest generations of horse-breeders, in their first home on the Asiatic steppe, had no choice in the matter of sires; they had to take what foals the local variety of wild horse would get off their mares. At this stage it was quite beyond the capabilities of early pastoralists to cope with adult stallions, and their domesticated mares were covered by their being tethered, when they came into season, where it was hoped the wild stallion would come out of the steppe and cover them. The pastoralists' economy was based on mares' milk and the meat of the year's colts; only the filly foals were allowed to grow on. The breeding would have attempted to produce the milkiest mare and the most meaty colt; two qualities inherent, luckily for the owners, in pony type II which was the type of horse most common in north-central Asia.

When it came to breeding horses for work in a later age, the desired type was a packhorse. The use of animals for pack arises sooner or later among all nomadic pastoralists who have been living quite well on meat and milk but are getting tired of carrying their own bedding, tents and cooking pots. The analogy of primitive camel-breeders, donkey-herds, even of ancient cattlemen and yak-breeders, bears this out.

Only after this stage does any significant divergence in the desired type of horse arise. Now the option is between a saddle horse and a occurred in Persia in remote antiquity, and started a chain reaction in striving for the Great Horse which lasted many centuries, culminating in the international, almost uniform west European charger of the late Middle Ages, from which many heavy breeds still seen today derive. But for many centuries Persia had a

A rare complete German Gothic armour, made at Landshut in the fifteenth century. It is an excellent example of the workmanship of the period and is now a prized item in the Wallace Collection in London.

head start because horse type III, the tallest of the primitive breeds, was native to that region. The discovery that crossing type III with type II, very common in Europe north of the Alps, would produce a hybrid bigger than either parent, was a fortunate one. Most of the mediaeval records of horse-breeding that survive in Europe west of a line drawn roughly between Venice and Stettin are concerned with the product of this cross, because it was ridden in war and on ceremonial occasions by kings and nobles who, even if they could not actually write themselves, employed people who could.

harness horse. In the ancient world, one might almost say in the pre-Biblical world, those who bred for riding lived on the Mongolian or north-eastern side of the primeval horse-rearing region, while the breeders of harness horses inhabited the south-western or Iranian side. The distribution of primitive types, either in their wild state or recently domesticated, did not fit in with the requirements of the two groups and their preferences led to the beginnings of horse trading on an international scale, and also to wars of conquest with almost the sole object of acquiring by force working horses and breeding stock.

The invention of body-armour suitable for horsemen and the adoption of the lance both

With perhaps the single exception of the Andalusian, few breeds of horse extant today can be traced back with convincing continuity to the era of classical antiquity, but many can be shown to have their roots in well-known mediaeval breeds. This is because of the great technical gulf that separates, say, the world of Julius Caesar from that of William the Conqueror. The ancient world was a stirrupless world, a fact which greatly influenced the choice of types of horse for riding. By Roman standards the mediaeval world was a roadless world, and this in turn had a profound effect on mediaeval

MEDIAEVAL HORSES

ideas of what could be expected of a harness horse. When European road surfaces later achieved the standard, on some stretches of expensive turnpike, that had been commonplace throughout Roman Europe, the whole nature of harness and methods of harnessing had undergone such a radical change, making different demands on equine anatomy, that comparisons between the harness horse of antiquity and that of eighteenth-century England, for example, are simply not valid. It is perhaps enough just to point out that the Graeco-Roman horse could not 'throw its weight into the collar' because there was no collar!

Modern harness horses are derived from what were originally riding types. Hackney horses, and most carriage types most effective at a trot, have a foundation in breeds such as the Norfolk Roadster, which had a reputation in the pre-macadam era as a fast *ridden* trotting horse. Throughout Britain and the whole of the continent of Europe a profound modification overtook light horse breeds when parish roads became good enough for the farmer to go to church and market in a gig. As long as he had to travel in the saddle, with his wife riding pillion behind him, he both demanded and bred for himself an 'easy going' sort of horse that would amble or pace. Once the gig was a practical means of transport, however, whether in Norfolk or in Normandy, he went in for a 'hard trotting' horse. Only in peripheral areas such as Iceland, Portugal and Spain is the utility light horse of local pony blood still a pacer, ambler or 'racker'.

During the eighteenth century the packhorse, which had carried most of the goods traffic of Europe since the days of antiquity, also became obsolete except in certain obstinate mountain regions such as the Pyrenees, and was converted into a harness horse of about 'vanner' size. This is true of, for instance, the Fjord pony, the now extinct Devon packhorse, of Fell and Dales ponies and the Arriegeois of south-eastern France, among many others.

A group of young Thoroughbreds at a stud farm near Ocala in Florida.

The picture in eastern Europe, and over a large part of Scandinavia, is a rather different one. There the harness horse had its chance to excel, not on road surfaces (which were no better than elsewhere) but on frozen snow over several months of the year. It was as much before the sleigh as before the trap that the light horse breeds of Hungary, Poland, Sweden, etc., put up their most impressive records. Even the celebrated Dutch Harddraver performed at its fastest not on the roads but on frozen canals.

The establishment of a breed up to the point where it becomes 'fixed' and always breeds true to type takes longer than the lifetime of one man, which is after all only three times that of the horse. There are, therefore, very few individuals who have themselves accomplished this. Few families, too, maintain their social and economic position long enough to achieve it. Royal families sometimes succeed, and the country-wide network of studs established by Edward I of England for the breeding of heavyweight war horses continued to function, with a system of exchanges of stallions between all parts of the kingdom, until the time of Henry VIII. In the rest of western Europe the Church was a more durable establishment, and ecclesiastical studs, both secular and monastic, have brought into being and preserved some well-known breeds. The Einsiedler breed in Switzerland (the name means 'hermitage') had a monastic origin; and when the Andalusian horse was, at a late date in its long history, threatened with decadence and bastardisation in its native province the breed was saved and almost given a new life by a community of Carthusian monks.

Continuity of human effort and maintenance of the same objective over a long period are both essential if consistent characteristics are to be stamped on a breed. In the case of the Thoroughbred, the objective has remained the same for between two and three hundred years: to breed a horse which will run faster than any other horse over a range of distances that have not materially varied in all that time. The effort has not been put out by any single organisation but by the racing community as a whole all over the world, though it should be taken into account that for a long time the community consisted of a narrowly defined social class in one country, most of whom were known to each other.

It is worth noting in this context that the

old-style 'pedigree' handed over at the sale of an Arab horse to an outsider (any non-desert dweller) made sense. It simply stated that such-and-such a horse had been foaled among the tents of the Beni-so-and-so from one of their mares by one of their stallions. It went without saying that since the Bedouin clan concerned had been following the same way of life in the same general area, and keeping horses for the same purpose, since the time of Mahomet and earlier, any horse genuinely bred by them must conform to the standards of the traditional type.

Most European countries – Britain is an exception – breed horses as a state enterprise. Generally the state owns stallions and individuals own mares. Typically, in France a system perfected under the first Napoleon obtains whereby in each *département* (in practice, nowadays, each group of *départements*) there is a stallion depot, from which stallions are sent out to covering stations in the country. Their progeny by privately-owned mares is later inspected and the male progeny approved – or not – for breeding purposes by the responsible authority. Thus regional breeds like the Breton draught horse have conformed to the standards desired by the authority of the region concerned. There is no competitive element in breeding, and yet competition is the one factor that has moulded the type of the only universally recognised breed of horse, the Thoroughbred.

A scene from the past. Horses are now seldom seen engaged in tasks for which mechanised agricultural equipment has been developed.

PART TWO

The world's principal breeds and types

The first progenitor : the Arab

It is virtually impossible accurately to date the emergence of the Arab horse as a breed. All that can be said with certainty is that the Arab is the oldest of the world's recognised breeds and that he has had the greatest influence on the world's horse population.

The creation of the Thoroughbred is undoubtedly the greatest achievement of the Arabian horse, but it must also be recognised that all of the established warm-blood breeds of the world either originally contained Arab blood or have, at some time in their history, been improved by the introduction of that blood. Additionally, not a few of the heavy, cold-blood breeds of north-west Europe were at some time influenced by the Arab.

In Britain, the repository of the world's finest ponies, most of the nine indigenous breeds contain a percentage of Arab, particularly that most beautiful of all ponies, the Welsh Mountain, who is in so many respects a miniature Arab: the head, the dish of the face and the large, wide-set eyes are particularly Arabian in character.

The Arabian influence is generally acknowledged all over the world. In France there is the prime example of the carefully-bred French Anglo-Arab, and one of the great horses of France, the Percheron (cold-blood), was in early times the product of a cross between Oriental and Norman horses, and other heavy French horses have also been subject to outcrosses of Arab blood, the grey Boulonnais breed being a case in point. The Normandy breed of France, an excellent saddle horse, traces ancestry to the English Norfolk Trotter, Young Rattler, a descendant of the Godolphin, and it is also from Norfolk Trotter blood that the Demi-sang Trotter stems.

In Sweden, the Swedish warm-blood has been developed from East Prussian, Hanoverian and Thoroughbred horses, all originally based on Oriental ancestry. The breeds of the USSR similarly owe their existence to this source. The ancient Turkmene derives from the strain of Munighi Arabian; the Orlov Trotter, for example, is an amalgam of Arab, Thoroughbred, Dutch, Mecklenburg and Danish blood, the first two predominating.

In the American continents the breeds that have been developed also have the Arab at their base. They range from the Standardbred Trotters and Pacers, descended from the English Thoroughbred, Messenger, to the spectacular five-gaited American Saddle Horse, who has no European equivalent of note but certainly carries Thoroughbred and Trotter blood.

The unique position held by the Arab can be attributed to three principal factors: first, to its antiquity as a breed of fixed type and character; second, to the development and improvement effected by the desert Bedouin over a period of centuries, and especially to the Islamic religion, which virtually incorporated the keeping of horses as an article of faith; and finally, as a result of these two factors, the supremacy of the Arab is due to the inherent qualities of stamina, hereditary soundness, conformation, courage and speed with which the Arab characteristically stamps his stock. Furthermore, Arab blood has the capacity to mix with virtually any other and effect an improvement in the resulting progeny.

It is not surprising that the Arab, with such a record and being, as he is, an animal of undeniable beauty, should have induced an almost fanatical devotion in his admirers. Add to all his exceptional qualities the romance, real or imagined, of the East, and the Arab horse becomes irresistible in his appeal. There have been many authorities on, and historians of, the Arabian breed – King Solomon is among the former – and there are likely to be many more, but pride of place must surely be given to the late Lady Wentworth, author of *The Authentic Arabian* and other works of comparable note.

It is to Lady Wentworth that we are indebted for much of the existing background knowledge of the Arabian, but even that formidable scholar, faced with the confusion of myth, legend and reality that forms the enigma of Arab origin, found difficulty in remaining entirely factual. Lady Wentworth's bland assertion that the Arab horse was an established breed in 5000 BC, a sort of separate creation, 'having its origin solely in the Arabian peninsula where a separate race of wild horse having no connection with any other strain once lived and was domesticated', is not entirely acceptable. However, Lady Wentworth was not the only one to fall under the Arab spell, and while it is difficult to establish dates there is no doubt that the breed is, indeed, very old, existing as a warm-blood in Arabia when the greater part of Europe was marsh and swamp.

The Arab historian El Kelbi (AD 786) traced –most unusually, since the later desert Arabs kept no written records–pedigrees of Arabian horses from the time that Bax, great-great-great-grandson of Noah, captured and tamed the wild horses of the Yemen. His pedigrees show a line of descent from a mare called Baz and the stallion Hoshaba. Evidence of the breed's existence in ancient Egypt is supported, quite feasibly, in depictions of recognisable Arabian types on monuments dated around 1300 BC, and possibly the first representation of a ridden horse is a statuette of a very Arab-like mount of about 2000 BC, also found in Egypt. The same unmistakable types occur in the rock carvings of Nejd and Syria.

Nearer to our own time, evidence of the development of the Arab horse within what may be termed Arabia is provided by the French General Daumas (1803–1871). This officer spent many years in the desert with the object of securing Arab horses for military purposes, an object in which he was more successful than most Europeans. The General struck up a friendship with the Emir Abd-el-Kader (1808–1883) of Algeria, with whom he carried on a voluminous correspondence. Abd-el-Kader was an accepted authority on the desert-bred Arabian. He divided the history of the horse into four separate eras: Adam to Ishmael, Ishmael to Solomon, Solomon to Mohammed and Mohammed onwards. The first period must be largely discounted since the Arabs as a people did not exist at that time, but with Ishmael the personification of the Arab people, the close association of the man of the desert and the horse had its real beginnings. It later became the task of Solomon, undoubtedly the greatest horse dealer of all time, to maintain the Arab breed. This he did to some effect, having 1200 saddle horses and 40,000 chariot horses in his stables to complete disregard of the Israelite law forbidding the keeping of horses as idolatry.

Lady Anne Lytton's famous Polish-bred stallion, Grojec. He has the classical Arabian head, and displays all the essential characteristics of the top-quality Arab.

The next era, that of Mohammed, is the most significant in the development of the Arab horse. As well as being an outstanding religious leader and a skilful soldier, Mohammed was also an astute politician, and it was the combination of these qualities that ensured the survival and extension of militant Islam for centuries after his death; it also resulted in a significant improvement in the status of the Arab horse. To the religious-minded Arab people, the promise of attaining paradise after death by obeying Mohammed's injunction to pay particular attention to the care of horses, while at the same time being able to enjoy on earth the benefits of doing so, was an irresistible inducement to contribute towards the creation of a superlative breed.

It is no wonder that the combination of assured spiritual and temporal reward should have resulted in the hordes of superbly competent Moslem cavalry bursting out of their desert homelands to conquer first Egypt and north Africa, and then penetrating in force through Spain into France. This happened after the death of Mohammed in AD 632, but the achievement was his and also, of course, that of the Arab horse which he had recognised as an essential element in the expansion of the faith. It is really from this point that the breed, by being brought into contact with those horses flourishing in the conquered territories, began to have a significant effect upon the world's equine development.

The supreme qualities of the Arab horse, in particular his stamina and soundness and his ability to carry weight out of proportion to his size for extended periods and at speed, are almost entirely due to the selective breeding practised by the desert Bedouins and to the methods of training and conditioning they used.

In the breeding of the horses five superlative strains were recognised, and these have persisted up to the present day. From these five stem all the hundreds of sub-strains which now exist. Written pedigrees, other than those compiled by El Kelbi, did not exist among the desert people, the breeding history of horses being passed down by word of mouth. To the European this practice seems open to enormous possibilities for error, and confusion and inaccuracy undoubtedly occurred in the early purchases of Arabian horses by Europeans. Even up to the time of General Daumas many Moslems adhered to the law forbidding the sale of horses to Christians, and even when the opportunity of pecuniary gain overcame their religious scruples horses were frequently sold unnamed and with the vaguest of backgrounds, such as: 'His mother drank of the desert wind and his sire outstripped the gazelle'! Nevertheless, those with knowledge of the tribes and of the strains favoured by each would have a pretty good idea of the particular bloodline involved even if this was not mentioned in the flow of adulatory verbiage accompanying a sale.

Today the Arab horse is bred all over the world, and he retains all the characteristics so highly valued by his original masters. All horses in competitive sport derive from the Arab to some extent and as a pure-bred he is still supreme in the long-distance endurance rides, though in his pure form he plays hardly any part in show-jumping or three-day events. Arab blood, outside strictly Thoroughbred breeding, continues to play a vital part in the development and up-grading of other breeds.

The danger facing the future of the Arab horse today, when his breeding has long since passed out of the hands of the occupants of his original desert habitat, is that the breed will become too popular for its own good. Already Arab 'studs' proliferate throughout Europe, America and Australia and exist in some numbers in Scandinavia. Careless overbreeding could result in the production of inferior specimens lacking the true Arabian characteristics. This would be a disaster, not only for the breed but for the whole world of horses, which still depends to a very considerable extent on the maintenance of these properties in the Arab.

In the light of what in certain areas may be termed promiscuous breeding it may be of interest to consider the perfect Arabian horse that breeders ought to be trying to produce. The head should be very short and of great refinement, and the face pronouncedly concave or 'dished'. A straight face is not, under any circumstances, acceptable. The muzzle is tapered and very small – so small, indeed, that it should fit into a half-cupped hand. The nostrils are exceptionally large in comparison and of great elasticity, being flared in excitement. The texture of the muzzle skin is especially soft in the high-class Arabian; any coarse feeling that it gives to the fingers denotes lack of breeding.

The eyes are set lower than in other horses and are widely spaced as well as being very large. R. S. Summerhays, an authority of standing, says that 'in appearance they must be dark and deep in colour, very soulful in the mare, in the stallion capable in an instant of showing great alertness, with enormous challenging dignity'. In many respects this typifies the Arabian: no horse is as masculine as the Arab stallion; none as feminine as the Arab mare. The ears are shapely, small, well defined, set apart and always carried alertly, exhibiting 'a most exquisite curve' (Lady Wentworth). Ears in the mare are usually longer than in the male. A further feature of the head, differing from that of other breeds, is the generous rounding of the jaws. It should be possible to insert a closed fist between the spread of the jaw-bones.

The neck should be long enough but not excessively so. In the Arab it is the *mitbah*, the angle at which the neck and head meet, which is important. There must always be a distinctive arched curve formed by the angle, peculiar to the Arab, at which neck and head join. The more accentuated this feature, the greater the degree to which the head may move freely in all directions. The texture of the mane is another distinguishing feature: fine, silky and soft.

The body must next be considered. The shoulder, a particularly important feature in the riding horse, must be sloped and long, with well-defined withers. An excessive slope, though possibly desirable in the Thoroughbred, is untypical of the Arab and the withers, although pronounced, are not so high. The back is short, slightly concave, very strong across the loin and carried on a level croup. Ribs are rounded and the girth deep. The chest, of which Lady Wentworth once wrote that 'it cannot be too broad' in fact certainly *can* be too broad, just as it can be too narrow. Too broad, with legs consequently set far apart, it is a 'harness' chest, whereas it should be normal in appearance and of usual riding type.

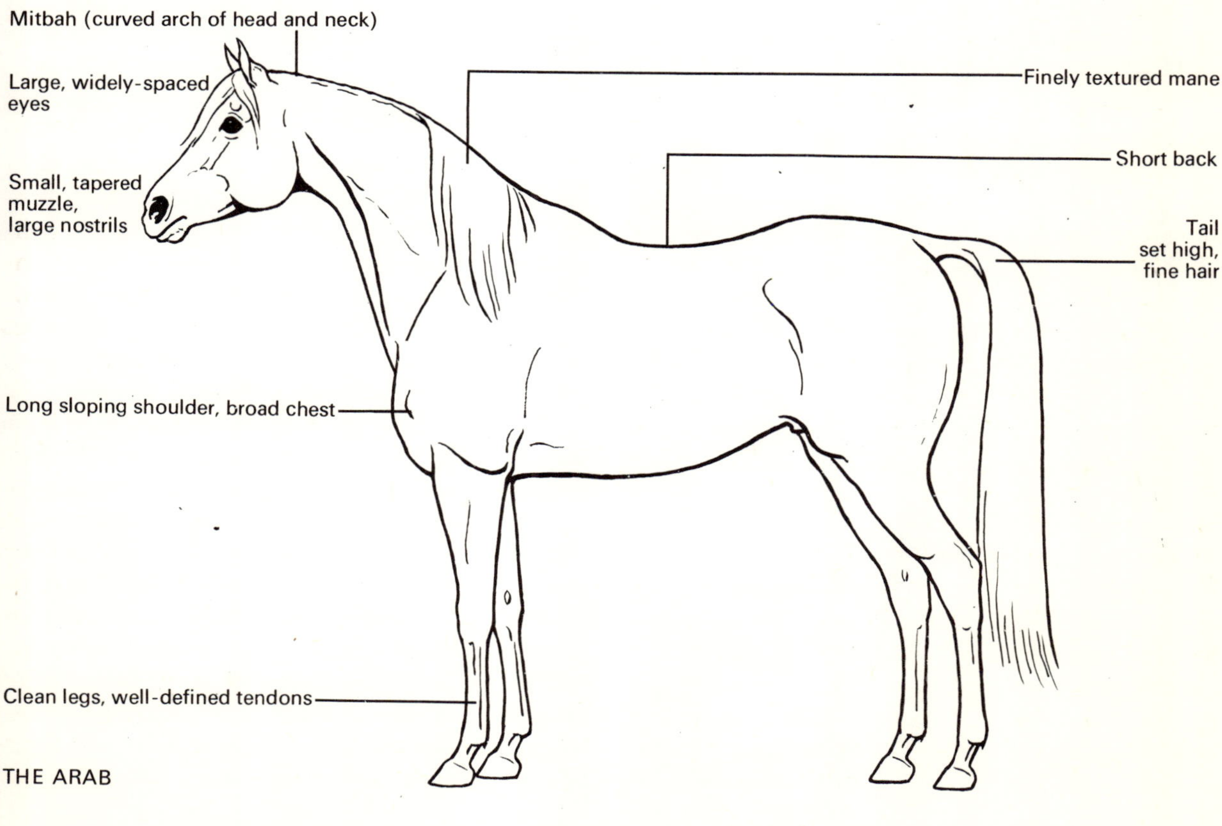

THE ARAB

The quarters are generously proportioned with the croup level, long and wide. The Arabian differs from other horses in the number of its ribs and lumbar bones, and this is responsible for the particular and unmistakable formation of its back and quarters. The root of the tail sets into the croup and its uppermost line and is carried arched and high in movement. Extreme length from point of hip to point of buttock is expected in the Arab, though it is not always found except in the highest class. The gaskins should be strong and prominent, and the hair of the tail should be similar to that of the mane.

Arab limbs are strong and hard and with clean, well-defined tendons. Knees are flat and low to the ground and the cannon bones short in consequence. Since the bone of the Arab is of greater density than that of other breeds the measurement below the knee need not be as great. Pasterns must be neither too upright nor too sloping and the joints should be flat and well formed. The hind leg is a weakness in the majority of Arabian show class entries. In the ideal Arab it is straight and not in any way sickle-shaped, bent or cow-hocked. Hocks must follow the line, turning neither in nor out, and they too must be set low to the ground. The feet of the Arab must be near-perfect; this is a fundamental characteristic of the breed, and any weakness or malformation is a serious fault.

Like so many other things Arabian the action is unique. It is characterised by a floating appearance, the horse moving as on springs and with the greatest of freedom in all his paces. The tail is always held high and bannered. The trot pace is not, as executed by the Arab, anything at all like the exaggerated, daisy-cutting toe-pointing of the show animal. Rather it is full, free and generous but with, nonetheless, little or no knee action.

Perhaps the most vivid description of the Arab is that written by Homer Davenport in his authoritative book, *My Quest of the Arabian Horse*, published in New York well over half a century ago. He concluded a detailed description of the breed by declaring: 'The build of the Arab is perfect. It is essentially that of utility. The space for the seat of the rider at once fixes his true position and his weight is carried on that part of the frame most adapted for it. If he be carefully examined it will be found that all the muscles and limbs of progression are better placed and longer in him than in any other horse. Nature, when she made the Arab, made no mistake.'

Arabs make superb riding horses. This horse, Saffron Sky, was a winner under saddle at the Royal International Horse Show at Wembley.

Dargee, a champion Arab stallion owned by Lady Wentworth, one of the greatest authorities on the breed.

THE ANGLO-ARAB

The Anglo-Arab horse, as its name implies, is the product of the Arab and Thoroughbred. Although in the strictest sense the Anglo-Arab cannot be regarded as a *pure* breed, since new blood, either Arab or Thoroughbred, is continually added, it is nevertheless recognised throughout the world as an established one, having its own well-defined characteristics.

The fusion of the two premier world breeds should in theory, and indeed frequently does in practice, produce a riding horse combining the best qualities of both. The modern pure-bred Thoroughbred is not always entirely suited temperamentally to present-day activities; he may be too excitable and highly couraged for the disciplines involved in dressage training or for the precision required in show-jumping arenas. Moreover, the Thoroughbred cannot be said to have retained the inherent soundness of his Arab progenitor. He has, however, speed

A French Anglo-Arab stallion.

and jumping ability far greater than that of the much smaller pure-bred Arabian horse. These qualities, together with the Thoroughbred's natural balance and good riding conformation, when combined – not necessarily a straight first cross – with the intelligence, soundness and stamina of the Arab can result in a riding horse of the finest type, even if it is not quite as fast as the Thoroughbred.

In Britain the Arab Horse Society maintains three stud books, for Arabs, Anglo-Arabs and part-bred Arabs respectively. To qualify for inclusion in the Anglo-Arab book a horse must be able to claim not less than 12½ per cent Arab blood. Since the establishment of the English Thoroughbred as a distinct breed during the last century, from the point where re-crossing to the pure Arab ceased, there have been numerous examples of English Anglo-Arabs that have proved their excellence in every equestrian field outside the racecourse. In recent years Anglo-Arabs have been successful in hack show classes, as dressage, show-jumping and event horses and also as hunters. Nonetheless the development of the Anglo-Arab in Britain is insignificant when compared with the standing attained by the French Anglo-Arab in the country of his origin. That the French Anglo-Arab should be pre-eminent is due to the encouragement given to its breeding by the long-established, state-owned national studs which have influenced horse-breeding in France ever since they were first created by Louis XIV's minister, Colbert.

No such situation has ever existed in Britain, where horse-breeding is, and has always been, in the hands of a number of individuals usually operating on a comparatively small scale. It is true, of course, that the breed societies are extremely active, but membership, although necessary, is voluntary and arbitrary control quite out of the question. The conditions of registration in the various stud books, therefore, and the definitions of type and characteristics which are laid down for guidance, are the only governing factors. In France, Colbert's early national studs served mainly as a source of supply for the royal stables and it was not until the arrival of the Emperor, Napoleon I, that the studs began to take the form that they have today.

Napoleon, increasingly prodigal in his use of horses and of men, was in constant need of cavalry replacements for his ambitious campaigns. He favoured the small, tough Arabs for use as his personal chargers, and realised that a crossing of Arab blood with the horses existing in the south-west of France would produce ideal cavalry remounts. As a result the Haras National de Tarbes was founded in 1806 together with the studs at Pompadour and Pau.

The English Thoroughbred element was introduced around 1830 when 'English blood', adding size and speed, became fashionable in France. To begin with, the French were over-enthusiastic, using so much Thoroughbred blood that the original qualities of hardiness, stamina and a level head, so desirable in a cavalry horse, were lost. However, the lesson was learnt and subsequent generations practised a rigorous system of Thoroughbred selection based on performance, stamina and conformation, that has persisted to the present time.

The renowned horseman, Henry Wynmalen, with his Anglo-Arab stallion of the Shagya Arabian strain.

The French Anglo-Arab, subject to the academic reservation made at the opening of this article, must now be regarded as a definite breed, and as such it is far more firmly established than its English counterpart. Pedigrees, indeed, usually show either one or both parents to be Anglo-Arabs. Domollan, for instance, an Anglo-Arab (25 per cent) standing at Tarbes, is by Tetouan (Anglo-Arab 25 per cent) out of Domerede, who had the same blood percentage. The best products of the French studs show between the minimum 25 per cent and 45 per cent Arab blood, and many of the Olympic medals gained by France in the equestrian events, as well as numerous honours in international show-jumping, have been achieved by Anglo-Arabs within these percentage limits.

Few pure-bred Arabs for use as stallions (it is the practice to put an Arab stallion on a Thoroughbred mare) are bred at the national studs, it being though preferable to import these from Syria, and particularly from Tunisia, where climate and soil conditions encourage the necessary and much-prized hardiness and typical Arabian character.

As an additional and most important factor in the breeding of the French Anglo-Arab the racing programme confined to the breed provides an excellent means of selection or rejection according to performance. Thirty-one race meetings for the Anglo-Arab are held each year and there are also jumping and cross-country tests (some of the latter over the formidable bank country of Pau).

The present-day Anglo-Arab is usually about 16 hands and rarely exceeds 16·2 hands. Its appearance is marked by a distinctive head, well-set and clearly defined withers, sloping shoulders, roomy girth, good limbs and joints and excellent hard feet. Endowed with intelligence, sobriety, scope, endurance, balance, weight-carrying capacity and no mean turn of speed this is one of the world's greatest all-round riding horses.

An Anglo-Arab in peak condition for being shown in hand.

Chapter two

The English Thoroughbred

The evolution of the horse, from the very earliest times when the ancestors of our present equine species began to lose their toes and to walk upon a recognisable hoof, has been marked by numerous specific and significant developments.

The early changes in form and habit before the domestication of the horse between 3000–2000 BC, were dictated by environment and were part of a gradual process spreading over thousands of years.

Thereafter, however, the evolutionary process accelerated as the human race, seeing the horse as a means of assisting its own progression, took an increasingly large part in the development of animals best suited for the purpose of transport, communication and, most particularly, of making war upon its fellows.

Over the centuries that followed the first recorded domestication of the horse by the Indo-European tribes, which inhabited the steppe lands north of the mountain ranges bordering what we now know as the Black and Caspian Seas, specific types of horses, selectively bred for one use or another, began to appear. The apotheosis was reached in the establishment of the warm-blood Oriental horses, which we call the Arabian horse and which must be acknowledged as the fountainhead of the world's breeds, since there are few if any that do not owe their existence to the prepotency of these desert-bred horses.

It was due to the importations of these Oriental horses into England that the next great development, that of the emergence of a super-horse, the English Thoroughbred, was achieved.

The exact origin of the world's premier breed of horses is a matter of some controversy, and there are authorities who disagree with the late Lady Wentworth's positive and sweeping assertion that the Thoroughbred is entirely and exclusively the product of Arabian blood.

Racing and racehorses were in fact an integral part of the English sporting scene for centuries even before the formative period of the Thoroughbred during the hundred or so years following the Stuart Restoration in 1660. It would therefore be reasonable to assume that native elements already existed which, when crossed with the Arab, contributed to the combination of qualities characteristic of today's Thoroughbred.

This theory is based upon the belief that there was a well-established breed of native 'running horses' in Britain before the large importations of Oriental blood (known variously as Barb, Turk or Arabian) that took place in the latter part of the seventeenth century and continued well into the eighteenth. On the other hand, it is known that efforts were made to improve the native horses by the importation of Oriental sires and mares well back into English history, and so it can be argued that the running horses might well have carried proportions of Oriental blood. At this distance in time, however, no certain proof is available.

What is fact, however, is that James I had Arabians, among them the Markham Arabian, in his stud of racehorses, these being brought to England by Sir Thomas Esmond.

Sir John Fenwick, Master of Horse to Charles II, is said to have furthered the influx of Oriental blood by bringing back, at the command of his master, the Royal Mares, as well as some stallions from the Levant. That, at any rate, is the General Stud Book explanation of the Royal Mares found in so many early Thoroughbred pedigrees. It is more probable, however, that these mares were supplied from various sources by James D'Arcy, Master of the Royal Stud, who had a contract with King Charles to supply 'twelve extraordinary good colts' each year for the royal stud at Sedbury, in Yorkshire, for an annual payment of £800.

During the period 1721–59 some 200 Oriental horses are listed in Volume II of the General Stud Book. Of these 176 were stallions, of whom three exerted a particular influence on the subsequent development of Thoroughbred stock, all modern Thoroughbreds being descended from them in the male line. These were the Darley and Godolphin Arabians and the Byerley Turk.

The Darley was the founder of the Eclipse line, the horse that inspired Dennis O'Kelly, his owner, to make the now-famous remark, 'Eclipse first, the rest nowhere', and how right he was, for this son of Marske was never beaten up to the time he retired from the racecourse in 1760. From Eclipse, and so from the Darley Arabian, descend today's important male lines of Blandford, Phalaris, Gainsborough, Son-in-Law, Boss, Teddy and St Simon. The Darley was also responsible for another famous son, Flying Childers, the first great racehorse out of an Oriental mare, Betty Leedes, and so exclusively of Eastern descent. The Darley also heads the direct sire lines of Sun Chariot and Big Game.

The Godolphin Arabian came to Britain in 1728 and lived to the ripe old age of twenty-nine, dying in 1753, in Cambridgeshire, at the Gogmagog estate of his owner Lord Godolphin. The Godolphin founded the Matchem line; Matchem, foaled in 1748, being by the Godolphin's son, Cade. Matchem, although not as successful on the racecourse as his half-brother, Gimcrack, who is perpetuated by the Gimcrack Stakes for two-year-olds held at York in August, far exceeded the latter as a stallion and was the leading sire in the north of England, his fee starting at five guineas and rising to the then astronomical sum of fifty guineas as his fame spread.

The Byerley Turk was the first of the three to come to England. Byerley's Turk sired Jigg, to found the Herod line from whence came the Tetrarch, among other great horses of our own century.

There were, of course, other Oriental stallions whose influence still remains, although not in the male top line. The Leedes Arabian, for instance, appears in more pedigrees than any other horse and every grey Thoroughbred traces back to Alcock's Arabian.

Few, if any, of these horses ever ran in a race, and one wonders why the breeders of the seventeenth and eighteenth centuries made so much use of horses that had not been subjected to the process of selection and rejection imposed by racing performance. Certainly it was not because of the speed of the Arabian, which would have been of little account even in those days. It has to be concluded that the Arab was used because of its inherited quality and its ability to breed true to type, an essential in any breeding enterprise, and particularly so in circumstances where the native stock had been debased by repeated recourse to foreign blood.

On the female side the all-pervading influence in the early Thoroughbred is the Arabian foundation mare, Old Bald Peg, by the Unknown Arabian. Repeat crosses of this mare appear 367,162 times in the pedigree of Big Game; 233,579 in that of Sun Chariot; 138,827 in Hyperion; 113,740 in Fairway and 184,091 in the pedigree of Windsor Lad – according to Lady Wentworth's table of repeat crosses to be found in *The Authentic Arabian.*

In order to establish and improve a breed it is imperative to keep reliable records of pedigrees and matings, as well as records of performance in the case of racehorses. Early records of the Thoroughbred are naturally enough not entirely comprehensive, but by 1791, when the first of Weatherby's General Stud Books appeared, a definite pattern had been established. Today's GSB, still published by the same family, includes all pure-bred mares and their progeny together with pedigrees of both mares and sires. Only animals entered in the GSB are eligible to compete on licensed courses in Britain. Loosely, therefore, the English Thoroughbred can be defined as a horse of proved pedigree eligible for entry in the GSB.

The word Thoroughbred, as applied to the racehorse, was not used until 1821 when it appeared in the second volume of Weatherby's GSB, and it was not until much later that Arabian outcrosses ceased and the English Thoroughbred was established as a breed in its own right. Indeed, it is only in the last hundred years that the Thoroughbred has achieved its remarkable increase in numbers and has spread to all parts of the world.

In 1962 the world Thoroughbred population, within the racing and breeding world, was

Horsemen round up a herd in the Hortobágy region of Hungary.

Horses in the Camargue.

Threshing wheat by traditional methods in Greece.

A police horse demonstrates its courage and obedience.

The Godolphin Arabian. (*Fores Ltd*)

The Darley Arabian. (*Fores Ltd*)

Quality Fair, an outstanding modern Thoroughbred stallion.

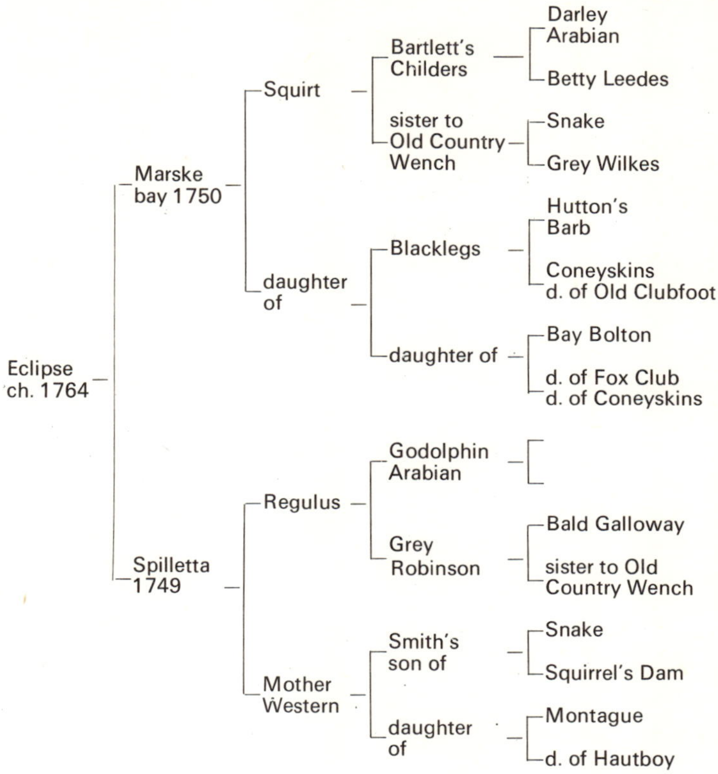

estimated by Franco Varolo in the magazine *Courses et Elevages* as 233,000. Now, some ten years later, it is probably approaching a number in excess of 400,000. In addition, of course, the Thoroughbred exists in quite large numbers outside racing. That a near-perfect racing machine has evolved over so comparatively short a period of time is tribute to the knowledge, judgement and enthusiasm of the generations concerned with its production.

Apart from the necessary quality of speed the Thoroughbred is the ideal riding horse. In movement the action is long, free, easy and fast at all paces and since action is dependent upon conformation that of the Thoroughbred is the nearest to perfection. Add to this the elusive factor of quality combined with balance, and true symmetry of proportions, and the Thoroughbred is without doubt the aristocrat of the equine species.

In all these respects he far exceeds his progenitor the Arab, and still retains the latter's fire and courage. Generations of confined breeding, however, as well as other factors, have resulted in a loss of much of the inherent soundness that is a characteristic of the Arab horse. Many authorities would hold the practice of two-year-old racing responsible for so much wastage. Despite the precocity of the modern Thoroughbred, there is inevitably risk of overstrain or breakdown when weight has to be carried at speed on an immature structure.

This apart, the Thoroughbred is regarded as the ideal cross to produce hunters and jumpers when mated with either half-bred mares or with Arabs and Arab crosses.

Chapter three

British native breeds

Great Britain has a unique heritage in her famous native breeds. Apart from the heavy horses, most of these are ponies which, while they share some characteristics common to all true ponies, have developed along different lines according to the conditions of their particular environment, the addition of foreign blood in their breeding patterns and the different uses to which they have been put.

The characteristics that all real ponies share have been developed for several reasons. In general, horse-breeding has been influenced by men to a greater extent than the breeding of ponies. Until recently, with a few exceptions such as the introduction of Arab blood into the island strain of Scotland's Highland breed, ponies have bred naturally, and nature's process of selection has led to superlative quality in each breed for the needs of that breed. Perhaps the most difficult pony quality to define, and that which essentially differentiates ponies from horses, is 'ponyness'. True pony character is the result of an independence and intelligence fostered by life in natural and frequently harsh environments, combined with an equable, sensible temperament. Far fewer ponies than horses 'hot up'; far more ponies than horses will find their way out of an apparently impenetrably-fenced field. There are, of course, plenty of horses with positive and individualistic personalities, but ponies do seem generally to have more character than their larger and better-bred relations. Part of the elusive pony character is to be observed in their expressions. Typically, ponies have short, trim ears on neat heads, giving them considerable alertness and a particularly intelligent and endearing expression. Even among the carefully-bred show ponies, some of them in many respects very different from the native pony stock on which they are founded, this essential quality of pony character is regarded as all-important. Without it, a pony can become just a smaller version of the horse.

There are also some physical features shared by pony breeds which are not always found in horses. They are tougher than many of the horse breeds, able to thrive in harsh conditions and to get themselves out of trouble. Centuries of having to fend entirely for themselves in order to survive at all in the mountain and moorland areas of Britain have sharpened their faculties and increased their hardiness, stamina and sure-footedness. Ponies are rarely ill, seldom go lame if properly looked after, are remarkably adaptable, and seem to have a tremendous sense of fun. Curiosity, a necessary part of survival in the wild, is just one characteristic typical of the pony in domestic surroundings which endears him to his human companions.

It is hardly surprising, in view of how much they have to offer, that ponies are now widely used in the breeding of horses, and that crossed with a quality blood horse – such as a Thoroughbred or an Arab – they provide the best of all foundation breeding stock. The quality of the British native pony has been widely recognised, and examples of all the breeds – in particular those from Wales – are now exported all over the world. The establishment of breed societies for each of the native breeds during this century has ensured the protection of the quality of the ponies and the maintenance of their traditional attributes.

Exmoor

The Exmoor, oldest of all the native pony breeds, is known to have existed in the Exmoor area of Devon and Somerset in prehistoric times. The moors are wild and solitary, the climate is harsh; there is little protection from the elements here. In order to survive, the ponies had to develop what is still one of their prime attributes – hardiness. This is one of the characteristics that has led to their wide use as foundation stock for breeding, and Exmoors are exported for this purpose to many countries;

they are specially popular in Canada and Denmark. Crossed with Thoroughbred bloodlines, the Exmoor produces superb stock.

In addition, the Exmoor is known for its agility, quick-witted intelligence and considerable powers of endurance. The ponies are strong out of proportion to their size, and it is not unusual to see a diminutive though sturdy pony of about 12 hands carrying a large West Country farmer both during his shepherding work and behind hounds. Properly handled, the Exmoor is a good pony for children, too, though its primary value is considered to be for breeding.

Exmoor ponies should not exceed 12·2 hands, though the stallions may reach 12·3 hands. The short pointed ears of the breed enhance the kindly look typical of these ponies, with their wide foreheads and nostrils and large eyes. Exmoors, which are bay, brown or mouse-dun in colour, have the attractive mealy muzzle and 'toad eye' as well as being lighter in colour on their bellies and the inside of the thighs. The coat colour may not include any white.

The strength of this breed is displayed in the deep, wide chest, powerful loins and clean legs, and the hardness of its feet ensures its surefootedness even in difficult country. The winter coat of the Exmoor is extra thick and has no bloom, for it is unusually wiry in texture in order to protect the pony from the rigours of the winter climate on the moors. The foals are similarly protected, with a thick woolly undercoat covered by a waterproof top layer of long hairs.

The sturdy Exmoor pony.

Dartmoor mares and foals at a stud in Devon.

Dartmoor

The Dartmoor pony, neighbour of the Exmoor, has had to learn to withstand similar climatic conditions, and has developed some of the same characteristics as a result. The moorland environment has not only produced hardiness and surefootedness in the breed; the Dartmoor is a valuable pony for many other reasons as well.

Infusions of outside blood meant that for some time there were great variations in type of the Dartmoor, but since the formation of the breed society and the registration requirements for quality animals formulated at the end of the last century, considerable efforts have been made to stabilise the Dartmoor's breeding and to produce a really first-class children's riding pony.

One of the advantages of the Dartmoor is its versatility. These ponies are remarkable jumpers for their size – they do not exceed 12·2 hands – and are much appreciated as foundation stock in the breeding of larger all-round ponies and top-class quality riding ponies. Their value in this respect has been shown by the fact that a part-bred Dartmoor register has been formed.

The Dartmoor pony is strong, well-made and good-looking. It should have a small, well-bred head and an alert expression, strong shoulders, back and loins, a full mane and tail and particularly well-shaped feet. The typically good front and relatively high head carriage makes Dartmoor ponies feel specially safe to children. To all these physical advantages the Dartmoor adds its intelligence, an ability to think for itself, a kind, sensible character and a particular liking for children.

New Forest

The New Forest pony is to be found all over the extensive woodland and open heath of the New Forest in Hampshire. The Forest has been the home of this breed since the days of King Canute in the eleventh century. As stock of other breeds was turned out to run in the Forest from time to time, the blood of the native ponies has become rather mixed. The breeding was affected by deliberate attempts, made at the end of the nineteenth century, to improve the breed by the introduction of other bloodlines; when these efforts were limited to the introduction of other native pony breeds all went well, but alien blood was also brought in and at one time the breed was in danger of losing its hardiness.

The ultimate result has been that there are now great variations in type of the New Forest pony. Its size varies between 12 and 14 hands; with the increase in height one finds an accompanying increase in sturdiness and overall size. The larger ponies are ideal as all-round family riding ponies. The New Forest pony is intelligent and quick to learn, has an aptitude for sporting events, is very useful in harness, and is thoroughly reliable.

The ideal New Forest pony should have a conformation enabling it to perform as the best type of riding pony. The head should be well set on, there should be a good length of rein, a short back, strong loins and quarters. These ponies must carry a substantial amount of bone, a good forearm and second thigh, short cannons and healthy feet.

The majority of New Forest ponies still run loose on the Forest. The stallions are inspected for quality every year and a careful watch is now kept on breeding in order to maintain a high standard of quality. There are some private studs also breeding these ponies, and cross-breeding them produces excellent small hunters. There is a considerable demand for New Forest ponies from abroad, principally from Europe.

A New Forest mare and foal on open grazing near Lyndhurst, Hampshire.

Welsh breeds

The ponies of Wales are widely recognised as being among the most beautiful in the world. In addition to this, they possess to such a remarkable degree all the other desirable pony qualities that almost all children's riding ponies today carry a percentage of Welsh blood in them.

The Welsh Stud Book caters for all four of the distinctive Welsh breeds. The first and most important of these is the Welsh Mountain pony (section A). The Welsh Mountain pony may not exceed 12 hands, and is at best one of the supreme examples of all that a pony should be. It combines in its character intelligence, courage, gentleness and endurance. Its exquisite, concave face is set on a graceful neck, it has a deep, sloping shoulder, short, strong back, a tail set high and proudly carried, short limbs with dense, flat bone, and hard, small feet.

The Welsh Mountain pony has always been bred out among the hills of Wales, and as a result has the most hardy constitution. The action is typical of a quality pony accustomed for centuries to moving over rough ground. It is sure-footed, quick and full of action, moving with the utmost freedom at all paces. These ponies are natural jumpers, and very adaptable, being equally at home ridden or in harness.

Welsh Mountain ponies are now being bred all over the world, but the exceptional quality of the herds still roaming over the Welsh hills is such that breeders frequently go back to Wales to improve their own stock.

Section B of the stud book caters for the Welsh pony, which inherits many of the characteristics of the Welsh Mountain but is bred more specifically as a child's riding pony. The emphasis is placed on quality, bone and substance, hardiness, constitution and pony character. The Welsh pony may be up to 13·2 hands in height, and makes an ideal show and performance pony. Like all the other Welsh strains, these ponies may be any colour other than piebald or skewbald.

Welsh ponies of cob type and the Welsh Cob, covered in sections C and D respectively, are similar, though the former may not exceed a height of 13·2 hands, while the latter are over that height, averaging 14·2 to 15·1 hands. The cob ponies are sturdy, active and strong, with considerable substance; a quality head is important, as they should still display true pony character. Free, true, forceful action is essential in both types, accompanied by considerable presence and zest, two typical characteristics of the breed. The larger type makes an excellent hunter, being a natural and clever jumper and with considerable staying power, while the smaller one is unsurpassed for trekking, now a very popular holiday activity in Wales and elsewhere in the British Isles.

The Welsh Section C combines pony characteristics with robustness and substance.

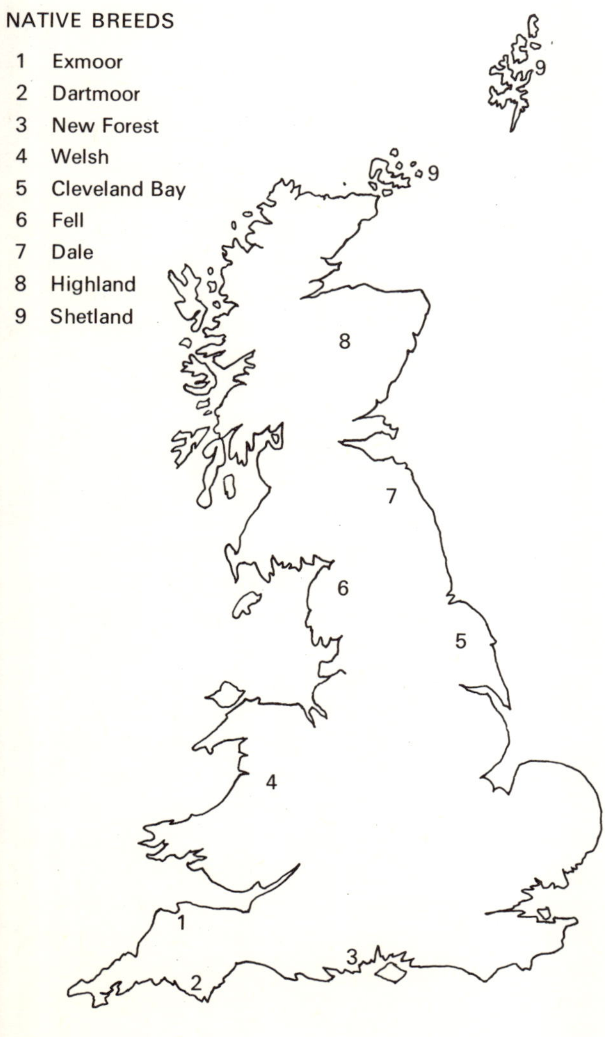

Fell

In the northern counties of England some rather different breeds have developed. The Fell pony, native to Cumberland and Westmorland, is descended from the Celtic pony. During the seventeenth, eighteenth and early part of the nineteenth century it was greatly valued as a pack pony, principally being used to carry lead from the mines inland to the port towns on the north-east coast. To be able to carry a sizeable load the Fell pony had to be powerfully built and with considerable stamina; it is a graceful as well as strong pony, with a compact, muscular body, strong, squarely-set clean legs with plenty of bone and open, hard, round feet. It has the typical charming head and prick ears of the pony, carries its tail high and has an overall air of gaiety and liveliness. Great emphasis is placed on the importance of straight and active movement at all paces. Fell ponies should not exceed 14 hands; they are usually black, dark brown and bay in colour, though an occasional grey can be found.

In the past the local farmers organised trotting

races between their ponies when the pack work had been finished for the week. The Fell also used to be the all-round work pony on the north country farms, being used for riding, shepherding, pack work and light haulage as well as being called on to pull the family trap. Now the ponies, which stand at between 13 and 14 hands and are still very strong, are used for all sorts of different activities. They make excellent riding ponies, are still used for shepherding in the more remote areas, and their stamina stands them in good stead in competitive events such as long distance rides and combined training events.

The Fell pony is highly valued as breeding stock. Little alien blood has been introduced since the beginning of this century, and great care is now taken to protect the purity of the breed by controlling the requirements for eligibility to the breed register. A Fell pony crossed with hunter stallion blood produces a tough, dependable, big-hearted small hunter; when crossed with a Thoroughbred a good jumper should result. The substance of the Fell has been responsible for the current demand for the breed from abroad.

The versatile Fell, though a pony in stature, can easily carry an adult.

Dales

The Dales pony is in some respects similar to the Fell, for it originally sprang from the same stock. It is found principally in County Durham and Northumberland, to the east of the Pennine range. Like the Fell, it has great hardiness and can thrive in spite of the cold, harsh climate of this part of the country.

The Dales pony is the largest of the native pony breeds; it is also the heaviest and strongest of them, capable of carrying a full-grown man all day on the fells. In the past, Dales ponies were used for all kinds of farm work, and after suffering a decline when motorised transport and machinery became generally available, they are now increasing in popularity again with the growing demand for ponies suitable for trekking. They are ideally suited for this work, being hardy, quiet, sure-footed and above all sensible.

Great emphasis is again placed on the importance of sound limbs and clean, healthy feet where the Dales is concerned. Straight, true action is also demanded–action which should give an impression of rhythmic movement and power. In spite of its size and strength the Dales should show no sign of coarseness. A neat pony head with no suspicion of thickness of jaw or throat is expected; the body should be compact, with a short back and powerful loins, well-sprung ribs, depth through the girth and well-developed quarters. The ponies should not exceed 14·2 hands, and their colouring is again limited to blacks, bays, browns and a few greys.

The combination of active movement and substance of bone makes the Dales pony excellent stock to cross with Thoroughbred blood in the breeding of hunters, jumpers and eventers.

Highland

The Highland ponies of Scotland used to be divided into two types, though the mainland strain has achieved general recognition throughout the Western Isles as the older and more pure-bred strain. The mainland ponies are larger than their island cousins, standing at about 14·2 hands, while the island type can be as small as 12·2 hands.

Highland ponies have traditionally shared the life of the crofters, being used as general utility ponies in the craggy uplands of Scotland. They are capable of carrying great weights, and combine this strength with remarkable agility and sure-footedness. They are still mostly found in their native land, and are often used, for example, to carry the carcasses of deer down over difficult hillside terrain.

The famous strength of the breed is coupled with an attractive appearance and true pony character. The remarkable strength is to be found in the quality of bone: stallions should carry between nine and ten inches and mares well over eight inches. Neat heads and kindly eyes, a solid, deep and powerful body, muscularly well developed, and short strong legs are all combined in Highland ponies. Their natural paces are the walk and trot, for a turn of speed has never been required and has seldom even been practicable in this sort of country.

Highland ponies are most frequently found in one of the many shades of dun colour–ranging from golden-brown to silvery-blue. It is usual to find the dark dorsal eel-stripe in almost all pure-breds except for those predominantly black, grey-black and brown.

The Highland pony still plays an active part in the life of its native Scotland.

Shetland

The charming and lovable Shetland pony is also a native of Scotland. It comes from the Shetland and Orkney Islands off the north coast. This diminutive pony is of ancient origin and until very recently has been the only means of transport for people and for goods in the islands. Strongest of all the native breeds for its size–the Shetland is a tiny 10·2 hands–it has a small, refined head which is frequently almost hidden beneath its abundantly protective mane. The weather in the northernmost parts of Britain is extremely unpleasant, wet and cold, and the Shetland needs all the protection it can get from its double winter coat, profuse mane and thick, long tail.

The Shetland gets its strength from its short back, depth of girth, sloping shoulder and well-sprung ribs. The hard, short legs and small open feet give a surprisingly airy action for one apparently so stocky in appearance. Of an independent and sometimes headstrong mind, the size of the Shetland makes it ideal as a pet and a child's pony, but it needs firm, though kindly, handling if it is not to get above itself.

Cleveland Bay

The Cleveland Bay is another British native to be found in the north of England. The Cleveland is the only native horse–as opposed to pony–that does not belong to the heavy horse group. It has been known since mediaeval times in the part of Yorkshire from which it takes its name. Cleveland is the centre of an area in the eastern part of the North Riding, where there has always been a demand for large, strong packhorses. Until recently mining work has been carried out in this part of the country, and the endurance and weight-carrying capacity of the Clevelands has ensured their participation in the industry.

Before ironstone mining became general in Yorkshire and thus provided pack and haulage work for the horses, there was plenty of work for them to do in other ways. Until the mechanisation of agriculture, the Cleveland Bay was a real all-purpose horse. The long back and strong loins typical of the breed made it ideally suited to all kinds of pack work, for it was able to carry larger and heavier loads than the neighbouring Dales pony. These same qualities came in useful, too, for carrying the farmer's family – to market or to church – before the improvement in road surfaces led to the introduction of the gig.

As well as facilitating the family's movements, the Cleveland was used for agricultural work, was ridden to hounds and, when carriages became fashionable, was adopted as the most popular carriage horse. Breeders of Clevelands began, about a hundred years ago, to distinquish between the carriage horse and the agricultural type, though this versatile breed could in fact perform in either capacity.

The Cleveland adds other characteristics to those which have earned it a reputation as a valuable work horse. It has excellent hunter qualities, frequently being a natural jumper, up to a great deal of weight, with the courage to go all day and a cleanness of leg and strength enabling it to move 'over the top' rather than through heavy going, as most horses up to weight – themselves of the heavyweight sort – must do.

The Cleveland is usually a bright bay in colour with black points, though brownish bays are also seen. Standing at between 15·2 and 16 hands, the Cleveland has a large, well-made body with hard, short, clean legs and some nine inches of bone. Its back is still relatively long and its withers unpronounced, and it has a fairly large head on a long neck. Stamina and longevity should be added to those qualities of character already discussed.

Where breeding is concerned the Cleveland has several assets. Its greatest is its prepotency when crossed with other breeds; in particular it passes on its hardiness and strength of bone. Much favoured for cross-breeding with Thoroughbreds to produce hunters and lighter, faster carriage horses, and in considerable demand for export, the pure-bred Cleveland was at one time in danger of extinction. This situation has now been remedied. Though a lighter stamp of hunter is now often preferred in Britain, Clevelands are still in demand, and are also widely exported, for example to Germany, the United States and Canada, South Africa, Japan, Pakistan, Czechoslovakia and Australia.

An elegant pair of Cleveland Bay carriage-horses parading in the show ring.

Chapter four

British heavy horses

The heavy horse breeds in Britain today fulfil a rather different role from those of their forebears. Originally horses of considerable size and strength were needed by mediaeval horsemen as chargers to carry them and their weighty armour into battle. In days when agricultural work and transport relied entirely on the horse the heavy breeds to be found in Britain flourished in most parts of the country. With the advent of increasingly mechanised equipment for farming during the present century there has been less and less demand for these horses, but recently interest in them has been increasing again, and splendid examples of these noble breeds can be seen at shows all over the country. There is also a lively international interest in the British heavy horse breeds; Clydesdales have for a long time been exported to countries both throughout the Commonwealth and also to others where their value for haulage has been recognised; and the equally famous Shire horse is also appreciated, particularly in the North and South American continents.

Clydesdales, named after the area of Scotland now centred in Lanarkshire – the Clyde valley – have a long history. The breed comes from a region always regarded as the best horse-breeding land in Scotland. It was here that the majority of horses required to carry Scottish knights were bred. Edward I of England was the first of many English kings to impose a ban on the export of horses of military value north of the border, and as a result stock was mostly drawn from Flanders and also from Denmark. The rural horse stock of the south-west area of Scotland therefore probably carried a fairly high proportion of the famous mediaeval 'Great Horse' blood. The 'gig mares' of those parts were covered by Flanders stallions in the late eighteenth century to produce the modern Clydesdale horse.

This Clydesdale mare is a good example of a breed with great strength but no hint of coarseness.

A Clydesdale mare and foal at Stirling in Scotland.

With the coming of the Industrial Revolution, demand favoured draught animals of great substance and strength, for the production of goods preceded modern methods of transport. Demand has caused, perhaps, the perfection of the breed as we see it today. The stabilisation of breed characteristics in native mares showing qualities of size, strength, quality and above all activity has meant that still, today, the Clydesdale is the most active quality draught horse in the world.

The Clydesdale is a handsome, powerful animal standing at about 17 hands, and combines great stamina with strength and gaiety. The amenable disposition of the Clydesdale makes it an easy horse to manage, while the emphasis placed on the quality of its legs and feet has made it an outstanding horse for both farm work and haulage in town conditions. The typical well-built Clydesdale is an attractive and almost elegant horse as well as a strong and energetic one, with an open, flat-profiled head, and straight legs planted well under its close-coupled body. Clydesdales are dark brown or black in colour, often with a white-striped face and white hind shanks. They carry a good deal of shaggy feather, which shows off their active paces to great advantage.

Shire horses were founded by mares supplied from the stock of the Old English Black horse bred on the east coast to the north of East Anglia. Like all the heavy breeds it was originally used for military purposes, and its origins can be traced back at least as far as Tudor times. Originally, as its name implies, the Shire was to be found in the Shires, in particular the counties of Leicestershire, Staffordshire and Derbyshire, though it is now more often found in the northern counties, in East Anglia and in Wales.

In both height and weight the Shire is the greatest of the heavy horses, and it has played a major part both in achieving more efficient agricultural cultivation and in helping the movement of heavy loads from one part of the country to another.

The Shire has all the qualities needed in a draught animal: strength, constitution and stamina. It is an immensely powerful horse of commanding appearance, the stallion sometimes standing at as much as 18 hands, and weighing anything over a ton. Wide-chested, with legs well set under the broad, muscular body, the Shire, like the Clydesdale, is remarkably docile. It is characterised by white markings and heavy feather against a body colour of black, bay or grey.

A splendid Shire horse.

The Clydesdale and Shire areas of England met traditionally along a line running from Scarborough to Preston. Thus in Yorkshire, farmers in the North Riding who did not use Cleveland Bays in the plough used Clydesdales, while those in the East Riding ploughed with Shires. A cross between the two breeds was also popular. Further south, in the industrial region of Lancashire and the West Riding, heavy urban haulage was mostly entrusted to Clydesdales. As a plough and wagon horse the Shire expanded westwards across the Midlands, and it was more usual to see Shires than the native Suffolks in use in the southern parts of the country.

As the name indicates, the Suffolk Punch breed originated in Suffolk, though it has for a long time been considered native to the whole of East Anglia. The breed has been mentioned as far back as 1506. It is possible that the foundation stock may be a legacy of the Viking invasions hundreds of years before this time: the Suffolk Punch bears a marked resemblance to the Jutland horse which faces it across the North Sea. In the sixteenth and seventeenth centuries stallions of the Flanders breed were used on Suffolk mares.

Bred since the sixteenth century, the Suffolk Punch is particularly valued for its powers of endurance.

The Suffolk Punch shares qualities of stamina, strength and docility with the other heavy horses of Britain. To these it adds unusual longevity and an ability to thrive under poor feeding conditions. The average height of the

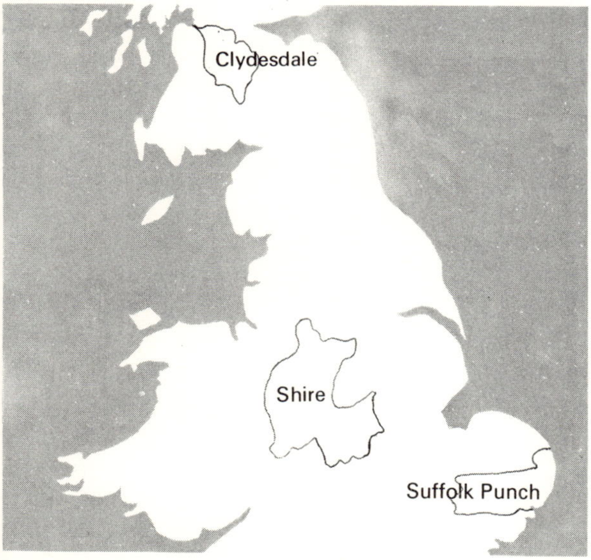

Suffolk is 16 hands – less tall than the Shire horse, but its equal in weight, for many Suffolks also weigh over a ton. Rather large-headed, the Suffolk is characterised by the depth of its neck in collar, its deep, round-ribbed body and short cannon bones. Typically, it too is tremendously well-muscled and impressively powerful. Suffolks are always chestnut, though there are seven acknowledged shades ranging from a dark to a bright chestnut. They carry noticeably less feather than the other heavy breeds, and less white is found on the face.

The Percheron, a native of the La Perche region of France, is also a familiar sight in England. It was introduced into the latter country in 1916; it is perhaps the most widely dispersed heavy breed in the world. Its qualities are discussed in the chapter on French horses.

The maintenance of quality has always been of prime importance to breeders of heavy horses, and each breed has its own society to encourage breeding standards. Heavy horses are much admired even in this mechanised age, and they can also still be found at work both on the land and, particularly, in use with drays of the big brewery companies. The efforts of the breed societies have been largely responsible for the recent renewal of interest in these magnificent animals, and their breeding is encouraged. They are a familiar sight at all the major shows, beautifully decked out and carrying themselves proudly on parade.

The Percheron, originally native to France, has been successfully bred in England for nearly sixty years.

Chapter five

British types

THE HUNTER

The word 'hunter' could with justification be applied to any horse used for the purpose of hunting. Like show-jumpers, hunters may come in all shapes, sizes and, indeed, colours, since they are not a specific breed.

In practice, however, the word refers to the *type* of horse best calculated to carry riders of various weights safely, expeditiously and comfortably over a country during the course of a season's hunting. That is the criterion by which a hunter is judged, but it has to be borne in mind, also, that the type will vary according to the country in which it is to be used. For instance, in the Shires – that part of the Midlands including Leicestershire, Rutland, Warwickshire, Northamptonshire and a part of Lincolnshire, which is regarded as the best English hunting country – the type best suited to the big, strongly-fenced pastures is the Thoroughbred or near-Thoroughbred. Size, speed and scope in full measure are needed if a horse is to remain in touch with hounds over this sort of grass country where scenting conditions will nearly always be good.

Conversely, such a horse might be a liability in a rough, hilly country, in ploughland or an area of small enclosures and big woodlands such as are to be found in the 'provinces' – a disparaging term originally used by those fortunate enough to hunt in the Shires when referring to hunting countries outside their own. In such countries the big-striding, bold Thoroughbred horse, so well suited to galloping over grass, is inhibited and likely to become increasingly frustrated when asked to 'creep' a trappy place or to pick his way up and down rough, stone-strewn moorland combes. His courage is then a disadvantage, his speed an unimportant factor, while the slow business of finding and hunting a fox in and over poor scenting land may be altogether too much for his volatile temperament.

In these circumstances a short-legged, compact sort of little horse, possibly carrying a percentage of native pony blood and the temperament that goes with it is the type needed. Many good hunters of this stamp are the result of a Thoroughbred cross with one of the larger native breeds, the Connemara, Highland or Welsh Cob, the last two, particularly, producing an animal of good substance and bone, capable of carrying considerable weight. A second cross to the Throughbred will result in greater quality, more freedom of action, increased size and so on but will usually retain the characteristics of the native pony: sure-footedness, sagacity and a strongly developed sense of self-preservation that will keep both horse and rider out of serious trouble. Many of the hunters used in the West Country will have a proportion of Exmoor or Dartmoor blood. Both these breeds are, of course, small (12·2 hands) but like all the native ponies are immensely strong and are capable of carrying weight out of proportion to their size. The first cross with Thoroughbred blood is unlikely to produce a horse big enough for a large adult, but the second cross often results in a horse entirely suited to the demands made by the varied country over which the West Country packs hunt.

A champion hunter stallion shown in hand.

In plough countries, like those of East Anglia, speed is again not a vital factor, since the arable lands constitute 'cold' scenting conditions. It is frequently thought that a big, strong, common-bred horse is the sort to be used here; in fact, this is not so. The old maxim that 'an ounce of blood is worth a pound of bone' is just as applicable to plough countries as elsewhere. The heavy commoner, often having an uneconomical, pronounced knee action, expends far more energy, to less avail, ploughing through the deep than the better-bred horse, whose action allows him to 'go over the top' and who, in general, has more stamina. A Thoroughbred would not usually be considered a good choice here, although there will always be the exceptions, and the ideal is possibly the good stamp of active, short-legged half- or even three-quarter bred with enough scope to tackle the wide ditches and enough common sense to cope with them when they are 'blind' in the early part of the season.

No discussion of the hunter could exclude the Irish variety, bred on the limestone and rich pastureland that are eminently suitable for breeding stock, and having the advantage of the mild Irish climate.

Ireland is traditionally a horse-breeding country, and even today many Irish farmers continue to keep mares and breed youngstock. At one time the market for their produce was almost entirely confined to England. Today, while the English are still substantial buyers, the market is international and the prices, for successful entries in the hunter classes at the Royal Dublin Society Horse Show, are little short of astronomical. Many of the best types of Irish hunters are bought for show-jumping and eventing by the representatives of continental countries and not a few have dominated the international arenas, particularly those bought by the Italians – The Rock, for example.

The Irish hunter was frequently bred out of an Irish Draught mare – a big, roomy sort, not exactly beautiful, but possessing excellent limbs and great bone – and by a Thoroughbred stallion, whose pedigree was often not exactly definitive. The resulting progeny were big, galloping horses, up to weight, admirably balanced, usually temperate and as clever as cats over any kind of country. Their ability to jump was virtually in-bred by the practice of leading or long-reining youngstock over banks and ditches at an early age, and then riding them quietly to hounds in a snaffle and on a long, even loose, rein. In this way the young horses learnt to look after themselves and most of them developed that invaluable 'fifth leg' which keeps a good hunter on his feet in situations where the more impetuous land on their expensive noses.

Although the Irish Draught is now not so numerous, the Irish hunter is still very much in evidence, even if the eye of his breeder is fixed more firmly on the bulging wallets of the eager continental buyers who are responsible for most of the high prices obtained.

In England the encouragement of hunter breeding is largely in the hands of the Hunters' Improvement and National Light Horse Breeding Society, founded with the object of improving and promoting the breeding of hunters, as its title implies, as long ago as 1885. This Society maintains a list of recommended hunter judges, publishes a Hunter Stud Book, and puts on two shows a year, one for stallions and the other for youngstock. The principle activity of the Society, however, is to make available suitable Thoroughbred stallions for the purpose of producing the most desirable type of hunter, an animal which in these days may also be successful in point-to-points, National Hunt racing and eventing.

At its annual Thoroughbred Stallion Show held at Newmarket in March of each year the Society awards sixty-seven premiums and fourteen super-premiums to the owners of the stallions that in competition are judged to be deserving of these awards. The premiums are worth £445 each and the super-premiums range from £270 for the show champion and reserve to £80 for the last two of the fourteen horses in the final line-up. The stallions are then allocated to 'routes' covering every part of the country, and are available to members of the Society at a service fee representing only a small portion of their actual worth. Non-members may also take advantage of the premium stallions at a higher service fee, although the amount is still far less than would be usual for Thoroughbred stallions outside the scheme.

Before competing for a premium these stallions, the majority of whom have acceptable racing careers, are required to pass a veterinary inspection that ensures complete freedom from hereditary disease. Breeders are therefore assured, provided that the mare is similarly sound, of producing youngstock free from the transmittable diseases and defects of cataract, roaring, whistling, ringbone, sidebone, bone spavin, navicular disease, shivering, defective genital organs and parrot mouth.

Clearly, the HIS Premium stallion has a considerable influence on type, but the greatest factor in setting the standards that should be aimed for by breeders is the show-ring, or more specifically the show hunter classes and the judges appointed to officiate in them. The show hunter, which is the ideal, is Thoroughbred or nearly so, and is judged on conformation, action and on the ride he gives. That such horses may never go hunting, most of them being too valuable to run the risk of disfiguring blemishes, is immaterial. The show classes are there to provide the pattern of horse who, by virtue of his make and shape and the action that should accompany near-perfect conformation, will fulfil the requirements of a top-class hunter–namely, to carry a rider throughout a season safely and comfortably at a good hunting gallop with the minimum of effort and the least risk of incurring physical strain.

The show hunter must, therefore, be well proportioned; in fact, nearly perfectly so, since good conformation not only results in a balanced ride under saddle but is also less likely to lead to the unsoundnesses that are inherent in a badly made horse. The structure of back, loins and quarters must be powerful and the quarters capable of great propulsive effort, while the shoulder must be well sloped and in no way straight or loaded. The horse should give an impression of substance and power ('blood weeds', often pretty and elegant, are not hunters) and the body, showing a deep girth to allow unrestricted use of the lungs, needs to be set on good limbs, having clearly defined, clean joints, short cannon bones and a good measurement of 'bone' beneath the knee. It is this last measurement, combined with the quality of bone, that together with the general proportions of the animal determines its weight-carrying capacity; an important point in any hunter.

Few hunters of any worth will have less than eight inches of bone, while those intended for carrying heavyweight riders (above 14 st. 7 lb.) may well have more than ten inches. Measurement of bone by itself is not, however, sufficient; its quality also matters. Thoroughbred bone is of greater density than that found in common or cart-bred horses and will carry, inch for inch, a greater weight. Twelve inches of poor quality bone with a wide central core encompassed by only a thin wall is obviously a structure of less

A ladies' hunter, shown to great advantage by being ridden side-saddle.

This novice has considerable promise, but still has room for development.

strength than bone of a lesser overall circumference with a smaller central core surrounded by a thicker proportion of denser bone.

However, while conformation is of the utmost importance it would be of no value unless the action of the horse were equally good. A hunter must walk and trot freely and straight, but the gallop is the pace at which he must excel. The stride should be long, low and effortless and devour the ground. Wasteful knee action and going 'into the ground' instead of over it are faults to be heavily penalised.

Good manners, an equable temperament, a kindly disposition and a good mouth are also necessary requirements in the show-ring. Gassy, tearaway horses that career away, fighting and fretting and almost out of control, do not make for comfortable or safe rides, however fast they may gallop.

Hunter classes are divided into those for heavyweights (horses capable of carrying 14 st. 7 lb., or 203 lb., and over); middleweights (13 st.–14 st. 7 lb.; or 182–203 lb.); lightweights (not more than 13 st. or 182 lb.); and small hunters, horses over 14·2 hands and not exceeding 15·2 hands. There are also classes for ladies' hunters, to be ridden side-saddle, and for novices (four-year-olds).

Outside what might be termed the 'beauty' classes are the very popular competitions for working hunters, in which considerable emphasis is laid on performance. Judging is carried out on the basis of 40 per cent for jumping ability, 30 per cent for conformation and presence and 30 per cent for ride and action. The course, in accordance with the recommendations made by the HIS, the body responsible for initiating this system of judging, consists of six natural-looking fences of various sorts and allows for ten marks to be awarded for the style of jumping. Style is not judged on the performance of the rider but on whether the horse, going at a fair and smooth hunting pace, jumps freely as a hunter should, instead of being 'hooked back' for each fence and jumping the course in a series of stops and starts after the manner of the less accomplished show-jumper. Blemishes, honourable scars and the like are not considered to be so detrimental in working hunter classes as in those for show horses proper.

A winning show hunter, mature, fully developed and in the peak of condition. Compare the horse in the preceding photograph with this one.

THE HACK

What is a hack? Today one thinks of a hack only in terms of the show ring, but to get it in its proper perspective one should really go back a hundred years, to before the days of motor cars. This was the age of the hack.

There were, in fact, two types of hack: the park hack and the covert hack. The former was the kind of horse on which ladies and gentlemen of leisure went riding, either in their park at home or in the public parks in the towns. In the latter case the horseman or horsewoman was exposed to the gaze of the public, and so it was essential that the whole appearance should be pleasing to the eye. Consequently, the park hack had to look very elegant, move beautifully and be perfectly turned out. In addition, of course, it had to have perfect manners, as did the hack used in the country; and also be a really comfortable ride. The emphasis, therefore, was on elegance, movement, manners and comfort. It is exactly the same today, or should be.

The covert hack was slightly different. This was the animal on which a man rode to the hunt while his hunters were hacked on at a steady jog by road. On his covert hack the hunting man tittuped across country at a fair pace, only getting onto his hunter at the meet. It was, of course, again essential that the horse should be thoroughly comfortable and well-mannered, but looks were of slightly less importance, while in conformation the covert hack was probably rather tougher than the more elegant park hack.

In considering the hack now one should never lose sight of its prototype, the hack of a hundred years ago. And so a judge today will look first and foremost for quality. The show hack should be a delight to look at and, of course, a pleasure to ride. The hack exhibitor is always striving to find the animal with these two attributes.

To me, personally, real quality means Thoroughbred. There can, perhaps, be a little Arab blood creeping in. Small hacks tend to have it more than large hacks, because obviously it is more difficult to find a small Thoroughbred horse. Often pony blood is also present, but on the whole one finds that the champions are pure Thoroughbred with just a touch of Arab.

The name of one consistently successful exhibitor that springs to mind is Miss de Beaumont, who has had such a famous line of hacks, all bred by herself. First there was June, then Honeysuckle, Juniper, Jupiter, all by the same stallion, Basa. The last in the line was Last Waltz, a son of Honeysuckle. All these have dominated the hack world, and each had two outstanding qualities: superb action and movement, and wonderful self-balance. In addition they each had a marvellous front, with exceptionally well set on head and neck.

If one stops for a moment to consider these priorities one realises, I think, that it is these that are essential when one is setting out to buy a hack. But how extremely difficult to find horses with these qualities, especially in the Thoroughbred, when most of the best horses are bred for racing.

Once one has bought a hack, young and with good potential, the next step is to produce it.

The production of a hack is an extremely specialised art, because although one might have the best-looking horse in the class it does not follow that it will win if it does not go correctly, or if it does not give the judge a good ride. To produce a hack is a more specialised job than producing either a hunter or a pony, as perfect manners and a good ride for the judge are so absolutely essential.

A charming example of a show hack.

An informal race in progress in the Cook Islands in the South Pacific.

Magnificently attired, tribesmen from all over northern Nigeria gather for the *Durbar* at Kaduna.

A *Csikó* horseman of Hungary.

Lippizaners from the Spanish Riding School in Vienna giving a display of *haute école*.

Suppleness and obedience mark the quality hack. This Anglo-Arab became a successful dressage horse.

Recently there has been a most encouraging increase in the popularity of hacks. A few years ago they seemed to be dwindling. Many people thought that the hack class was being replaced by dressage, and that the dressage-trained horse would soon swamp the hack altogether.

In fact, there is a great difference between the two. The hack may, today, be considered a little old-fashioned, but basically it is the elegant riding horse, to be enjoyed both as a ride and in appearance. What more can one ask? It is really not surprising that each year it is increasingly popular.

In the production of a hack showmanship is all-important: a judge will always be influenced by first-class showmanship. Not surprisingly, therefore, the history of the hack is full of the names of great artists who brought the production and presentation of a hack back to a fine art: Major Harry Faudel-Phillips; Mrs Dinah Kent, who as Dinah Heasman used to ride horses produced by Harry Faudel-Phillips in the Olympic days; the one and only Sam Marsh; Count Robert Orsich–has there ever been a more elegant showman? Harry Tatlow and now his remarkable, versatile son David; Jane McHugh, Dinah Kent's daughter; Jenny Lauriston-Clarke, who as Jenny Bullen was such an outstanding rider of children's ponies; Irene MacIntosh.

To watch these great artists in the ring was an inspiration. Nor would it be an exaggeration to say that, because of their artistry, many far from outstanding horses have stood at the top of the line. But then that is what showing is all about–certainly in the world of hacks.

Second only to the presentation is, of course, the ride: particularly the ride from the judge's point of view. The judge looks for a well-balanced, free-going ride, and perfect manners. As horses can change from day to day it is up to the rider or trainer to see that their exhibit keeps up to the mark. All too quickly a hack, of all show horses, can deteriorate in the way it goes and the ride it gives. There can be no relaxing if one is to stay at the top.

Finally a word about the show–that is, the individual show that the judge will ask each competitor to give. Often there is controversy over what is required. Most judges are satisfied with a simple show–walk, trot, canter, a couple of simple changes at the canter, an extended canter, perhaps, halt, rein back and then stand still.

Such a show, carried out smoothly and without fuss, is much more attractive and indeed more impressive than a niggly show where half-passes and so on are attempted, often incorrectly.

There is only time for a short show. Often this is a problem: but done correctly, with forethought and practice, even a couple of minutes is long enough to show that your hack is elegant, moves well, is well-balanced and has good manners. And this is exactly what a hack should be.

THE COB

In days gone by the cob was as much used in harness as under saddle, and was ideally suited to this dual role. Essentially, the cob was the mount of heavy, elderly riders who wanted a comfortable ride without any fireworks and who appreciated his lack of inches when mounting and dismounting. This does not mean that a cob is necessarily a slug. In the show ring he is expected to walk, trot, canter and gallop, and much emphasis is laid on his ability to cover the ground comfortably and expeditiously at all these paces. Very necessary attributes of the cob, however, are his temperament and good manners. He should be able to gallop flat out without hotting up and at all times he should come easily back to hand. Most cobs have equable temperaments, are inherently sensible, and nearly all are great characters. They are in fact intensely interesting horses, generally having exceptionally good balance and being anything but stodgy rides, even though they are often classed as 'confidential' and designated by clever dealers as 'gentlemen's gentlemen'.

The Welsh Cobs are an established breed; there are, however, many cobs that are not Welsh but are nonetheless of a well-defined type. The word cob refers to a strong, stocky animal, large in the body and carrying himself on powerful, short legs. Ideally, the head should show as much 'quality' (fineness) as possible and be well set on to an elegant but powerful neck. The body must be of exceptional depth at the girth and the back short and strong. In height the cob does not exceed 15·3 hands, though the limit set in the British show ring is two inches less than that, i.e. 15·1 hands. In the show ring it is also stipulated that cobs should be capable of carrying 14 stone.

At one time the best sort of cobs came from Ireland and, indeed, some very good ones are still to be found there. In general these were the progeny of Irish Draught and vanner mares put to a Thoroughbred horse or polo pony, and the majority fulfilled the dictum 'a head like a ladies' maid and a bottom like a cook'. Otherwise, riding cobs may come from all sorts of liaisons. Welsh Cobs are frequently used as a foundation, being put to Thoroughbred riding pony sires and even to Arabs. Similarly, a good cob can be got from Highland pony mares.

Until the nicking and docking of tails was stopped by Act of Parliament in 1949, cobs were always shown docked in the ring, and very jaunty they looked. However, it was a cruel practice and today cobs are shown with full tails but always, of course, with the hogged mane which suits so well the particular conformation of the cob's neck.

Today, when there is no demand, or very little, for harness cobs of the sort that were once a familiar sight in the big cities drawing milk floats and tradesmen's vans, really good riding cobs are not so easily found; even so, good classes can still be seen at major British shows.

The cob–sturdy and reliable.

A demonstration of the extravagant, showy action for which the hackney is famed.

THE HACKNEY

A good deal is known about the origin of the modern Hackney, but it would be extremely difficult for anyone to attempt to decide exactly how the breed was evolved. It can, however, be proved that the ancestry of the Hackney on the sire's side dates back to the Darley Arabian and about 1704, though horses known as Hackneys were used in this country for centuries before the Darley Arabian arrived on these shores, and the trotting horse, as distinguished from the ambler, pacer or the galloper, was recognised as a valuable breed. These trotting horses were probably improved by animals brought in at the time of the Danish invasion, as in those days the Scandinavian horses were celebrated for their trotting action.

By the nineteenth century the Hackney had become firmly established as a riding horse and pack horse, being the means of transport used by many farmers, and it was as a riding horse that the first Hackney made its name.

The introduction of the railway system reduced the demand for the Hackney as a riding horse, as it became customary for farmers to use the railway to go to market. But in spite of this, its popularity as a harness horse increased. The improved roads and much improved carriages in the second half of the nineteenth century demanded a lighter, faster type of horse than was used up to that time and the Hackney soon became the most fashionable harness horse as it combined good looks, action and courage.

The popularity of the Hackney encouraged a number of breeders to establish a register of the breed and, as a result, the Hackney Horse Society was formed in 1883. The first stud book was printed in 1884, at a time when the breed was beginning to attract increasing notice from horse breeders far outside its native eastern England. At that time there was a rapidly growing demand for light horses for many purposes and Hackney stallions were particularly sought by overseas breeders to produce the types of horses most needed.

By the turn of the century the Hackney had firmly established itself in Britain and abroad as the finest harness horse in the world, and has since maintained its world-wide popularity. This overseas demand for Hackneys for breeding purposes reached a peak in about 1906, when the Society's records show that horses were shipped to many foreign countries, with Canada and the United States taking the largest number. Today there is a considerable market for Hackneys to Argentina, Australia, Brazil, Canada, Denmark, France, Holland, Italy, Japan, Mexico, New Zealand, Portugal, South Africa, Spain and the United States.

The Hackney Horse Society has guarded jealously the interests of the breed and its type. This and other contributing factors have resulted in the development of this unique harness horse which has a reputation reaching far beyond the British Isles.

Here is a brief description of the breed. The Hackney should have a small convex head, with a small muzzle; large eyes and small ears; longish and well-formed neck; powerful shoulders and low withers; compact body with great depth of chest; tail set and carried high; short legs and strong hocks, well let down; well-shaped feet; fine silky coat. The most usual colours are dark brown, black bay and chestnut.

Both in action and at rest the Hackney has highly distinctive and readily observable characteristics. Shoulder action is free and progressive, with a high, ground-covering knee action, the foreleg being well thrown forward, not just up and down. Action of the hind legs is the same to a lesser degree. In a good Hackney the action must be straight and true. At rest the Hackney stands firm and foursquare, forelegs straight, hind legs well back, so that it covers the maximum of ground; the head is held high, ears pricked, with a general impression of alertness.

With its extremely high action and almost volcanic personality, the Hackney stands alone in the show ring as the personification of beauty and elegance. At the same time it continues to prove its stamina, courage and versatility. It is outstanding for private driving purposes, as can be seen in many of the marathons organised by the British Driving Society, and today Hackneys compete with great success in the new FEI Combined Driving Competitions, which consist of a dressage test, a marathon of well over twenty miles and an obstacle driving test.

RIDING PONIES

Working hunter pony classes test a pony's all-round usefulness.

The high standard of English ponies is known the world over, and what is more no other country in the world breeds such a great variety of ponies. Not only do we have native breeds with outstanding characteristics, but also the development of the crossed strains of riding pony. A cross between one of the native pony breeds and a small Thoroughbred or Arab sire will, at best, produce a superlative pony, full of quality while retaining its native hardiness, intelligence and pony character.

Buyers from overseas are on the increase each year, mainly for the pure native breeds, the most popular definitely being the Welsh Mountain. Large Welsh pony studs have been started in Australia, America, Canada, South Africa, Holland, Sweden, Belgium, Germany and elsewhere, and all obtain their foundation stock from the leading English studs. All this overseas trade has encouraged more people to start a pony stud; prices have been good for the good ponies and there does always seem to be an excellent trade for the best. Over the last few years, however, I think this country has reached a bursting point and now people have been made to stop and think more carefully.

There are more Welsh Mountain stallions and Arabs than any other type standing in England. This is probably due to the fact that these are the two most popular breeds for the overseas buyer.

The National Pony Society, founded in 1893, is the oldest of the pony societies. Its job is to look after the registrations of the mares, stallions and youngstock of all the native British breeds. A great job they do, with the individual breed societies taking care of their respective stud books. The British Show Pony Society, founded in 1949, looks after the ridden ponies. They have greatly lifted the standard of these and have increased their popularity over the years. Most shows in Britain that have pony classes are now affiliated to the BSPS, thus ensuring a panel judge and encouraging all ponies to become registered.

Last year they started working pony classes for ponies of 13 hands, 14 hands and 15 hands. These classes are catching on fast in popularity: children enjoy them and they do encourage the type of pony a child can really have fun on – a good performer rather than a pony that just looks beautiful.

In 1952 Mrs Glenda Spooner, with her great love of ponies and genius for organising, started the now famous Ponies of Britain Club. This, in my opinion, gave the pony breeding world its biggest boost and must be largely responsible for the fantastic boom in ponies. If one looks back before 1952 there were very few shows other than the NPS Show that put on breeding classes. We now take for granted the inclusion in a schedule of plenty of classes for youngstock.

Pollyanna, a famous show pony of the 1960s and perhaps one of the most beautiful ponies ever bred.

There is no doubt that the enthusiasm Mrs Spooner created by introducing her three club shows encouraged people to breed and improve our English ponies, so demanding more and more shows to put breeding classes in their schedules. Her stallion show was the first of its kind, held purely for pony stallions at the beginning of the summer, when people are busy looking for suitable sires for their mares. Then there are two summer shows, one in the north and one in the centre of the country, each with classes to suit all members of the family. There is no denying that pony people owe a great deal to Mrs Spooner for all her hard work in helping to raise the standard of the English pony.

Riding in itself must also be mentioned as responsible for the increasing demand for ponies. It is now – to soccer – the second most popular sport in England. It is helped by television and also by the Pony Club, which is tremendously beneficial to children and encourages many to take up riding.

An exquisite pony yearling with a great future.

The kind of people who breed ponies are interesting. They can be put into four categories. There are large studs run by experienced people; the experienced person who runs a smaller stud; the individual who breeds a few ponies; and last of all those who like to do the 'in' thing and send their perhaps unrideable mare to stud as they don't know what else to do with her.

Can it go on? Pony breeding over the last ten years has almost been overdone. One feels there are too many ponies about – not enough good ones, but too many indifferent ones. The first two types of my pony breeders one need not worry too much about, but the last two are perhaps responsible for swamping the market with the wrong type of pony. I am not now thinking of a quality show pony, but of an ordinary child's pony.

The NPS is doing all it can to control the licensing of pony stallions and, indeed, of mares. This is the fundamental point in our future pony. Too many people, bitten with breeding, have used untried, unshown and badly bred two-year-old colts, thinking that doing so saves money. This is what must stop, for it is spoiling the market for people who carefully plan their mares' breeding programme. People are finding it quite hard to sell their youngstock, having thought that the whole thing was too easy.

I do also feel, though, that people must be prepared to pay far more than they do now for the ordinary young pony with potential for the Pony Club, jumping or eventing. The cost of getting a mare in foal, her keep and the foal's keep until it is two or three years old, is quite considerable. As things are now one has to turn out several champions to begin to cover costs. Breeding is a fascinating hobby and most people do it for the pure love of it – but at the moment the breeder tends to be the person who comes off worst. It is most often the middleman who asks for large prices. Prices are getting out of proportion and are, I feel, bad for the future of ponies. The top ponies must be worth a good deal of money as they are hard to find.

If the boom for ponies is to continue, we must not forget that ponies are meant to be for children; secondly, the breeding aspect must be given far more careful thought by owners. One must think carefully before keeping a colt, and before putting a mare in foal be quite sure that she is good enough – one must know just what one is hoping to get.

The standard of our English pony is extraordinarily high, and any overseas person visiting England for the first time is usually – and highly – amazed. It is essential that the standard is not allowed to deteriorate.

Chapter six

Ireland

Ireland is a country which can be justly proud of its horses. The Irish-bred Thoroughbred, the Irish hunter, the Irish Draught horse and the Connemara pony together provide the country with a major product for export to all parts of the world, worth many hundreds of thousands of pounds.

The influence of the Irish Thoroughbred on racing stock is enormous. Nearly all the big overseas bloodstock breeders rely on the importation of European-bred stallions to maintain and improve their stock. Many of these horses are of Irish origin. Perhaps one of the most famous of these is the horse Nasrullah, whose influence in North America has been quite outstanding during the last twenty years. Breeders in South America, Australia, South Africa and recently in Japan are also major importers of top-class breeding stock, while there are of course many Irish-bred stallions standing in Europe.

Because of the favourable conditions, many of the greatest names in the bloodstock world own stud farms in Ireland, and there are also many old-established farms which have been in the same hands for several generations. The big Thoroughbred studs are mostly grouped in either the Golden Vale area of Tipperary and in adjoining Co. Limerick, or close to the Curragh in Co. Kildare and nearby Co. Meath. In the breeding areas the limestone qualities in the water and pastures make the young horses develop good bone and substance, while the mild climate also has its advantages. This environment is of course as advantageous for the Irish hunter as it is for the Thoroughbred.

It is fair to say that the Thoroughbred horse suitable for flat racing has become an international breed, with the recent influx of French and American bloodlines. However, the Irish-bred steeplechase horse is definitely a speciality. These horses are bred from families of mares whose ancestors have for many generations bred 'chasers. They are bigger and slower to mature than their flat racing cousins, and are more usually bred by farmers and breeders with just a few mares. The young horses are allowed to grow and mature until they are about three or four years old before being broken in. They still gain much of their early experience in the hunting field.

In the pedigrees of outstanding steeplechase horses, Irish blood is very frequently found. The famous Arkle, for example, was sired by an Irish stallion out of a dam of good 'chasing stock.

The Grand National is something of an Irish preserve. It is a race particularly suited to these bold, courageous stayers, whose balance and athletic ability has been bred into them over generations. The three biggest events in many Irish horsemen's calendar would undoubtedly be the Grand National, the National Hunt meeting at Cheltenham and the Dublin Horse Show.

An Irish hunter at the Dublin Summer Show.

The Irish hunter, or Irish half-bred, also has a solid claim to international fame. The demand for high-class horses is increasing with the growing popularity all over the world of show-jumping and three-day events. This change of emphasis, from the days when the hunting field and the mounted armies of Europe were the biggest outlet for half-bred horses, has been a tremendous asset to breeders. Irish hunters are still, of course, to be found in the hunting field both in England and in the United States, but it is with horses bought for the competitive sports that the biggest demand–and the most money–is now to be found. Ambassador, the German-owned grey show-jumper, winner of the gold medal at the Munich Olympics, is just one example of an Irish half-bred at the top. Ambassador's dam must carry a good deal of Irish Draught blood. This is the breed which puts the bone into the weight-carrying hunters. They have quiet temperaments, tremendous limbs and practically no feather, and above all are noted for their wonderful action. There are unfortunately not all that many pure-bred Irish Draught horses to be found now, as an enormous number were exported earlier this century, but the Irish government is doing what it can to stimulate the breed, particularly with a view to preserving and increasing the number of good-quality mares suitable for breeding. The best cross for the half-bred is to use a Thoroughbred sire on an Irish Draught mare, which is one reason why the shortage of Irish Draught horses is so badly felt.

For three-day eventing and in fast hunting country a horse carrying three-quarters Thoroughbred blood is generally preferred, for while the Irish Draught has unbeatable weight-carrying ability, marvellous action and staying power, a greater proportion of Thoroughbred blood does enhance quality and produce a faster horse. These horses have dominated the Olympic scene for some time. The Italians have always been good customers for Irish horses, and their methods of training suit these high-couraged horses. Recently, the Swiss, Dutch, French and Spanish have also been big buyers, and there is always a considerable number of high-class hunter-type horses exported each year to the United States, particularly to Virginia and Maryland.

The annual show held by the Royal Dublin Society in August is the shop window for Irish-bred horses, and has for a long time been considered the best place to find potential champions. In particular, the youngstock classes are always hotly contested, and winners will fetch very high prices at subsequent sales.

There is one native breed of pony in Ireland, the Connemara. As its name suggests, it comes from the area of western Ireland known as Connemara. These ponies are a very old breed and a very hardy one. They should stand under 14 hands and are usually dun-coloured or grey, though occasional brown ponies are to be found. The mountains of Galway and Mayo have ensured that they are remarkably sure-footed and with an ability to survive on very little food. They also show a marked ability to jump. There is a very active breed society for Connemara, which maintains records and a stud book. Connemaras are primarily used as children's ponies, though they are also used for farm work. Breeding experiments to improve the quality of the breed by crossing with small Thoroughbreds and Arabs have been notably successful.

Because of the importance to Ireland of its horses, there are several bodies responsible for maintaining standards and encouraging breeders. The Department of Agriculture looks after the allocation of premiums for qualified mares and also owns a number of Thoroughbred stallions, which are located throughout the country. The Irish National Stud at the Curragh is a state-owned Thoroughbred stud, where a number of very high-class Thoroughbred stallions stand at reasonable fees. The Irish Horse Board is the co-ordinating body for the various breed societies and specialists, and promotes equestrian teams to represent Ireland abroad; the Army Equitation School produces the Irish Army jumping team; the Royal Dublin Society holds a position of importance; and there are associations catering for both show-jumping and eventing.

Chapter seven

The Americas

NORTH AMERICA

Man and his progress through the centuries owe much to the horse. It is on his back that armies travelled; that conquests were made; that different cultures and civilisations met and mingled. The horse has been valued as friend and companion; used as transport in quests for new territory; even revered as a god by the Aztecs when, in the sixteenth century, Cortez and his *Conquistadores* rode into old Mexico bringing Spanish dominion with them, destroying old cultures and imposing their rule and religion on the New World of the Americas. Without the horse, the *Conquistadores*' ravages into the interior would have been well-nigh impossible, a fact fully realised by Cortez when preparing for his expedition. Sixteen horses, eleven stallions and five mares, left Spain for the New World. It is from the survivors of this band, and the many that followed, that the vast herds of America grew, and from the subsequent importations of bloodstock from England, Europe and the East, that the breeds of America took shape.

The Quarter Horse

Formally recognised as a breed in 1941, the Quarter Horse has a long history dating back to Colonial days in America, when horses descended from old Spanish stock were crossed with blood imports from England. Among the most famous of these was a Thoroughbred imported in 1756. Named Janus, he was a grandson of the Godolphin Arabian, and is considered the ancient foundation sire of the present Quarter Horse breed. Today his name figures as head of nine of the twenty-four main 'families' of Quarter Horse. However, the horse who holds the distinction of being the first registered Quarter Horse is Wimpy, Grand Champion Stallion at the 1941 Southwestern Exposition and Fat Stock Show in Fort Worth, Texas.

Now to be found all over the USA, in Colonial days the breed was centred mainly in Virginia and the Carolinas, and it was here the breed name originated. Freed from weekday work, farmers, land and plantation owners indulged in their favourite Sunday pastime—match racing their fleetest horses down the Southern towns' main streets. It was a rare town that could boast a street longer than a quarter of a mile, and from this the stocky, fast-starting little horses took their name.

Naturally, those fortunate in possessing match race winners found their horses in demand as breeding stock, thus ensuring transmittal of the Quarter Horse's prime asset—early speed. Nowadays, through selective breeding, the Quarter Horse is the fastest horse on earth over his distance, the quarter-mile.

However, speed alone is not the only factor ensuring his popularity. He has a more than equable disposition, is generally easily handled and broken, and when trained lives up to the typical but descriptive American phrase of 'a kid-broke horse'. He is also rated supreme as a working cowhorse, combining intelligence with muscular build and superb conformation that enables him to perform with ease manoeuvres calling for incredible agility. Watch a Quarter Horse leap into a full gallop from a standing start—the muscles in those powerful hindquarters bunch as he explodes into action. Or as he slides to a stop try discerning the outline of his quarters from the dustcloud spurting around him—you can't. Yet again, nothing is more agile than the Quarter Horse cutting cattle. He snakes in low, mind and body tuned fine, watching, out-thinking, out-flanking each and every move of the calf.

The average Quarter Horse stands at between 14·3 and 15·1 hands, and weighs between 1100 and 1300 pounds at maturity, which with this breed tends to be somewhat earlier than with some others. A good Quarter Horse should possess a short head with small muzzle, foxy ears and the well-developed jaw so characteristic of the breed. His neck is of medium length, joining the shoulder at an angle of approximately 45 degrees, neither heavily crested nor too light of muscle. Good shoulders slope also at about 45 degrees, making for a smooth ride. The chest is wide and deep, allowing for the generous heart room so vital for this active horse. Forelegs are heavily muscled but without coarseness. The back is short-coupled and very strong, especially across the loins. Ribs are well sprung and very deep through the girth. Hindquarters are very broad and heavy with powerful muscles, especially in the thigh and gaskin. Unlike other breeds, who tend to narrow here, the Quarter Horse's stifle is wider than the hips. It is from these tremendously powerful quarters that he gets his quick acceleration and the ability to stop rapidly, throwing all his weight onto his hindquarters. Hocks are well let down, and cannon bones short. Medium length pasterns and hooves are set at an angle of 45 degrees which, coupled with a good shoulder, increases the horse's durability as leg concussion is thereby minimised.

Within the breed there are distinct types, raised with a view to the work they will be called upon to perform, yet every good Quarter Horse retains the breed's hallmark of muscular strength coupled with refinement. Hence we find that the stockier 'bulldogger' type, for all his muscle and greater body size and weight, retains the refinement so apparent in his elegant brother, the 'racing' Quarter Horse, and the agility of a cutting horse. In this sphere, where inborn cow-sense and agility are called for, the Quarter Horse reigns supreme, and it is in racing that he shows his lightning burst of speed, scorching the track with fantastic twenty-second times for the quarter-mile.

Year by year the number of horses registered outnumbers that of all other breeds in the USA. The uses to which the Quarter Horse is put are legion, but first and foremost he excels as a working horse, having taken the highest honours wherever he competes in cutting and roping. He is also in great demand in every other Western sphere. Throughout the USA there are Quarter Horse shows, where the cream of the breed competes for points in pleasure, reining, trail, barrel racing, pole bending and working cowhorse classes. In addition he is also used under English saddle in pleasure, jumping, working hunter and polo classes. At the year's end the horses accumulating the highest number of points are awarded the American Quarter Horse Association's Championships and Register of Merits. Quarter Horses also figure largely and successfully in open shows, where many breeds show in the same classes.

Since the Quarter Horse is by far and away America's most popular breed, being versatile, easy-going and tough, horsemen in other countries have become eager to buy that blood. Exports of Quarter Horses have been made to Canada, Old Mexico, South America, South Africa, Australia and England. In particular there is a big trade in Quarter Horses with Australia, with England serving as a quarantine base for horses in transit. Many Quarter Horses have found homes in England, where the breed is currently gaining favour. Thus the Quarter Horse cannot only claim numerical superiority in his country of origin, but now also ranks as a truly international breed.

The well-developed muscles and powerful hindquarters of the Quarter Horse enable it to change speed and direction with particular agility.

The Paint Horse

Closely allied to the Quarter Horse is his colourful cousin, the American Paint. Formed as recently as 1962, the American Paint Horse Association is dedicated to upgrading the quality of this increasingly popular breed. From the initial year's 250 registrations numbers have now leapt past the 22,000 mark, amply substantiating the APHA's claim to be the fastest growing breed registry.

Not to be confused with the Pinto Horse Association, which incorporates coloured horses of all breeds and types, the APHA was formed solely to further the interests of the Western Stock type of horse. To ensure consistent top quality and conformation any horse up for registration must have been sired by either an APHA stallion, a registered Quarter Horse or a Thoroughbred. Paint mares and geldings foaled prior to 1964 may be registered even though of unknown breeding. However, after 31 December 1969 the registry became 'closed', and all horses foaled after that date must have at least one parent in the regular or breeding stock registry, the other being either a registered Paint, a Quarter Horse, or a Thoroughbred. The Association's policy of inspecting stallions for soundness and conformation at two years of age is steadily upgrading the quality of the American Paint. Any stallion rejected by the Association inspectors cannot be included in the Regular Registry, and his foals are therefore ineligible for registry. By these stringent rules the APHA has been able in a remarkably short time to fix the standards sufficiently so that future generations will conform to a breed as well as a colour type. Consequently Paints, from often being denigrated in the past when many of them were either coarse or weedy with poor conformation, have made giant strides in the equine industry.

Other rules of the Association are that in order to be eligible for registry animals must be a minimum of 14 hands. Gaited horses that 'singlefoot' or pace are not eligible, whereas a foxtrot or running walk in a horse is acceptable. However, the greater majority of registered horses are not gaited in any way, performing only at the ordinary walk, trot and canter.

Modelled along stockhorse lines, and because of heavy infusions of Quarter Horse blood, the ideal conformation of the Paint Horse is closely allied to that of the Quarter Horse. The Paint breed falls into two basic categories of colour pattern: the Tobiano – a predominantly white animal with large patches of colour, especially on head, flanks and chest, with white often spreading unbroken across the back; and the Overo – a solid-coloured horse with white splashed mainly on the mid-section of his body, and rarely with white spreading across the back. The Overo's legs are frequently dark-coloured, and in contrast to the Tobiano he will often be bald-, apron- or bonnet-faced according to the amount of white. Overos frequently lack the extra clear-cut demarcation lines between colours that are to be found in Tobianos – there is often a blurring of shades where the colours meet. In addition to the white, the accepted body colours of Paints are black, bay, sorrel, dun, palomino and roan. Some Paint characteristics are blue eyes, white above the hocks and knees, variegated manes and tails, a pink skin under the white hair, and a bluish tinge where the dark and light skins meet.

The APHA has improved and stabilised the characteristics of the American Paint in a remarkably short time.

In breeding for colour Paint genes are very strong, and it is unusual, though not impossible, to have a solid colour foal where either or both parents are true Paints. I once owned a very strongly marked Tobiano Paint mare who, when put to my chestnut Arabian, Royal Command, produced a solid bay filly. The next mating between these two resulted in a Tobiano paint colt who looked exactly like his dam. Should a solid colour foal result, it is highly likely that foals bred from this will in turn be coloured, thus puzzling an owner if he has no knowledge of his horse's ancestry. A solid-colour foal from Paint parents is eligible for the Breeding Stock Registry of the APHA, but it may not compete for any Paint horse awards given at breed shows.

One facet of the breed is that no horse is so highly individualistic as a Paint. Though they will be either Tobiano or Overo, no two horses carry the same markings, each animal being distinguished by his own Paint pattern–a real safeguard these days when horse-stealing still happens, as reference to horse journals proves.

One of the reasons for the breed's rapid growth in America is the Paint's versatility. Predominantly a Western horse, he figures largely in open shows where he is certainly an eyeful of colour. With his legacy from Quarter Horse blood he competes on level terms in the many events open to Western horses–pleasure, reining, roping, cutting, speed events, trail–as well as the English divisions in jumping and pleasure. He also earns his share of the prize money in the action-packed rodeo arena, where it is not the style he displays that counts but the sheer speed and co-operation with which he places his rider ready for a throw in calf-roping, or moves up close to the steer for the bulldogger rider.

Currently gaining favour and becoming more numerous and valuable are races for Paint horses. These are run on the same lines as the Quarter Horse races, and the infusion of Thoroughbred blood lends enough speed to make the colours blur down the straightway.

The Standardbred

The foundation stock of many American breeds was imported from Britain, and the Standardbred breed is no exception. It was a grey Thoroughbred stallion named Messenger, foaled in 1780 and imported into the United States in 1788, who was the breed's foundation sire.

In those days, particularly in the eastern states, harness racing was popular at country fairs. It was at these gatherings that owners with fast horses competed. In the early days race winners were determined by holding several heats, the horse winning most being the overall winner. This tough method resulted in the elimination of weaker animals, only the horses 'in the money' being rated best breeding stock. Although Messenger was used extensively and successfully as a sire of Thoroughbreds, it was when crossed on these harness horses that the ability to trot at racing speed resulted in his off-spring. From these small beginnings the Standardbred breed took hold, receiving a tremendous impetus with Messenger's famous greatgrandson Rysdyk's Hambletonian, who gained such lasting fame as a racer that the name Hambletonian became synonymous with Standardbred. All present-day Standardbreds trace back to Rysdyk's Hambletonian, and through him to Messenger.

The breed name originated because when the breed type was being fixed and the American Trotting Register was started in 1871, horses were accepted for permanent registry only if they could trot or pace a mile in a standard time. For the trot it was 2·30 minutes, and for the pace 2·25 minutes. However, the first recorded trotted mile under that speed was made in 1845 by the mare Lady Suffolk, in 2·29½ minutes. As the years passed and Standardbreds, through selective breeding for speed, became faster, the time for the mile diminished.

One of the greatest of all Standardbreds was Greyhound, a grey gelding who, in 1938, became the fastest trotter, setting a record which held for many years with a time of 1·55¼. Recently the star speedster, with a new mile record of 1·54⅕ was Nevele Pride. Pacers, too, reach fantastic speeds, and of the two gaits the

This Standardbred mare, ridden by the author of this chapter, has just completed one of the gruelling endurance rides held in the United States.

pace just slightly edges the trot in the record tables, the fastest ever recorded being 1·52 by Steady Star. For identification purposes all horses racing are now tatooed.

At first glance the Standardbred horse reminds one of the English Thoroughbred, to whom he owes so much, but comparing the two one sees that the Standardbred is a heavier-boned animal, longer in the body and with a flatter rib-cage, and quarters sloping to a greater degree. The muscles of the quarters, particularly in the thigh, are powerful and long, and this, combined with the extra slope to the quarters, gives the Standardbred such impetus in his racing gait. In keeping with the heavier bone, the head is not so refined as that of a Thoroughbred, often being straight or very slightly convex, but with a good generous eye. The nostrils are capable of immense flaring, necessary for oxygen intake when at speed. The ears tend to be rather on the long side. The neck is somewhat shorter and straighter. The most predominant colour is bay, followed by browns, chestnuts, blacks and a few greys. The manes and tails grow profusely and very long, with the mane being extremely fine in texture. The average height ranges between 15 and 15·3 hands, though some horses go over 16 hands.

Although Standardbreds started their career as country fair racers, going on to increase in numbers and achieve a phenomenal popularity after the second world war on recognised race-tracks mainly in the eastern states, Pennsylvania, and on the west coast, they also appear in many other guises. Top-class shows always include classes for roadsters, which are 'driven to bike' and with the driver wearing racing colours. The winner here is determined by several factors, speed being only one, albeit a very important one. In championship classes the horses are judged on performance, speed, presence and manners, and will show at jog trot and at speed. When the drivers 'turn them on' they immediately flash out in long, ground-eating strides, excitement at the ringside building as the speed increases. It needs a skilled driver to handle horses going at near-racing speed, especially on tight corners, considering that show arenas are almost one constant bend. Their manners are really impressive: when the announcer calls 'jog them down', they decelerate with amazing ease. Most horses travelling at speed take a while to slow down but I have never seen a show roadster pull against restraint.

I lived for many years in North Carolina, fifty miles from Wilmington, where several trainers had winter quarters for their trotters and pacers. Upwards of a hundred horses were stabled, coming south when the cold weather set in up north. In the mild southern winter the horses could maintain condition and be fully prepared and ready for the spring season's opening at the northern tracks. Most grooms had charge of three horses, and it was during the winter that the young trotters and pacers were tried and tested. Those not making the grade were sold, and I did buy one brown gelding off the track for a fraction of what his winter training bill amounted to. He made a good saddle horse once over the transition from harness to saddle, though at first he alarmed himself when cantering, doing so in a manner often shown in old-fashioned prints.

In order to facilitate speed the training track was of an exceptionally hard composition, mainly baked clay, and it did not surprise me to find that leg ailments among Standardbreds seemed more prevalent than among Thoroughbreds.

Many of the horses not fast enough for the track returned to Pennsylvania, where there was a ready market amongst the Amish people, or as they are also known the Pennsylvania Dutch. These farming people, whose religion forbids them to use motor-driven machinery, still drive the old-fashioned buggies. They take great pride in their horses, being only too pleased to have the fine-blooded quality of a Standardbred in their shafts.

Although used primarily as harness horses, many Standardbreds make excellent mounts. I have owned three, all bays, two pacer geldings and one trotting mare, which I still have with me in England. This mare, Magnet Regent, by Prince Regent out of Red Magic, bred by Harvey B. Hunter of Charlotte, North Carolina, I bought as a long yearling before she had gone to the track. Green-broken as a two-year-old, she was ridden lightly until the age of four. I showed her quite successfully as a pleasure horse under English tack and jumped her a little with moderate success, though I think the Standardbred conformation does not lend itself to jumping. As a six- and seven-year-old I tried her in endurance riding, at distances ranging from thirty to a hundred miles. Here the breed's known toughness paid off, as she was absolutely indefatigable, winning and being placed in many competitions. She once completed the Cape Fear Trail Riders Endurance Ride, of close on fifty-six miles, in four hours forty-six minutes, on a day with temperatures in the eighties. She even reverted to the old-time Standardbred forte of match racing, but under saddle rather than in harness. Since being in England she has hunted regularly, produced two filly foals, completed two British Horse Society Qualifying Rides for the Golden Horse-shoe, and is currently in foal again, to the same stallion, Nizzolan, a grey Arabian.

Having ridden both a Standardbred and Arabians on endurance rides, I believe that the cross should give the best possible horse for this activity. The Standardbred whose legs have not had to take a persistent hammering on unyielding going is equal to the Arabian for stamina. As far as temperament is concerned I find from my own experience that the Standardbred is honest, very keen and very stable, channelling all his energies into producing a first-rate effort and wasting none on flighty nonsense, as is often the case with other very highly-bred horses.

The Standardbred, while most numerous in the USA, is also found in Australia, South Africa, Rhodesia, New Zealand, Japan and several European countries. Britain, however, has very few, as harness racing is only supported by a tiny minority of horsemen.

The Walking Horse

America has developed many breeds, and one of the most distinctive of these is the Tennessee Walking Horse. It gained official status when leading breeders met in May 1935 and formed the Tennessee Walking Horse Breeders Association.

The beginnings of the Walking Horse breed trace back almost two centuries, when the rich farmlands of central Tennessee started attracting settlers who brought horses of Standardbred, Thoroughbred, Morgan and Saddlebred breeding with them. From crosses of these four breeds came the foundation stock of the present-day Walker. Each breed transmitted its prime asset – Morgan tractability; Saddlebred elegance; Thoroughbred quality; and the substance of the Standardbred, which also strongly influenced the development of the Walker's gaits.

In early days the Walker was bred primarily as an all-round working horse, equally at home in harness, under saddle, or earning his keep on the land, but it was as a saddle horse that he really excelled. Because of his remarkably smooth gaits, the Walker was in demand on the vast cotton and tobacco plantations of the southern states, where landowners and their overseers, who spent hours in the saddle, needed a horse capable of giving a supremely comfortable ride. Hand in hand with comfort went looks, as southerners took immense pride in the quality of the horseflesh they raised. Before long these smooth, gliding horses became known as 'plantation horses'.

Many prepotent sires figured in the breed's formation, but it was not until the early 1900s that the horse considered as the foundation sire of the modern Walking Horse began to have such an impact on the breed. Black Allan, a direct descendant of the great Standardbred, Hambletonian, was foaled in Lexington, Kentucky in 1886 and was originally intended for the trotting tracks. However, he showed a predilection to pace and remained unraced. Many of today's Walkers are inclined to pace or trot early in life, a tendency which is curtailed early on as the trainer induces the correct walking gaits. Black Allan was later sold to Mr J. R. Brantley of Manchester, Tennessee, and it was here that, among other good offspring, he sired three of the most famous foundation Walkers – the stallions Roan Allan and Hunter's Allan and the mare Merry Legs.

Some breeds have many characteristics in common, but the Walking Horse's conformation is unique and unmistakeable. Average height ranges from 15·2 to 16 hands, and his frame is very compact and full of substance.

The head is rather long, with the profile straight and with narrow pointed ears. Carried high, the head joins the neck at approximately 90 degrees. The neck is fairly long and powerful, moulding into a sloping shoulder that contributes so much to that smooth ride. The back is extremely short, while the quarters slope slightly. The profuse tail is carried high. The legs are fine, with the forelegs set slightly forward and the hind legs set with the hocks well away from the body. Colours are sorrel, chestnut, black, roan, white, bay-brown and occasionally yellow, and white markings are often very prominent.

The Walking Horse is synonymous with comfort. In all his gaits–gaits possessed by no other breed–he gives a ride completely devoid of any jarring. These gaits are the flatfoot walk, a true four-beat gait in which the horse seems to glide over the ground, all the while exhibiting a slight nod of the head; the running walk, a much faster and very spectacular version of the flat-foot walk in which the hind feet overstride the front–with exceptional Walkers by as much as fifty inches. The running walk is again accompanied by the nodding head, only now much more pronounced. The Walking Horse's canter, often referred to as a 'rocking chair' canter, is very distinctive, as the horse elevates his fore-hand in a rolling, forward motion, while the action of the hindquarters remains relatively level. The hind legs in all three gaits are carried well underneath the horse, creating impulsion and drive and also enabling the horse to achieve his remarkable smoothness. I have heard it said, and can quite believe it, that a good Walker in a canter can balance a glass of water on his quarters and not spill a drop.

When the Breed Association was formed in 1935, registration rules were drawn up. Research into bloodlines helped establish the foundation lines, and horses of known Walker ancestry were admitted to the Registry. Since 1947 the Walking Horse Registry has been 'closed', which means that both mares and stallions must be fully registered in order for their foals to be eligible. However, any gelding broken to ride that has 50 per cent Walking blood may be registered, provided he does true Walking Horse gaits. These rulings serve to preserve intact the bloodlines of producing animals.

Walking Horses come in two categories–show Walkers and pleasure Walkers. The show Walker is always shown with a set tail, and his gaits show more defined action, particularly in the running walk. To encourage greater action the front hooves are often weighted, mercury sometimes being inserted in the tip of the shoe, and the hooves themselves are built up with several layers of leather pads. In fact the horn of the hoof is pared very little in comparison with that of other breeds, and it is a matter of great concern to the owners of show Walkers if any part of this extra long hoof breaks away. A farrier shoeing Walking Horses has to be exceptionally skilled and the cost of shoeing a Walker is about twice that of ordinary saddle horses.

Pleasure Walkers are bred from the same stock but usually lack the potential of their more showy brothers. Their training is not taken to such a high degree, and though they perform in basically the same manner there is none of the high-powered drive and action of the show Walker. At all good-sized shows there are classes for show Walkers and others for pleasure Walkers. Show Walkers are invariably trained by professionals, sometimes being shown by their owners but more often by their trainers. Pleasure Walkers are usually owner-trained and ridden.

During the last week of August and the first week of September the top show Walkers in America converge on Shelbyville, Tennessee, for the annual Tennessee Walking Horse National Celebration. Here the world championship is decided, as well as classes being held for every category in which Walkers are shown. During these weeks there are also many auctions and sales of top Walkers in Shelbyville and on breeding farms nearby.

The Morgan

Durability and stamina are just two of the characteristics that hallmark the Morgan breed. Now rated as one of the foremost of America's many horses, the Morgan breed began in Springfield, Massachusetts, with the birth of a bay colt named Figure in 1789.

Most breeds evolve over a period of time, as

various horses, bloodlines and existing breeds are crossed and recrossed before a breed type is established, but with the Morgan it happened in just one generation–the time it took for Figure to get his first crop of foals on the ground. Coming into the ownership of the local schoolmaster, Justin Morgan, in Randolph, Vermont in 1793 in payment of a debt. Figure stood at stud to local mares as well as working as an all-round saddle, harness, racing and draught horse. In those early days much land still had to be cleared for homesteads, and Justin Morgan's tough bay stallion excelled in snaking heavy logs out from clearings. Today's Morgans still prove their strength in the show ring pulling a heavy stone boat–the only breed that has to show both as harness, show and draught horse.

Taking his new name from that of his owner, Justin Morgan's successes in match-racing, pulling contests, and as a trotter in harness, when pitted against all comers, meant he was in steady demand as a sire. Even when put to quite ordinary mares, the resulting foals were born inheriting their versatile sire's compact and very robust frame as well as his elegance and kind disposition. All these qualities, combined with inherited versatility, made him one of the most successful sires in history. Over the ensuing years, until Justin Morgan died at the ripe old age of thirty-two, he continued to stamp all his offspring in the same way, so that well before his death the breed had become established and very popular.

Justin Morgan's own breeding is something of a mystery. That of his dam is not recorded, but it is generally believed that his sire was the racehorse True Briton, a song of Traveller, who was imported from England. This would account for the strongly marked Arabian influence, as at that period the English Thoroughbred was closely allied to the desert stallions who greatly improved British bloodstock.

As well as being a famous breed in its own right, the Morgan has largely influenced the formation of three of America's best-known breeds–the Standardbred, the Saddlebred and the Tennessee Walking Horse. Many present-day Standardbreds have Morgan blood in their ancestry. The Saddlebred, too, owes much to this tough breed, gaining its short coupling and rounded barrel from Morgan conformation as well as its docility of temperament. The Tennessee Walking Horse has been directly and strongly influenced by the Morgan. Black Allan, the Tennessee Walking Horse foundation sire, was a direct descendant on his dam's side from Ethan Allan. This famous great-grandson of Justin Morgan was rated world champion trotter, sweeping all opposition out of the way in northern races.

Predominantly bay, the average Morgan stands between 14 and 15 hands. His conformation is robust but very refined–a rare combination. The head is small and dry, with neat ears and a large expressive eye. The neck is of medium length and carries a proud, heavily-

This Morgan stallion is being driven to a Meadowbrook cart in Rhode Island.

maned crest. Well sloped shoulders enhance his smooth ride. The back is short, and the barrel rounded and deep. Loins are strong and the croup level, with the long and profuse tail carried high. The limbs are very fine and strong. The general carriage of the Morgan is alert and gay. He is also a thrifty keeper – a great asset in these days of high-priced feedstuffs.

In the past the Morgan was used mainly as a utility or working horse on the farms, in the family buggy, and as an all-around saddle horse. Today he still shows his innate versatility, but more along the lines of the family pleasure horse, taking readily to any aspect of equine activity that is asked of him.

The Morgan Horse Association holds regular breed shows throughout the country. The main categories in which the Morgan is shown are as an English pleasure horse, a Western pleasure horse and a fine harness horse. Morgans also compete as roadsters, successfully utilising their great trotting capabilities; working hunters; jumpers; and equitation horses for junior riders. One event that shows off his elegance and fire to great advantage is the 'park horse' class, where the horse moves very collectedly, his gaits, particularly the trot, showing tremendous animation.

Outside the competitive atmosphere of the show ring the Morgan is the ideal family horse, having a docile but lively temperament and a size that makes him suitable for both children and adults.

In the modern day's horse world there is seldom any need to prove the endurance of our mounts, but many riders do enjoy the challenge of endurance riding and choose the Morgan for his proven toughness and durability. One I do recall was Soxaaral, a light bay Morgan/Arab cross mare who was on the 1967 Asheville, North Carolina, 100-mile competitive ride. Not only did she win the heavyweight division for her owner, Ira Gordon of Hunter, New York, but she also became Grand Champion over all the other entries, with a superb score of ninety-six out of a possible hundred fitness marks – a great achievement when one considers that judging is exceptionally stiff on American rides. A part-bred Morgan I used to board at livery in North Carolina for his owner, Mrs Diane Kallio, was Panic, a bay gelding by the Morgan stallion Panez. Panic was shown in English pleasure and working hunter classes at local shows, crowning a successful season by winning the open pleasure class for horses of hack type at the 1965 North Carolina State Championships. Morgans are also making quite an impact on one- and three-day events on the west Coast. Here the Morgan's powerful quarters are a real asset in jumping, hunter, and eventing competitions, and their lack of size, in comparison to the usual jumper of Thoroughbred breeding, is more than offset by their willingness and courage.

Set on the road to popularity by its prepotent founder, this very versatile breed gained swift recognition in the 1800s, followed by a tremendous increase in numbers, particularly in the northern and eastern states, although Morgans were to be found all over the country. Later on, they were used largely in New York when tram-cars relied on horses rather than horsepower. With mechanisation their use, and therefore the numbers bred, declined. However, many still see service in the north as police horses, as their tractability, courage and ease of handling make them ideal in the welter of city activity. Now that the horse is enjoying a popularity boom there is a resurgence of interest in the Morgan, his particular attraction being that he can honestly be said to be an all-round performer.

The American Saddlebred

Most of America's breeds of horses originated east of the Mississippi, and the American Saddlebred is no exception. Hailing from Kentucky, this elegant animal was yet another of the breeds developed by the settlers staking homesteads in the early days of American independence. As people drifted further west and deeper into the south looking for new land tracts to support the increasing population, their horses played a vital part in the success of their enterprises. Once settled and becoming prosperous, Kentuckians began evolving a new breed particularly suited to their own needs from their original stock of Thoroughbreds, Morgans, Standardbreds and Carolinian Naragansett Pacers. By crossing the best of these breeds the American Saddlebred was born.

The Saddlebred foundation sire was a Thoroughbred named Denmark, and most modern Saddlebreds trace back to him through Gaines Denmark, the result of a mating between Denmark and a pacing mare. It is from the Gaines Denmark line that the easy gaits evolved.

Combining the best of all the foundation breeds, the Saddlebred presents a picture of extreme refinement and elegance. One of America's taller breeds, he stands between 15·2 and 16·2 hands, with occasional individuals over or under this height. The predominant colour is a rich coppery chestnut, often accompanied by a flaxen mane and tail. Other common colours are bay, black and grey, and lately there have been a considerable number of palomino-coloured horses registered.

The typical Saddlebred is a close-coupled, proudly alert horse with high head and tail carriage. The head is refined, dry and rather narrow, topped by narrow mobile ears. The eyes, set wide apart, are extremely large and lustrous, enhancing his intelligent appearance. The neck is elegantly long, running into moderately high withers that hold a saddle well. Well sloped and powerful shoulders make it possible for the horse to achieve smoothness, while his short back and rounded barrel give strength. A level croup and rounded quarters give thrust to his gaits. Limbs are long and fine with tendons clearly defined. Mane and tail hair is silky textured and very fine.

Originally a utility horse destined for general farm and saddle work, his ability to perform at five distinct and smooth gaits combined with overall beauty made the Saddlebred a natural for the show ring when easier times meant horsemen could ride for relaxation instead of out of necessity. Consequently the modern Saddlebred, which is the product of intensive breeding to enhance all the finer qualities of gait and conformation, is primarily a show and pleasure horse. As such he is eminently successful, being shown as either a three-gaited, fine harness or five-gaited horse, as well as introducing children to the show ring via the equitation classes.

The three-gaited horse is shown at the walk, trot and canter, all gaits being collected and animated. His mane is hogged, and the tail set and also clipped about a foot down the dock, leaving a banner of hair at the tip. At one time, setting a horse's tail involved nicking the muscles at root of the dock, breaking the tail and resetting it in a crook. Nowadays a tailset harness can achieve the elegant carriage without resorting to the reprehensible methods of the past—also once used on hackneys and Walking Horses.

The five-gaited horse is shown with a full mane and tail, and his gaits are the walk, trot and canter, all very animated, and the slow gait and rack. These two gaits are true four-beat baits with the hind foot striking the ground slightly before the forefoot on the same side. The slow gait is a slower version of the rack, which is performed with great style and speed and is very thrilling to watch.

Fine harness horses are also shown with full mane and tail at the walk and animated trot, and it is in this class that many owners start their young horses' show careers. After being trimmed some fine harness horses appear as three-gaited entries. All Saddlebred equitation horses are three-gaited and are shown trimmed, and ridden by a junior, great value being placed on a reliable, safe horse with which a child can start his show career.

Pleasure Saddlebreds are shown with natural, unset tail. They perform at the walk, trot and canter, these pleasure gaits and their whole attitude being much more relaxed than the firecracker animation of the horses reserved strictly for high-powered show classes.

The natural gaits of walk, trot and canter do not need as much concentrated effort as the slow gait and the rack. These two, while inherent to a degree in every Saddlebred, must be man-induced by careful training, and the rack especially takes a tremendous amount out of a horse.

Another sphere where the Saddlebred shines is as a parade horse, where flamboyant high-stepping is a must to be in the ribbons. Parade horses, quite frequently flashy palominos, are of the heavier type of Saddlebred, and the breed's natural *élan*, coupled with high head carriage and high-stepping hooves really catch the eye as horses parade in their heavily-ornamental silvered saddles.

Although capable of tremendous collection and animation, coupled with the eagle-proud head carriage brought on by intensive training, the Saddlebred not destined for the three- and five-gaited show events remains a much more relaxed horse that is eminently suitable for the more normal facets of pleasure riding. He is also being used quite considerably as a jumper, though high head carriage and action is a definite handicap, and some Saddlebreds even find their niche as working ranch horses.

Though acknowledging Denmark as foundation sire, there have been many other famous Saddlebreds in the breed's history and a large percentage of the modern successful horses come from the powerful Stonewall, Bourbon, Anacacho, Peavine and Kalarama lines. The most illustrious of all modern Saddlebreds is Mr and Mrs van Lennep's Wing Commander, a five-gaited stallion who was six times world champion in the sixties, and has since sired innumerable five-gaited and fine harness winners, many world champions among them.

One of the most famous Saddlebred families of recent years is that of horses owned by Mr J. L. Hutcheson, an owner who gathers in the best of the breed. Among aristocrats at his stable in Rossville, Georgia are a previous five-gaited world champion stallion, Indiana Peavine, his five-gaited daughter, Lily Merril, a 1965 world champion, and her dam Ensign's Fair Virginia, who has produced many champions since retiring to the brood mare band. These three horses, when bought by Mr Hutcheson, fetched the astronomical sum of $106,000.

The Appaloosa gained its name from the Palouse river in the lands of the Nez Percé Indians. The breed, which is now becoming increasingly popular throughout the United States, is descended from Spanish stock. The Appaloosa Horse Club was formed in 1937 to promote recognition of this special type of spotted horse.

LATIN AMERICA

Although there are a considerable number of breeds of horses to be found across the length and breadth of Central and South America, all but a few of them are descendants of one original breed – the Andalusian of Spain, which arrived in Latin America as the mount of the *conquistadores*.

The characteristics held in common by many of these breeds are the distinctive *paso* gait and the proud bearing inherited from the original Andalusians. While the Peruvian Paso has been the breed to make the biggest impact outside Latin America (principally in the United States), there are also paso-type breeds in Puerto Rico, Colombia, Cuba, Venezuela, Brazil and the Dominican Republic. And although the fiercely-nationalistic devotees of each breed fervently maintain its absolute supremacy, in fact all of them are remarkably similar both in appearance and in gait.

Interestingly enough, the North Americans have been the first to take a measured overall look at the situation, and accordingly have formed the American Paso Fino Horse Association, which takes in all the so-called Paso Fino breeds and is working towards blending the best of all of them to mould an ideal mount of this type.

The North American society employs the slogan 'The gait is the birthright of a Paso Fino', and this is, indeed, what really sets it apart. Basically the horse's movement is lateral, with the two legs on one side first being used, and then the two on the other. However, while somewhat similar to a normal pacing gait, the *paso* is actually performed in four steps, two by two, the two of each lateral pair being very close in timing. It could perhaps be described as a 'broken pace', with the hind foot touching the ground a fraction of a second before the front foot. This unique high-stepping action eliminates the jarring effect of a true pace, and so the Paso Finos offer a very smooth ride indeed which, allied to their proud bearing, amenable temperaments and considerable stamina, is why they are becoming so well known in the United States.

Representative of the *paso* breeds, and certainly the best-known, is the Peruvian Paso. Mature animals of this breed measure from 13·3 to 15 hands in height, with the average being around 14·2 hands, and they weigh 950 to 975 lb. Colours to be found are grey, white, chestnut, bay and black, and many of the horses have white markings on head and limbs. Palominos also occur. However, piebalds and skewbalds are rare, since they are most unpopular in Peru, and mares with these markings are normally taken well away from the band of brood mares and used for suckling mule foals.

Peruvian breeders have been steadily developing the breed for more than 300 years, from the time when the original imported Andalusians were putting their great hearts into covering vast distances over Peru's high and rugged mountains, through her parched deserts and in her humid, green jungles. These demanding conditions gradually produced a super-tough trail horse, chockful of stamina and able to perform on a minimum of forage. Since the Peruvian riders were in the saddle for many hours at a time over some really punishing terrain, they paid the closest attention to the smooth purity of their horses' natural *paso* gait, and they also looked for, and only bred from, animals with proud but very docile temperaments.

As a result of its remarkable temperament, the direct outcome of the most careful selective breeding during three centuries, the Peruvian Paso is famed as being very easy to break. One highly impressed US horseman, a well-known veteran Western rider, declared delightedly: 'Breaking them to drive or ride is really too easy to be called "breaking". Two saddlings and anyone who can ride at all can ride them. They don't shy, bolt or cut up, making them ideal for older, less active or timorous riders.'

This good nature and ease of handling also extends to the serving stallions of the breed, which is not really so surprising when one recalls that the parent Andalusian stock has exactly the same attractive attributes.

However, while the Peruvian Paso still retains much of the great pride, almost amounting to arrogance, of the Andalusian, the many hundreds of years in a very different environment have wrought considerable alterations to his conformation. He has become smaller and more wiry, and his legs in particular have altered to become so fine that a good Peruvian Paso looks very light of bone. However, the breed's bone is remarkably dense and tough, and the Peruvian Paso's legs stand up to an enormous amount of hard use, gliding him over the steep, rocky trails of the soaring Andes all day or eating up miles through burning sand in the featureless deserts.

The Peruvian Pasos have steeply-sloping quarters and low-set tails, and the cannons of their hind legs are often sloped forward so that they are somewhat sickle-hocked. The rear legs move with a minimum expenditure of effort in a smooth gliding action which imparts very little motion to the hindquarters. In the natural *paso* gait the front legs of the Peruvian Paso, moved from the shoulder, reach far forward and are lifted to between a foot and sixteen inches above the ground. The knee and fetlock joints are flexed at the same time, and the feet are thrown six inches or more to the side in a unique 'dishing' motion which is greatly prized by the South American horsemen, and which is in fact essential to the smooth, fast regularity

The Paso Fino is noted for its characteristic *paso* gait. This stallion is performing the *paso corto*.

of the horse's forward progress.

As well as the normal *paso* gait, the Peruvian Paso also performs the pace and the so-called 'marching *paso*' and can trot, canter and gallop as well. However, no self-respecting Peruvian *caballero* will ever allow his mount to perform anything but one or other of its *paso*-type gaits. The 'marching *paso*' is very similar to the pace, but the supports last longer than the suspensions, giving an impression that the horse is 'marching'. The print of the back hoof falls ahead of the corresponding front hoof but less than it does in the pace.

The other best-known of Latin America's breeds, Argentina's Criollo, also descends from the redoubtable Andalusian. This very strong, stocky, tough and agile horse is the pride of his homeland, where he carries the picturesque *gauchos* after cattle on the wide-flung *estancias* of the Pampas.

The Criollo stands some 14 hands, is short-backed and has very strong legs with short cannons. Most Criollos are dun in colour, but there are also piebalds and skewbalds. They are remarkably tough and long-lived, and two famous Criollos which were ridden the 13,350 miles from Buenos Aires to New York, crossing Andean mountain passes and Ecuadorian deserts in the process, lived until they were in their mid-thirties!

Today's Criollos owe their origin to the importation in 1535 by Pedro de Mendoza, founder of Buenos Aires, of some hundred Andalusians from Spain. When Argentina's fledgling capital city was attacked and sacked by Indians, many of the horses escaped into the Pampas. There they multiplied so quickly that within fifty years their wild descendants were numbered in thousands.

Harsh winters and scorching summer droughts ensured the survival of only the fittest, however, as did determined hunting by the Indians, who had acquired a taste for horseflesh. When taken from the wild herds by the moustachied *gauchos*, the small, tough, agile Criollos soon proved their worth as cattle horses, and also as strong, enduring pack animals.

However, about a century ago the breed was nearly ruined by crossing many of its members with stallions from Europe and the United States. The use of these imported sires produced speedier and more elegant animals, but it quickly whittled away the inherent toughness and stamina of the Criollo, not to mention his hard-won native sagacity, which could often be very useful on the large unfenced ranges of the time. Fortunately, before it was too late and the good qualities of the Criollo had gone for ever, a number of alert breeders became aware of the dangers that existed in crossing the native cattle horses with imported stallions, and they made a forceful and concerted effort to produce the old-style Criollo. Severely selective breeding from the best remaining specimens of the breed was a crash programme that worked, and in 1918 an association was formed to foster and to promote the Criollo breed.

Argentina is also famed for the excellent polo ponies it produces, which are not only models of speed, agility, balance and courage, but are often schooled to a very high standard by their expert Argentinian trainers. A great number of polo ponies are exported from the Argentine each year to Europe and North America, and many of their trainers journey with them to play as professionals in some of the world's best polo teams. The polo pony of Argentina is a specialised Thoroughbred type whose origins lie in the crossing, in the early years of this century, of Thoroughbred stallions on native mares. A continuous effort to upgrade the polo pony stock was made over many years, with today's splendid results.

Racing is also very popular in the Argentine, and to supply the considerable requirements of the racing industry there are a number of large and excellent Thoroughbred studs.

Chapter eight

Western Europe

FRANCE

The history of the horse in France has been greatly affected by changes in breeding policy and classification, and the appearance of most breeds has altered accordingly. There are two notable exceptions: the English Thoroughbred and the pure-bred Arabian.

The Thoroughbred breeding centres in Normandy and the Bearn possess records of racing as far back as the sixth century. The first regular races, however, took place during the reign of Louis XIV, and began with private bets and a series of dares, as they did elsewhere. Towards the end of the eighteenth century Thoroughbreds were imported from England, and in 1780 a set of racing rules was drawn up. The French Revolution of 1789 proved to be only a temporary setback to racing, and after 1805 the studs and racing were re-established and the French stud book created.

In 1833 the Société d'Encouragement, which still rules French racing, was founded, and two years later it created the Prix du Jockey Club – the French Derby. In 1865 Gladiateur's victories in England proved that the French Thoroughbred could compete successfully against its English counterpart. The breed is officially known as the Pur-sang Anglais – an English Thoroughbred bred in France.

The Pur-sang Arabe, pure-bred Arabian, has as yet been little developed in France, its chief purpose so far having been to furnish the Arab strain for breeding Anglo-Arabs. Till a few years ago the National Stud Commissions purchased most of their Arabian stallions in the Middle East, and only recently turned elsewhere and acquired three stallions in Poland.

The Anglo-Arab originates in the cross-breeding of English Thoroughbreds and pure-bred Arabians, though under special conditions the stud book has been opened to the progeny of part-bred Anglo-Arab mares. The Anglo-Arabs currently registered there result from breeding Anglo-Arabs to English Thoroughbreds or pure-bred Arabians, English Thoroughbreds to pure-bred Arabians or Anglo-Arabs to Anglo-Arabs. The name of any Anglo-Arab included must indicate the percentage of Arab blood the horse contains, the minimum allowed being 25 per cent.

Thus the modern French Anglo-Arab springs from two previously separate sources: the Demi-sang Anglo-Arabe (part-bred Anglo-Arab), also known as the Demi-sang du Midi (Southern part-bred) and Cheval Tarbais (Tarbes horse); and the Pur-sang Anglo-Arabe (pure-bred Anglo-Arab), descending from a direct cross of pure-bred Arabian with English Thoroughbred – that is, from a 50 per cent Anglo-Arab.

The Southern part-bred, resulting from the cross-breeding of pure-bred Arabians or English Thoroughbreds with native mares already strongly impregnated with Arab blood and whose ancestry could be traced, was eventually also bred to Anglo-Arab stallions. In 1941 and 1958 the National Stud Administration opened the pure-bred Arabian and Anglo-Arab stud books to the progeny of these part-breds provided that no strain alien to Arabian or Thoroughbred had been introduced for six generations. As part-bred Anglo-Arabs are in type, temperament and qualities no different from pure-bred Anglo-Arabs the meaningless distinctions between them were finally abolished, and now all qualifying Anglo-Arabs are registered in the same stud book.

The principal breeding country of France has always been the south-west, particularly the Plain of Tarbes and the Limousin around Limoges. Anglo-Arab stallions have for some years been used similarly to English Thoroughbreds for the upgrading of the quality of other breeds. For example, the progeny of Selle Français mares bred to Anglo-Arab stallions

A Selle-Français stallion from Normandy.

are classified under the denomination of their dams or, if the mare is without pedigree, simply under 'Selle'. The general description of Selle Français has replaced that of Demi-sang, which was formerly used to describe any horse other than an English Thoroughbred, a pure-bred Arabian or an Anglo-Arab.

The name of Selle Français, or French riding horse, can be applied to any horse with a pedigreed sire and dam. If a horse's conformation and speed make him suitable for racing he is called Autre que pur-sang (other than pure-bred or Thoroughbred). There are two further classifications: Selle (saddle), grouping the offspring of any non-Thoroughbred sire out of a pedigreeless dam; and Non-constaté (origin unestablished), which describes the pedigreeless mares themselves.

The description of the Selle Français is best broken down into the principal members of former part-bred 'families'.

The Demi-sang Normand or Anglo-Normand is probably the best known of these, as Normandy has always been famous horse country, even before the seventeenth century, when Louis XIV's enterprising minister, Colbert, imported Arabians to improve the native breed. Systematic breeding, however, had to wait until 1830, when English part-bred and Thoroughbred stallions were imported. The Normandy horses, bred specifically as carriage horses for the rich, saw their heyday at the turn of this century. Having more recently concentrated on breeding horses suitable for the saddle, Normandy can now boast a number of jumping celebrities.

Another Selle-Français stallion; this horse is from Charolais.

The Vendée, Charente, Loire-Atlantique and Deux Sèvres, all départements in western France, bred horses in the nineteenth century of a popular riding horse type to Anglo-Norman English Thoroughbred and Anglo-Arab stallions, producing the prototype of a highly-bred, rather tall, powerful horse of fine conformation known for its jumping ability.

The Charollais came from the centre of France, an ancient native breed up-graded by Anglo-Norman and English Thoroughbred blood. Raised on excellent grassland, it had plenty of bone and quality and produced fine hunters.

The Limousin was a remarkable horse: full of quality, averaging about 15·2 hands, extremely refined, supple and light in its paces, ideally suited to dressage work. Additions of English Thoroughbred and Anglo-Norman strains made it an even more excellent ride.

The Camarguais lives, as is well known, in the Camargue. The climate and soil in and around the Rhône delta combined with Arab and Berber origins to produce a small (13·2 to 14·2 hands), extremely robust horse used by the 'gardians' for herding cattle. Nature continues to be responsible for the horses in this area, and their breeding is not controlled.

The French Trotter originates in Normandy, where native mares were bred to English Thoroughbreds and a few American sires. In 1836 trotting races were instituted, and since 1858 only those stallions who have proved themselves on the racetrack have been recognised by the authorities. The stud book, opened in 1922 and closed in 1941, is reserved for the offspring of previously registered trotter sires. Some of them, in spite of their conformation (their straight shoulders, for example) have been successful in show-jumping.

There are many draught horses in France, the four principal breeds being the Boulonnais, the Percheron, the Ardennais and the Breton. The Boulonnais is raised in the northern départements of Pas-de-Calais, Somme and Oise, with a division into large and small horses. The large one can reach as much as 17 hands, and is sturdy with a short head and wide forehead, massive arched neck, straight back, short loins and double croup. It is used for heavy draught work in farming and transport. The smaller type, averaging 15 to 16 hands, is a fast trotter and was popular with city tradesmen. Some large black horses can be found, but the breed is typically grey in colour. It has been improved by the addition of Arab blood.

The Percheron, also infused with Arab blood, is raised in the area west and south-west of Paris. The breed originated with mares of the Parisian basin bred to Norman, Danish, English and German stallions. The large horses stand well over 17 hands; the smaller type, more harmonious in build, is known as the Percheron Postier.

The Ardennais, bred in north-eastern France, was formerly a highly-strung, agile and resilient animal, making an excellent heavy artillery horse. After early nineteenth-century cross-breeding with pure-bred Arabs, English Thoroughbreds and part-breds this breed was modified to a horse some 16 hands, today used principally in farming. Very deep-chested and showing little daylight, its colours are chestnut, bay and roan.

The Breton, though it has its own history going back to the Middle Ages, has received since the eighteenth century English, Arab and Norman blood, resulting in the two principal types seen today: the heavy one, with a wide, short, dish-faced head, barrel-shaped body,

The small, tough horses of the Camargue are an indispensable part of the life of this region of France.

double croup, a lot of bone and feather; and the light draught horse, better known as the Postier Breton, as robust and sober as his heavier cousin but somewhat more elegant, standing from 15 to 16 hands, with a nice shoulder and good withers. This lighter type, formerly used extensively for work with light artillery, is now a useful farm horse. The most usual colours for the breed are roan and grey.

The French Percheron, a heavier horse than its English cousin.

WEST GERMANY

In the Federal Republic of Germany there are today three principal breeds of warm-blood horses. Local provincial strains have been created by using stallions of these breeds, Holstein, Hanoverian and East Prussian (Trakehner) – principally the last two – plus the Thoroughbred. Some of them are also breeds of long standing, such as the East Friesian, with 1,360 registered brood mares, founded in 1715, and the Oldenburg. These strains share the same stallion lines and are based on the Dutch Friesian. Other breeds are the Westfalen, Kurhessen, Württemberg, etc. Altogether, in 1970 there were 252,000 horses in West Germany.

The Holstein is not as popular as it used to be, nor is it bred in such numbers. As well as Schleswig Holstein, Kurhessen has a small breeding centre, though in the seventeenth and eighteenth centuries stallions of the Holstein breed were used in many provinces. The horses were originally bred in the maritime climate of the marshes of the mouth of the Elbe and neighbouring rivers. They were then big strong horses with convex heads, high action and noble outlook, like their Neapolitan and Spanish ancestors. Later, Yorkshire Coach (Cleveland Bay) and Thoroughbred blood was introduced, and thus horses with jumping ability evolved for the show ring. The first stud orders were issued in 1680, and in the Royal Stud at Esserom the famous white horses were first bred. The Electress Sophia of Hanover, granddaughter of James I, is said to have been responsible for the creation of this white or cream breed of horses, of which the Holstein stallion Mignon was the founder. Her son, George I, introduced them to Great Britain, where they were used in the royal mews and at Windsor.

The Hanoverian is a true 'son of the soil' and his native land covered most of north-west Germany: Brunswick, Mecklenburg, Pomerania and Brandenburg. The original mares were native and were crossed with Mecklenburg, Thoroughbred, Cleveland Bay and East Prussian stallions. In the different districts coach horses, artillery and agricultural horses as well as saddle horses and warm-blood race-

A superbly matched team of high-class Holsteiners during a driving marathon.

This horse is a splendid example of the lighter type of Hanoverian now being bred in Germany.

horses were bred. The last two decades have seen a change in conformation to a lighter, more elegant saddle horse which commands a high price at the auction sales at Verden, held twice a year. In 1970 there were 6888 mares registered in the Hanoverian stud book, and at one show for brood mares held recently in the district of Lüneburg, 587 mares were shown. At the 1972 Olympic Games thirty-one Hanoverian riding horses represented eight nations, gaining two gold medals, one silver and one bronze. The countries were Holland, Sweden, Switzerland, Canada, the USA, Mexico, Japan, and, of course, West Germany.

The East Prussian or Trakehner horse is a refugee from East Prussia. The breed is now chiefly in private hands, though the Trakehner breed Society has a stud at Rantzau in Schleswig Holstein, and selected colts are raised, together with Hanoverian colts, at Hünnesruck stud in the Solling. During the war, this breed was almost decimated. Out of a total equine population in East Prussia of 50,000 horses only 700 East Prussian mares, most of them in foal and pulling farm waggons driven by their refugee owners, and with a small herd of entires, managed to make their way to the west under appalling conditions in the war winter of 1945. The first foaling season produced only aborted or dead foals and barren mares.

The Swedish warm-blood breed is based almost entirely on East Prussian blood, as is the Polish Masuren, or Wielkopolska as it is now called. There are now also studs in America and Canada and the East Prussian breed of horses is increasing in numbers.

The Dülmen herd of feral ponies, owned by the Duke of Cröy, is kept in a semi-wild state in the Merfelder Bruch. The ponies are known as Dülmen but are in fact of no particular breed, since a number of stallions belonging to other recognised pony breeds have been introduced. Ponies are culled and sold off each year.

The Haflinger of southern Germany is found in the Bavarian mountains as well as the Austrian Tirol and its native south Tirol. It is above all a sturdy mountain pack pony able to carry heavy loads on its back. The breed is also of extreme antiquity and may be descended from the small Noriker pack pony, a breed which was later improved by the introduction of Arab blood.

The Noriker or South German cold-blood horse is bred in Bavaria, Württemberg and Baden. It is a short-backed, strong horse and is used for draught work and in agriculture. In Austria the breed is divided into two types, the other being the famous Pinzgauer spotted horse. The evolution of the Noriker can be dated to the period of the former Roman province of Noricum and therefore to the Roman occupation of nearly two thousand years ago.

The Schleswig and Jutland cold-blood horses of north Germany and Denmark share the same family tree. The breed is partly descended from mediaeval war horses and was at one time used for draught work, especially for omnibuses and trams.

One of the best Arabian studs in Europe is Marbach an der Lauter which, in 1932, took over the breeding of the Arab horse from Weil stud. It is a state stud, the only one in the Federal Republic, founded by Duke Ludwig of Württemberg in the sixteenth century to breed hardy horses. This policy has now resulted in the modern Württemberg warm-blood horse. Nine Arabian mares and four filly foals with three stallions moved from Weil to Marbach. There are now eleven mares plus three in private ownership and three stallions.

A team of Jutland horses from North Germany.

ITALY AND SWITZERLAND

The Etruscans of 2500 years ago were probably the first people to produce for riding and driving a useful blood type of horse. The republican Romans were not noted for their horsemanship; during their rule and for several hundred years afterwards there does not appear to have existed any co-ordinated plan to breed a particular type or strain of horse apart from those of the local indigenous small animals. Horses were imported from Spain, Persia, Noricum and northern Europe. During the seventeenth century Arabs, Barbs and the Old Spanish horse breeds contributed to the evolution of the Neapolitan horse, which was bred around Naples and Sorrento. This breed was characterised by a markedly convex face, an over-bent neck, heavy hindquarters and not very good legs—the type of horse that one might expect to be created from a mixture of genes. However, the Neapolitan influenced other breeds throughout Europe in those studs where high-school and dressage horses were in demand. About a hundred years ago, the half-wild, strong and hardy Maremma could claim descent from the Neapolitan, and today the Maremma is used in the cavalry schools and for show-jumping, especially when crossed

The Italian Murgese, a versatile horse from the province of Taranto.

with the Thoroughbred.

Thoroughbred horses have for a long time played an important rôle in Italy, and many Italian-bred Thoroughbred horses have influenced and improved the breed generally. Equally, the American Standardbred trotter is enormously popular – the largest European stud for trotters is Le Budric, situated near Bologna. Other studs include the Karst stud Lipizza, and the two mountain breeds Avellino and Haflinger are also bred in Italy.

One of the earliest known engravings of the wild horse was discovered at Shaffhausen in Switzerland, and is dated to the Magdalenian period of the Early Stone Age, 1,000,000 to 50,000 BC. Two wild horses are engraved on reindeer horn. This horn is known as the 'commando baton'. Apart from this there are historically few references to wild horses or even to horse breeding in early historical times.

At the outset of knightly chivalry in the tenth century certain cloisters and abbeys were founded by the emperors of the Holy Roman Empire, which included modern Switzerland. The cloisters on Lake Constantine, Rheintal, Vorarlberg and Elsace were occupied by monks and young men aspiring to knighthood who, together with their patrons, presented property, corn and horses brought from their distant estates.

Thus one of the oldest studs in Europe is Cloister Einsiedeln, which was founded in 1064. Horses were also bred in the Schwyz province. The best years for the Einsiedeln Stud were from 1500 to 1800, and there was a considerable demand for horses in Italy. Their new owners called them the *cavalli della Madonna*. Turk, Spanish, Friesian and Italian stallions were used although generally the Einsiedeln strain was preferred.

The armies of the French Revolution destroyed the stud in 1798, but in 1801 the monks returned and began their task of rebuilding the breed of horses and their stables. By 1811 there were fifty-three horses once more in the cloister mews. In 1820 Abbot Conrad drove out for the first time in a carriage since the roads had been improved for wheeled vehicles. Thus the Einsiedeln horse became a horse used for riding and for driving, and was thenceforth much in demand for the Swiss army.

Federal studs such as Avenches and Freiberg and others in the Jura were founded about a hundred years ago to breed horses for defence purposes. In 1865 the first exhibition of Swiss horses was held in Aarau and the three principal breeds, the Freiberger, Erlenbacher and Schwyzer, were shown. Then, to improve these breeds, the Cleveland Bay stallion Bracken was purchased, as were two Anglo-Norman stallions, Corail (by a pure-bred Arab), who bred an outstanding line of trotters, and Egalité, who also produced a definite type. Thoroughbred and other Anglo-Norman stallions were later introduced.

Although the past decade has shown a decrease in demand for horses for agriculture and the army, a new generation of riders has grown up, and the Einsiedeln breed is versatile enough to ensure its popularity, producing good show-jumpers and saddle horses for all purposes.

Horses being watered in the plains of Hungary.

AUSTRIA AND HUNGARY

From the earliest historical times the vast *puszta* country that is now known as Hungary was the habitat of wild equine populations. It was also the home of one of the earliest of the horse-breeding peoples – the people of the Tripolje civilisation.

In the third century BC hordes of mounted nomadic Mongolians swept as far as southern Europe, and established their rule under Attila the Hun. In the ninth century AD there was a further horsed invasion of the Finn-Ugrit Magyars. Later, in its equestrian prime, the country was part of the great Austro-Hungarian Empire with close connections through the Court with Spain, another great equestrian country. Spain supplied stallions to most of the royal studs of Europe. These were commonly known as Spanish, though they were in fact of Andalusian breeding.

The native horse of Hungary was small, very fast and had plenty of stamina, and it formed the basis for several very hard warm-blood breeds.

The largest stud in Europe in 1774 was the Austro-Hungarian state stud Radautz in Bukowina, now part of the Ukraine in the USSR. Radautz extended for about 70 miles. It was situated in the valley of the river Suczawa, and contained within its demesne twelve studs and five estates, for Thoroughbred, Arabian and Oriental horse-breeding. These included one Huçul stud and depôts for yearling and two-year-old fillies, two-year-old colts and the stables for the Pepinière stallions and coaching entires. It was run on military lines.

The next largest Hungarian state stud is that of Mesohegyes, founded by Maria Theresa in 1785. In 1910 there were 4142 brood mares, divided into nineteen studs. Cattle were also bred on a large scale, and a population of 7000 cared for between 18,000 and 20,000 animals. This eighteenth-century industrialisation of animal breeding led to disease and pestilence; the limits of successful animal husbandry had been reached and the numbers had to be reduced.

The task of these huge state studs was to breed remounts and cavalry horses. In 1889 the stud Babolna was added, to breed entires for agricultural purposes. There were few pure-bred breeds at this time, and very little was known about genetics and the scientific practice of breeding any animals. For example, Mesohegyes bred light and heavy saddle and draught horses. Then the basis for the Nonius and Gidran strains was laid, and later the half-bred Furioso and North Star breeds were founded by stallions of these names. Nonius senior was born in 1810, in the French stud of La Rosières. Nonius was crossed with Spanish mares, and a strong dose of Siglavy Arabian blood was added to fix the breed. There are today both large and small Nonius (with more Arab blood). Gidran senior was an original Siglavy-Jidran Arabian. Furioso was a Thoroughbred by Privateer, foaled in 1836, and had a tremendous influence on the half-bred line which he founded. The black North Star, also a Thoroughbred, founded another half-bred line.

The state stud Babolna breeds some of the finest Arab horses in the world, although much valuable material was lost during the last war. A particularly successful cross of Arab stallions on native mares has resulted in the beautiful Shagya Arab breed.

The famous Lippizaner is bred not only in Austria but also in Yugoslavia and Bulgaria. These horses have made the Spanish School in Vienna famous with exhibitions of high-school dressage. The breed was founded in 1585 by the Archduke Karl, at Karst near Trieste. It was here that the old Karst horse, used by mediaeval knights, had been bred. The Spanish Jennet (descendant of Spanish-born Barb horse) the heavier Villanos from Castile, and the Italian Neapolitan with its high action, all helped to create the splendid Lippizaner. There are six bloodlines: Pluto (Fredericksborg), Conversano (Neapolitan); Neapolitano; Favory (Kladrub); Maestoso (Kladrub); and Siglavy (Arab). To these we may add a seventh, the unique Incitato from the Fogares stud in Transylvania. Although horses generally used for exhibition are grey, the Lippizaner may be any colour.

These Yugoslavian Lippizaners are competing in a team dressage event.

The chestnut Haflinger with its flaxen mane and tail is of old lineage and originated in the Tirol for mountain and pack transport. It still brings hay down the mountainside on its back. This is one of the breeds which is on the increase today, and Haflingers are bred in other countries as well.

BELGIUM AND HOLLAND

The equestrian Low Countries can be divided into several areas. The heavy Ardennes horse, bred in the Ardennes, has been known from Caesar's time, while Brabant is the home of the heavy Brabant horse which helped to sire several heavy breeds. Friesland is where the magnificent Dutch black Friesian horse is bred; and Gelderland is the home of the lighter saddle and harness Gelderland horse, which may be seen in many horse shows in Holland. In age and type peculiar to locality the Friesian runs the Ardennes a close second and since the Middle Ages has been the best known.

Both Holland and Belgium breed blood trotting horses for racing. Modern European trotters have evolved from three distinct breeds: the American Standardbred; the Russian Orlov; and the French Anglo-Norman. These three breeds have evolved on different lines but the Standardbred would seem to be the superior in speed. From the Standardbred crossed with the Orlov, a new, and it is said faster, horse called the Métis has evolved, while from the Standardbred crossed with the Norman, the Noram breed has been created. Trotting as a sport is immensely popular in most of the countries of Europe.

Two examples of working horses bred in the Low Countries. *Above,* the Dutch Friesian; *below,* the Belgian Ardennes.

The Dutch Gelderland horse is used both in harness and under saddle.

SPAIN AND PORTUGAL

The great contribution of the Iberian Peninsula to the world's horses has been the famous Andalusian of Spain. For many centuries this horse was the most sought-after mount in Europe; and at one time exportation of breeding stock from Spain was even forbidden, on pain of death and confiscation of property.

During the great years of Spain's 'golden age' the horse from the far south of the peninsula made history in both Europe and the Americas. In Austria in the sixteenth century the Andalusian laid the foundation of the Lippizaner of the Spanish Riding School of Vienna; the breed played a sterner role at this time across the Atlantic by providing the mounts of the *Conquistadores*, those few valiant mounted men who explored and annexed so much of North, Central and South America for the Spanish Crown. Hernan Cortes, who led a handful of riders to overthrow the Aztecs and capture Mexico for Spain, proudly proclaimed, 'After God, we owed our victory to the horses!' Andalusian blood lives on in the Americas in the Quarter Horse, the Appaloosa, the Saddle Horse, the Palomino, the Pinto and the Mustang in the United States and Mexico, the Peruvian Paso in Peru, the Criollo in the Argentine, the Colombian Paso in Colombia, and the Paso Fino in Puerto Rico.

The classic Andalusian continues to thrive in Spain–in fact its popularity is greatly on the increase. In latter years, good quality Andalusians have been exported to many countries, including Australia. In Spain itself there is also a growing demand for these fine horses from the newly-affluent in the country's developing economy.

The Andalusian is a most impressive sight, with his sculptural beauty, proud bearing and natural high action. The horse is strongly built and yet extremely elegant; naturally high-stepping and yet with catlike agility; while he presents a picture of spirited animation under saddle or led in hand, he is at all times perfectly amenable to the will of the person controlling him, and has a friendly, docile temperament. The Andalusian's beauty lies in the balanced symmetry of his noble proportions, and he was for centuries used as a model by sculptors of Europe. The head is majestic, with large, kind, well-set eyes, a broad forehead and well-placed ears. The neck is reasonably long, broad yet elegant, and well-crested in stallions. Well-defined withers precede a short back from which stems the Andalusian's notable agility; the

The Spanish Andalusian.

quarters are broad and strong. The croup is gently rounded, being neither horizontal as in the Arabian nor steeply sloped like that of many Quarter Horses. The tail setting is rather low, and both tail and mane are luxuriant and silky and worn long. The horse's shoulder is long and sloping, the chest splendidly broad and the body well ribbed-out. The legs are of medium length, clean-cut and elegant, yet more than strong enough to support the robustness of the body. Andalusians average around 15·2 hands, weigh something over 1,000 lb., and are white, grey or bay in colour. Blacks and roans also exist.

The Andalusian's temperament is exceptional; his calm, good temper and ease of handling is to be found even in serving stallions of the breed. But though the Andalusian is so docile he is no slug, and he moves with a tremendous amount of elegant animation.

Many Andalusians 'dish' extravagantly, an action very popular with the Spanish *caballeros*. High, wide and handsome is how they want their stallions to come down the decorated streets of Andalusia's white towns during carnival time; so they look for knees that come flashing right up in front of the horse's broad chest, and forelegs that rotate showily out to the side as the horse moves majestically forward. The Andalusian is also born with a tendency to perform a *paso* gait, and this is always developed by the Cordoban-hatted horseman of the south. In this gait, the horse moves forward proudly with a high, collected head carriage, and with extreme smoothness and vivacity, in four disjointed steps, two by two, the two of each lateral pair being closer in timing. His forelegs swing out elegantly to each side as he proceeds, his hind legs step straight forward and far under his body with the utmost economy of motion. This gait is the natural forerunner of the *paso* that has been so particularly developed—with a number of variations depending on pace—in some South American breeds, such as the Peruvian Paso and the Paso Fino, descendants of the Andalusian.

The pure-bred Andalusian is a great luxury even in his native country. He is bred (as a hobby) by the great aristocratic and landowning families, and is used only for the most glamorous activities: carrying the *caballeros* and their gaily-dressed girlfriends in street parades; giving private displays of *haute école*; and as the specially trained mount of the *rejoneadores*, the gentlemen bullfighters, when they lead the colourful parade into the bullring. Some of the speedier Andalusians are also used in the actual encounter with the *toro bravo*.

Andalusians are also crossed with Arabians and Thoroughbreds to produce excellent general riding horses; a successful blending of all three breeds, known as the Hispano-Anglo-Arab, is particularly highly prized. These crosses with the Andalusian produce the alert animals which are specially trained to work with the irritable, fast-starting, fast-turning and supremely dangerous fighting bulls in the huge pastures of Spain. The Andalusian cowboy riding with his left hand on the reins and his

A pony and caravan in Ireland.

Farm horses at the Plitvice lakes in Yugoslavia.

A drive through the Tierpark near Copenhagen.

An Italian *carrozza* takes tourists through Rome.

right hand holding a long, lance-like pole, needs a mount on which he can literally stake his life. When the big black bulls start to play up the *vaquero* must have a bold, fast horse reacting to his slightest aid in order to keep the irascible bulls under control. The Thoroughbred and the Arabian are used to supply speed in the cross-bred; the Andalusian provides the bold yet calm temperament and the turn-on-a-*peseta* agility.

The origin of the Andalusian is not definitely known, although most Spanish authorities adamantly maintain that the horse is a pure native of Spain owing not a single feature to any breed from outside the Iberian peninsula. Another theory suggests that the Moors developed and fixed the Andalusian during their stay of nearly eight centuries in *Al Andalus*, southern Spain, by crossing their Barbs with the light, agile horses they found in Spain. These original Spanish horses are also thought to have been of Barb type, since the theory postulates that they were descendants of horses which had crossed from north-west Africa centuries before when the Iberian peninsula was linked to Africa by a narrow strip of land.

Closely related to Spain's Andalusian is the Lusitano of neighbouring Portugal. However, while some strains of the breed are of very high quality, the Lusitano tends to be longer-backed and longer-legged than the Andalusian, and is consequently less agile. Their heads are often rather unattractive, too long, with Roman noses and too-large ears. It would appear that the Lusitano is in fact descended from a cross of the Spanish Andalusian on heavy, rather coarse mares.

Another horse containing a fair dash of Andalusian blood is Portugal's Altér, a tall, impressive bay horse with high action, powerful quarters and broad chest which is used for high-school displays. There are not a lot of these horses, and a really good individual is greatly prized.

Both Spain and Portugal have a number of Arabian studs, with some very good stallions and mares being owned by the government stud farms. Spain's Arabian population is much bigger than that of Portugal, with most of the animals concentrated in the southernmost

A Portuguese Lusitano performing the *haute école* Spanish Walk.

region of Andalusia. The environment there is said to be exactly the same as that of the best Arabian-breeding areas in the Middle East, and certainly Spain produces some magnificent-looking Arabians–lean-bodied and stylised, with excellent steel-tendoned limbs and hard feet, and with exquisite heads showing the much sought-after quality of 'dryness'.

The parent stock of Spain's Arabian population was drawn from a number of sources, principally the desert, Poland and Hungary. Some excellent animals also come from Britain in the form of Arabians purchased from Lady Wentworth by the Duke of Veragua. However, considerable confusion was caused in breeding during the Spanish Civil War, when the papers testifying to the authenticity of the pedigrees of many fine Spanish Arabians were destroyed.

There is a limited amount of racing in Spain, and only a few Thoroughbred studs exist, the quality of whose stock is modest. The north of the country, near the Bay of Biscay, is favoured for breeding and raising Thoroughbreds.

The small, tough Garrano ponies are also to be found in the north of Spain. They have close relatives across the border in Portugal known as Minhos. Their origin is obscure, though one rather colourful theory has it that they are direct descendants of the prehistoric horses portrayed in the Stone Age wall paintings in the caves of Altamira in Spain.

Andalusian stallions, harnessed in Jerezano style, being driven in *media potencia* to a four-wheeled Spanish brake.

Chapter nine

Eastern Europe and Scandinavia

THE USSR

Since the USSR consists of fifteen republics covering an area of over $8\frac{1}{2}$ million sq. miles (17·1 million sq. km.) of desert, steppe, mountain, forest and tundra, it is scarcely surprising that a great number of breeds is to be found within the USSR.

These breeds include, just as naturally, desert, steppe, mountain and heavy breeds of horses. In the far north-east, between the rivers Kolyma and Omolon, there are also said to be wild horses with pony characteristics.

Most of the pony and steppe breeds are descended from the Tarpan, *equus przewalski gmelini Antonius*, and are therefore counted among the native primitive stock of the district in which they are found. The Panje pony, which is also found in Poland, belongs to west Russia as does the Lithuanian Imud, which is both large and small in type, and the Zeimatuka, with its pronounced dorsal stripe, which has been crossed with other breeds. There is, too, the strong Klepper which is not so much a breed as a type, obtained by crossing Arab and Ardennes, and which in turn has been used to improve the Obwa draught breed.

The troika, a traditional driving style favoured by the Russians.

Going further east, there is the 14 hands Kirghiz pony, which has been used to form the slightly larger New Kirghiz by crossing with Don and Thoroughbred stallions. The Bashkirs of the Ural Mountain district have a strong all-purpose draught and saddle pony known as the Bashkir. In the north of the Urals around Archangel there is the 14 hands Viatka, which is often palomino or chestnut in colour with the dorsal stripe. The same strain is found in the Obwinski and in the Kasanski breeds.

In the Baltic states there are also some heavy-weight harness draught breeds of horses. These are the Toric, the Lithuanian and the Latvian. Very widely spread throughout the Ukraine, Udmut, Kirovograd, Archangel and Vologda provinces is the small but powerful Russian Heavy Draught horse whose average height is only $14{\cdot}2\frac{1}{2}$ hands, and the Vladimir Heavy Draught horse, bred up from Ardennes, Percherons, Suffolk Punches and Cleveland Bays.

Two breeds of mountain horses, separated by more than 1,500 miles but with a distinct similarity of conformation, are the Kabardin from the northern Caucasus and the Lokai from Tadzhik in central Asia.

The Moshua collective farm at Ashkabad breeds the Turkmene desert horse. Desert horses were bred at Ashkabad in 1000 BC and were even then mentioned as racehorses. The Turkmene is therefore one of the oldest *breeds* of horses known. King Darius of Persia, born in 522 BC, had a cavalry bodyguard of 30,000

Bactrian horsemen, mounted without doubt on the same type of Turkmene horse which is being bred today. The Turkmene, which has influenced almost every light horse breed throughout the world, is herded in *tabuns* on the steppe in the care of mounted herdsmen.

The Akhal-Teké is a direct descendant of the Turkmene (the Jomud is another) and is bred in Turkmene, Kazakh, Usbek and Kirghiz. This is a genuine desert horse of almost unbelievable stamina. One test of endurance covered 2,580 miles, of which 600 miles were over a trackless sandy desert without water. The distance was covered in eighty-four days. The Akhal-Teké is described as occasionally spotted or striped and of a golden dun colour. Like the Turkmene, these horses often have a metallic glint to their coats.

The Don breed of horse originated during the eighteenth and nineteenth centuries in the steppes and plains bordering the river Don. It was originally a small, tough, active oriental type of horse which was known as the Old Don breed. These horses were crossed with Orlov trotters and Orlov-Rostopschin, a saddle breed which originated in the Orlov and Thoroughbred. Don horses are very hard and are used for agriculture and riding. Their endurance test consisted of covering a distance of 160 miles in twenty-four hours.

The Bujonny, named after Marshall Bujonny, who improved the breed, was bred in the Rostow district from a cross with the Don and Thoroughbred. It rapidly became a fixed breed and stallions were subsequently used to improve the native horses of other republics. The Bujonny has real jumping ability and makes an excellent saddle, harness and draught horse. The champion stallion Sanos covered a distance of 185 miles in twenty-three hours excluding rest pauses.

Of all the Russian breeds perhaps the Orlov trotter is the best known. As so often happens, it was the genius of one breeder that was responsible. In 1775, Count Orlov-Tschmenski wanted a hard, strong horse capable of covering great distances in the Russian winter in front of a sleigh or troika. He imported to his stud at Ostrowo a number of Arab stallions, including the grey, Smetanka, who had remarkable

'Battle for the Tufts'—a national Uzbek competition.

An Akhal-Teké mare of the Komsomol Stud in the Turkmenian SSR.

trotting action. This stallion was mated with a blue-dun Danish mare (probably of the Fredericksborg breed). But the offspring of this mating, Polkan, showed both the good and bad points of the Oriental horse; his forehand action was particularly bad. However, Polkan's mating with a Dutch Harddraver mare with oriental blood produced the great stallion Bars 1, foaled in 1784. Bars 1 became the founder of the Orlov Trotter breed, standing at stud in Chrjenow for seventeen years. Count Orlov trained his horses himself in Moscow, but after his death inbreeding was practised, the horses lost quality and eventually the stud was taken over by the state. There are now twenty different bloodlines.

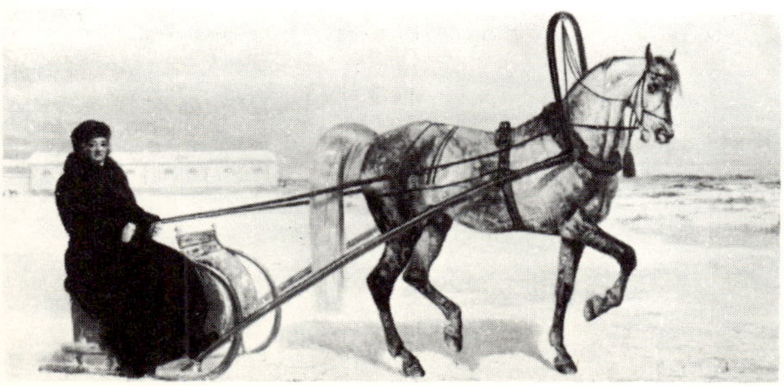

Count Orlov-Chesmenski, founder of the Orlov trotter.

POLAND

The very important place held by Poland among the horse-breeding nations of the world results first of all from sheer numbers. In spite of a considerable decline in recent years, there are still nearly three million horses in Poland, which places her second in Europe only to the Soviet Union.

The great historical traditions of breeding and horsemanship in Poland, and their military and sporting achievements, still pervade many aspects of Polish life today, from art to politics. Horsebreeding is fortunate to have a genuine basis, rooted firmly in agriculture, in what is mainly a country of small peasant farms and light soil, in which the horse, particularly the warm blooded type, still has an important part to play, and where modern machinery – designed for horse traction – is still being produced: for instance, the combine drill.

The Ministry of Agriculture is directly concerned in improving the general standard of horses in the country. As an instrument of this policy, it has adopted a system of regionalisation by which the various breeds and types of horse are bred within defined regions. All pedigree horses, both state-owned and privately owned, are registered in regional stud books. This system is designed to prevent haphazard cross-breeding. It does not apply to Thoroughbreds, pure Arabs or Anglo-Arabs, each of which has its own National Stud Books.

The largest regions are that of the 'Wielkopolski', a warm blooded half-bred hunter type with Trakhener blood, and that of the 'Malopolski' in the south, comprising rather lighter and smaller horses of traditional east European bloodlines of Arab half-bred and Anglo-Arab origins.

The modern Wielkopolski horse has emerged as a breed embracing several older breeds once prevalent in central and western Poland. These include the Poznan and the Masuren. The Wielkopolski is a big and handsome horse equally well suited to riding or driving work. The Malopolski is believed to be nearest in type to the original Polish horse. It displays several typical characteristics of Arabian bloodlines in its soundness, stamina, elegant movement and stamp of quality. Other regions in Poland specialise in heavy horses, such as the Slaski in the south-west, and ponies, for example the Hucul.

The Ministry runs its own government studs, each specialising in one or two breeds. At present there are over thirty of them, including seven for Thoroughbreds and two for Arabians. The purpose of these studs, equipped with a nucleus of élite breeding stock and excellent conditions, is firstly to breed stallions for the Ministry's stallion stations. They also produce the cream of Polish horses for performance and for export. The stallion stations themselves are the most important instrument of government breeding policies, for these horses are covering the ordinary working mares of the country. All stallions must pass not only the usual conformation and veterinary inspections, but also stringent performance and temperament tests suitable for each particular type. Throughout their lives at the depots they continue to work under saddle or in harness. The fees of these horses are purposely kept low, but many stallions are owned by private farmers and stand in competition with the government horses. To have a licence, these stallions must also pass a test of quality, and they are inspected every year.

Exports of horses, particularly those for hard currency, have a very important part to play in the Polish economy and, apart from the unfortunate but necessary meat trade, about 4,000 horses are exported every year, for riding, breeding, jumping, draught work, cavalry, etc. Of these, large numbers are bred by private farmers. Equitation is fostered on the studs and stallion stations and by government encouragement of riding clubs all over the country, while driving on light harness is very much a living skill.

Racing in Poland is popular and with a proportion of the money from the Tote Monopoly being channelled back into breeding, it is a vital link in the system. The pattern of races for Thoroughbreds is modelled on British traditions, and Britain is the main source of fresh blood through regular purchases of stallions and occasionally yearlings. There are seven races on the card three days a week at Warsaw during the season, with additional meetings at Poznan and Sopot. As well as the more usual races, these cards are filled by a complete programme of flat racing for Arabs, and by racing on the flat and over fences for 'non-Thoroughbreds' which are mainly Anglo-Arabs and the lighter types of half-bred. International race meetings are held between the east European countries at each capital in turn.

The Arab horse in Poland has a history of over 300 years, and is recognised as the great improver of other breeds. The pure-bred Arabs suffered severely in both world wars and the present stock owes its existence largely to the restoration to Poland of those horses which found their way into British hands in Germany. Numbers have been built up only very gradually, owing to a strict policy of quality control and culling. Particular attention is paid to the preservation of true desert type and to fertility and temperament. Racing is used as a test of soundness. The first post-war export of Arabs to a Western country was made to Britain in 1958, and Polish Arabs are now valued highly

Gerwazy, owned by the author of this chapter, is one of the most outstanding Polish Arabs.

A group of horses raised at the famous Janow Podlaksi stud in Poland.

in many countries. They have won important championships in Britain, Sweden, Holland and Canada, but it is in the United States that they have had the greatest success. This has been achieved by a large number of different horses among the very toughest competition. The greatest achievement was by the phenomenal stallion Bask, who not only won the American National Championship in hand, under saddle and in harness on three successive occasions, but followed this by his daughters twice winning the mare championship and reserve, and having at the same time several other progeny in the Top Ten. The 1972 stallion and mare championships were also won by horses imported from Poland.

The latest Polish Arab Stud Book shows that while brood mare numbers in the years 1966–70 were between 100 and 133, and there were fewer than 100 foals produced each year, the number of Arab exports during this four-year period (including geldings and foals at foot) reached the astonishing total of 150!

Poland is a comparatively poor country, but very rich in good horses and the horsemen to go with them.

SWEDEN, NORWAY AND DENMARK

The earliest historical representations of horses in Sweden were found at Kivik, and these are dated to around 2300 BC. The horses were engraved, together with a wagon or chariot, on the stone casing of a passage grave. From a moor in Seeland, Denmark, the famous horse and sun chariot was recovered, and dated at around 1500 BC. Returning to Sweden and the late Bronze Age at Järrestad, rock drawings of riders and horses were discovered; four of the six horsemen are using reins.

Interesting though these early historical discoveries are, they naturally give us very little indication of the type of horse then domesticated. From other sources, however, we may assume that it was the typical north European pony, *Equus celticus*, which was descended from the Tarpan.

The present-day Swedish warm-blood horse is descended from native mares whose ancestors had been brought to Sweden during earlier centuries, and perhaps also from an older native strain. To improve these horses as a breed, Hanoverian and East Prussian stallions were introduced with great success in 1916. At this time there were sixty-five imported Hanoverian stallions, and twenty-seven East Prussians. In 1956 there were two Hanoverians and three East Prussians, while the total number of warm-blood stallions was fifty-four. The Swedish warm-blood breed had been founded, and so successfully that for the past thirty years only home-bred horses have been used in the Olympic Games, most of them of the East Prussian bloodline. The state stud is at Flynge; a number of top-class horses have been exported from there under the direction of the well-known breeder, Dr Aaby-Erikson.

There are no small native ponies in Sweden itself. The native North Swedish horse is descended from an old Scandinavian strain and resembles a light draught or cold-blood horse. The breed was almost lost towards the end of the nineteenth century, but around 1890, with the help of the Norwegian Döle and Belgian stallions, the breed was revived. The best types

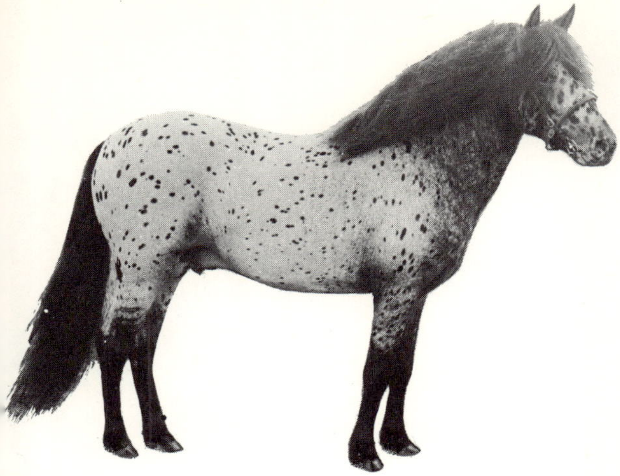

The Knabstrup coat markings can be highly individual.

The Knabstrups bred on this stud in Denmark are exported to circuses all over the world.

are found in Jämtland, Dalecarlia and Värmland, and along the Verboten coast. The colour is often dun with black points, dark brown and black, and although in conformation these horses may resemble draught horses, their lively temperament and energy are characteristics of a warm-blood breed. They are exceptional trotters and are raced, have strong hard legs and joints and considerable stamina, and are well-muscled and hard. Apart from trotting races, the North Swedish horse is used in forestry work and agriculture.

The island of Gotland, which belongs to Sweden, lies in the Baltic Sea. The native Gotland or 'Skogsruss' horse stands at about 12 hands and is the living representative of a very early primitive strain which lived wild in the forests. It is a typical descendant of the Tarpan, *equus przewalski gmelini Antonius*. The colours are brown, black, chestnut, dun, palomino and grey. These ponies are renowned for their stamina and speed and for a primitive contrariness.

Very similar in conformation and characteristics is the Norwegian Northlands pony, but of all the Scandinavian pony breeds the sturdy dun Norwegian Fjord or 'Westland' is best known. This breed is native to Rofaland, Hordaland, Sogn og Fjordane and More og Romsdal. The Fjord has been exported to the Canadian Rockies and many European countries, where it is used in harness, as a draught pony and sometimes as a riding pony.

The Fjord is thought to belong to an ancient primitive breed which was native to Norway before the period of the Vikings and was used by them, since they were able if rough horsemen. In colour they are always mouse dun or yellow dun, with black points, a pronounced dorsal stripe and silver and black mane and tail. The mane is often cut to an upright position. Some experts think that the ancestry of the Fjord can probably be traced to the Asiatic wild horse, *equus przewalskii przewalskii Poliakoff*, which is thought to have inhabited Europe in pre-historical days—and indeed, probably did so since it has been made immortal in a number of pre-historical paintings on cave walls in southern France and northern Spain. The Fjord certainly does look like a domesticated strain of the Przewalski horse. But several thousand years of selective breeding and domestication have changed their temperaments and characteristics. The Fjord is very hardy, long-lived, economical to keep, robust in disposition and fertile. It stands at 13·1 to 14·1 hands. They are genuine ponies capable of great effort and hard work.

A Norwegian pony stallion.

A display of skill given by the Royal Danish Hussars.

The Döle–Gudbrandsdal is a popular heavy draught horse and forms about two-thirds of the total equine population, which has decreased from 237,974 in 1946 to 142,247 in 1956 as a result of motorisation. This horse has hard legs and is very active. Its action has been created by crossing the native strain with Danish (possibly Fredericksborg) and Thoroughbred stallions. The black or brown Döle horse is used in city transport, in agriculture and forestry. Stallions of this breed, and all other Continental breeds of horses, have to undergo a severe test of endurance, stamina, health and willingness.

The Scandinavian countries are counted among the foremost countries to breed racing trotters based on American blood. The Standardbred crossed with Anglo-Norman and Norman is well known for its speed and stamina. In 1953 the famous Swedish-bred mare, Frances Bulward, created in Norway a European record.

The oldest breed of horse in Denmark is the Fredericksborg, founded in the sixteenth century. They were then regarded as the best parade and school horses in Europe, and were in great demand at all the courts of Europe. The Fredericksborg had basically Spanish Andalusian and Italian Neapolitan blood, though crosses of both Arabs and Thoroughbreds were introduced. The royal stud was disbanded about a hundred years ago and the Fredericksborg became a native breed of Seeland. The horse is now used for light harness work, as a draught horse, and for driving and riding. The spotted Knabstrup is a lighter type of the Fredericksborg strain, and these horses are very popular for circus work.

ICELAND

The island of Iceland was settled from Norway between AD 870 and 930; until then it had been uninhabited. These early settlers were Norsemen or Scandinavian Vikings led by Ingolf Arnarson. Generally speaking the Vikings, who were passable if rough horsemen, did not take ponies with them when they raided or colonised other lands – they were able to rely on native ponies for transport. In the case of Iceland, however, there were no native ponies available, so ponies had to be imported. There is no historical data to tell us what these ponies were but from a study of modern breeds it seems probable that the ponies belonged to a group from the Lofoten islands now extinct. A very similar breed is today's Northland of Norway.

The ponies that were brought to Iceland by Norse settlers were, according to skeletal remains found in Viking burials, rather larger than the modern Iceland pony, which stands at 12 to 13 hands. To carry ponies 600 miles across the North Atlantic in an open Norse longboat, a journey of some five to six days from Norway to Iceland, requires skill in handling animals. In fact the Norse settlers also carried cattle and sheep, and their boats frequently crossed these hazardous seas to bring extra families and household goods. By about AD 1000 there are said to have been some 60,000 people living in Iceland.

The Western Isles of Scotland provided some additional ponies, but very soon the Althing passed laws forbidding further imports of ponies. For 800 years the Icelandic pony has, as a result, changed very little except to become rather smaller than its ancestors. This is not unusual, since island ponies always do tend to diminish in size. As a breed, the ponies have become extremely hardy through natural selection caused by conditions of extreme cold and shortage of food. In very hard winters, when the ponies grow very thick coats, as many as 50 per cent of the herds of ponies may die. This was especially the case during the winters between the thirteenth and eighteenth centuries and also in 1784, when a great eruption destroyed a considerable number of them.

The ponies are kept in semi-wild herds and have developed both hardiness and stamina against weather conditions; they are also long-lived and fertile. The larger and stronger ponies are used for riding, and the performance expected of them is considerable. Until air and motor transport became available they were virtually the only means of transport as pack, riding and draught animals. We may think these ponies small to carry men, but in fact, except for the cob-sized pony seen in some representations of the period, the horse of Europe was of the same size and strain as the Iceland pony over a thousand years ago. Most of them are five-gaited and are especially valued for the easy gaits of the amble and the tölt, as well as the walk, trot and canter. The tölt is a very rapid walk which increases in speed to a canter. The ponies are very sure-footed and active over difficult mountainous going including rock, ice and snow, and riding ponies with tölt action are in great demand.

There are no predatory animals in Iceland so the semi-wild herds have never had to protect themselves, and once their first shyness is overcome they become friendly, kind and reliable. In addition to this they are economical to keep. They make general-utility ponies and have won the hearts of many adults and children not only in Iceland but especially in West Germany, where during recent years they have been imported in considerable numbers.

In 1962 there were said to be 42,000 ponies in Iceland. Breeding is based on descent from the female line. Stallions are privately owned and run out with the herds. Iceland ponies are to be seen in all standard colours, grey being the most common.

The shaggy coat of the Iceland pony is a valuable protection against the bitter winter.

Chapter ten

Australia and the East

AUSTRALIA

The Australian horse world is a vigorous and expanding one. Many of the world's breeds are represented and the country boasts several indigenous types as well.

Most famous of the Australian horses, an animal with a far-reaching and well-deserved reputation for toughness and endurance, is the Waler, which reached the peak of its worldwide fame during the last century and the earlier part of the present century. Its heir is the Australian Stockhorse. Mr R. M. Williams, Director of the Australian horse magazine *Hoofs and Horns*, and a very practical cattleman and horseman, has said of the latter: 'Many have commented on what they believe the breed to be, and where it all began; but it is sufficient to say that there is a lot of Thoroughbred and a dash of Arab, some cold blood from pony and plough, and plenty of individual preference.'

The origins of the Waler are rather obscure. It seems that the base stock came from South Africa in the form of Cape Horses, which were in turn the product of a mixture of Dutch, Spanish, Barb and Arabian blood. To this sturdy, tough and even-tempered horse was added blood from Britain – a considerable quantity of Thoroughbred, and possibly a fair dash of cob.

But while the beginnings of the Waler may present something of a mystery, the type that is called by this name, and its principal attributes, are easy to identify. The Waler is a hardy horse standing 15 to 15·2 hands, close-coupled, with an excellent sloping shoulder, a good length of rein, a sensible head, and very good limbs and feet. Carrying much blood from smaller Thoroughbreds, he has a deep chest and plenty of heart-room, and he is a speedy long-strider who grew up the hard way, ranging the huge spaces of the outback.

In the days of the British presence in India, the Waler was greatly in demand as a remount and artillery horse. During the first world war more than 120,000 of these iron-hard horses were sent to the Allied armies in Europe, the Middle East, India and Africa.

The Waler of today, recently designated the Australian Stockhorse, is very popular for working cattle on the huge stations of the outback, and has also shown up very well in endurance riding competitions, where his stamina, toughness and persistent will to win have all stood him in good stead. An official Register has been formed for the purpose of transforming a definite type of horse into a distinct breed.

The stocky, fast-starting, agile and very even-tempered Quarter Horse of the United States has been imported into Australia during the last few decades, and has become increasingly popular for certain types of cattle work. However, many ringers (Australian cowboys) feel that when it comes to racing for miles after fast wild cattle in heavily-timbered bush, there is nothing to equal the long, steady, ground-consuming stride and stamina of the Waler-type Australian Stockhorse.

The breeding of Stockhorses on the station properties is a very straightforward affair: the stallion is turned loose to run in the 'horse paddock' – which may well cover 10,000 acres! – with the band of mares. Often the entire will be a sound, well-built Thoroughbred who has

The Australian Waler, first bred by the early settlers.

been retired from racing in one of the coastal cities, and he soon adapts to the vigorous life of seeking out food and water for himself and his band wherever it can be found. In hard, dry times the blood horse and his band battle just like all other bush dwellers, and in the good, rainy times they thrive; and all the while they take an enormous amount of exercise, which keeps them hard and fit and also proves their soundness very effectively.

The Australian Stockhorse is often not broken in until he is perhaps four, or even five, years old and is fully developed. He is also likely to be fairly convinced of his preference for the wild, free life, but the skilful horse-breakers can soon turn a powerful, half-wild horse into a fairly reliable station mount without breaking his spirit, and these horses, who have been allowed to develop properly before being put into work, can be relied upon to labour hard and long.

Another horse of the Australian bush is the Brumby, a wild horse living in free-ranging bands. Descendants of domesticated horses which strayed, or former saddle stock gone wild, they too are tough, and very wily: they are as sharply-alert to danger as deer, and it is very difficult indeed to sneak up on them, let alone to catch them. However, there have always been hard-riding 'Brumby runners' who have made a living from capturing the wild horses by driving them into skilfully-concealed temporary stockyards, and then selling the best of them for saddle horses.

The American Quarter Horse, already mentioned, has arrived in Australia to stay, and there have also been a number of importations of the spotted, stockhorse-type Appaloosa from the United States. The high-stepping, peacock-necked American Saddlebred is another United States import, used mainly for breeding purposes.

The breeding of Thoroughbreds is an increasingly important activity in Australia, where racing is a major passion. There are now a number of high quality Thoroughbreds being bred in Australia, and several Australian sires have done well at stud in the United States during the last few decades. Virtually all the Thoroughbreds in Australia are descended from parent stock imported from the British Isles.

The most famous Australian Thoroughbred of all time, a racehorse of such renown that all subsequent Australian champions have been measured against him before their greatness can be acknowledged, was the legendary Phar Lap, popularly known as Big Red. Foaled in New Zealand in 1926, Phar Lap won thirty-seven of his fifty-one starts and collected £66,738 in stakes – an enormous amount of money at that time – during just over four years of racing, capturing the imagination and affection of the whole continent. He died in mysterious circumstances in California in 1932, soon after he had spreadeagled the field in his one and only American race, the Agua Caliente Handicap.

A herd of brumbies.

On the great majority of Australian stud farms brood mares are never stabled, not even for foaling. They roam outside all the year round, and many of them drop their foals unassisted in a quiet corner of a large paddock. Some of the better-equipped studs have special foaling paddocks which are floodlit at night so a watch can be kept for any mare in difficulties. Trotters and pacers are also bred in many parts of Australia in the same natural fashion.

Topping a new peak of popularity in Australia today is the Arabian. This breed is used mainly by the 'hobby rider', in many activities ranging from testing endurance contests to colourful costume classes in the show ring. Much of the parent stock of these Arabians came from Britain, which still supplies most of the imports of the breed, though increasing interest has recently been shown in American Arabians, generally of Egyptian bloodlines.

There are other popular imports from Britain. Hackney horses and ponies have been brought in for many years, and there are also general-purpose driving animals used in show ring classes. Heavy horses are not much in demand, for Australia's agriculture is highly mechanised, but a few enthusiasts do breed the big animals, mainly to show them; the attractive and active Clydesdale is particularly popular.

British native ponies have long been established in Australia, with the Welsh pony most in demand. Shetlands are popular, too, and recently a great deal of interest has developed in the Connemara, a number of people suggesting that its qualities would be useful additions to the Stockhorse.

Australian ponies are attractive and handy animals themselves, with their own stud book. They too are descended from various British native pony breeds, and probably also have a strong dash of Timor pony, from the Island of Timor in the Indian Ocean.

New Zealand has many of the breeds and types of horse found in Australia, and in fact much of New Zealand's original stock not surprisingly came from here. New Zealanders specialise in trotters and pacers, breeding them to a very high standard. The Thoroughbred is at home in the two islands as well, and a number of authorities think that New Zealand's excellent green pastures and temperate climate provide an almost ideal environment, superior to that of Australia for the breeding and raising of Thoroughbreds.

New Zealand horsemen are very interested in hunting, and as a result breed very good hunters. There is also a typical type of hardy cow pony to be found there. The product of a blending of various bloods, he stands about 14 to 14·2 hands, and is tough and agile, with exceptional stamina. New Zealand, too, with its great open spaces and timbered backblocks, has its wild horses. Large bands run free in the Lake Taupo district, and thrive there.

THE MIDDLE EAST

The Arab's Farewell to his Steed was a favourite recitation in Victorian drawing-rooms, but even if that sentimental picture of horses in the Middle East had approximated to the truth, poverty, modern warfare, mechanisation, and the riches that flow with oil would by now have changed it beyond recognition – as in many ways they have altered the true one.

There are still plenty of horses in those parts, varying from the bustling little jacks-of-all-trades in some of the Greek islands and the Peloponnese, and the indigenous Turkish ponies, to the frugal-living country-breds still found in quantity in most Arab states. And despite the speedy mechanisation of the Shah's 'white revolution' in Iran, horses are likely to remain essential to large sections of the population of that vast land for many years to come. In countries like Iran it is still cheaper to buy and run a horse than a lorry, and a fact that, however popular vehicles such as the Japanese mini-tractor may be in regions like the Caspian littoral, where small fields are enclosed by irrigation banks, a horse is still the better bet. In Jordan the Bedouin are proud to own a communal tractor for cultivating a few crops in the desert, but for tilling some pocket of plum-coloured earth on a crazily steep mountainside a horse has no equal.

These are the indispensible working horses of the Middle East, many of Arabian or Barb blood or a mixture of the two. In addition there

A quality Iranian horse.

are many well-bred horses, in Syria and Iraq, for instance, used for racing on the Beirut and other tracks in the Lebanon. In some of the sheikhdoms, where oil gushing from the barren sand has produced undreamed-of wealth, many of the sheikhs still own beautiful horses in addition to their Cadillacs. In Bahrein the stallions ridden by a section of mounted police are all bred at the ruler's palace, and Saudi Arabia retains much of its proud horse heritage. In Egypt, where for centuries 'dancing' horses have performed at the Feast of Luxor, the ancient studs, El Zahraa and Ein Shams, are now state-owned and still world-famous.

But while some of the Barbs, ridden in from the Moroccan desert by tribesmen come to honour their king with the wild charge – and last minute halt – of a ceremonial 'Fantassia', are true examples of a breed claimed to be as ancient as the desert Arabian, the majority carry Arabian blood. And throughout the Arab world there are relatively few desert Arabians that are truly what the Bedouin term 'asil', that is, pure-bred with a lineage direct to the few fanatically preserved bloodlines dating back to the eighth century.

In Jordan, however, where horses have been bred from very early times in regions that include much of the Old Testament's 'Promised Land', within a relatively few years a stud has been built up based on animals of the purest blood. King Hussein is acutely aware of anything that benefits or adds prestige to his country. For centuries much of the best stock had been sold abroad, and more recently poverty had forced the Bedouin to abandon the strictly selective horse-breeding of their ancestors. When Hussein realised the priceless heritage that was in danger of extinction, he asked an old friend, a noted horseman and horsemaster, to come to Jordan and remedy the situation.

Jordan is a poor country and there was no vast wealth with which to start a stud, only a tiny nucleus of mares, of the oldest bloodlines, that had belonged to the King's grandfather, the Emir Abdullah, plus the ability to find and trace the history of other suitable animals.

In some ways it was a romantic task. A mare of one of the rarest blood-lines was discovered drawing a Bedouin plough, another was found

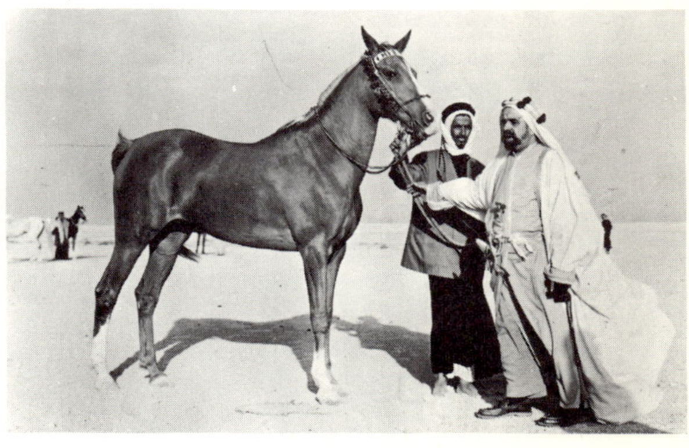

A successful racehorse in Bahrain, seen here with its owner, the sheikh.

among the horses of the mounted police; from his tents far out in the desert, a sheikh sent the king a present of a superb black stallion. A few horses came from the Beirut racetrack, and time and patience were required to check and recheck the lost records of their lineage; pure chance led to the acquisition of an impeccably bred old stallion whose stock is some of the finest.

Through the years the Jordanian Royal Stud, now housed permenently on the outskirts of Amman, has become known all over the world for the purity of its true desert Arabian horses –

animals that seldom gain the stature of those raised on the Continental or in American Arab studs, but which retain the original endurance, frugaility, soundness and beauty of horses exactly suited to conditions in the harsh environment to which they belong.

Iran is another country which has lately become aware of the danger of losing its age-old horse inheritance. The Persian Arab, taller and lacking the characteristic 'dished' profile of the smaller desert Arab, is by some considered to be the older breed, but without help of the fanatical in-breeding and culling practised by the Bedouin on a relatively small stock of horses for centuries by tribes living secluded in the desert, it was impossible to keep the Persian variant 'asil' in the same way. And for many years now horse-breeding, except of the Turkoman, has been sadly neglected. The Shah, however, is an accomplished horseman, and fine horses – Anglo-Arabs (a Persian Arab/Thoroughbred cross), desert horses such as the Darashouli from the south, and the Turkoman from the steppeland south-east of the Caspian Sea – have been kept and bred at the Imperial Stud at Farahabad, outside Teheran. And of late, when mechanisation and settlement of the Turkoman tribes, for centuries the traditional nomadic horse-raisers of this wonderful breed, threatened the country's legacy of horses further, a society was formed, under royal patronage, to encourage breeding on selective lines, and to promote racing and the new-found and increasing interest in other equine sports. A new breed, called Pahlavan after the dynasty and developed from proportions of Persian Plateau, Jadran Arab and Thoroughbred, is proving successful at the Imperial Stud, and an ambitious scheme, liable to take many years to complete, is under way to register all the different Persian breeds.

It is to be hoped that the forethought shown by both Jordan and Iran in preserving the fine horses that are part of their countries' ancient histories, will inspire other regions of the Middle East to follow suit.

THE FAR EAST

The Far East, and in particular the vast steppe, desert, forest and mountain area which now includes the countries of Mongolia, Siberia and China, has long been associated with the horse. The remains of *Equus przewalskii* (the primitive and early ancestor of the Asiatic Wild Horse) have been found together with the tools of palaeolithic Chinese man.

Equus przewalskii przewalskii Poliakoff, more generally known as the Asiatic Wild Horse, is a primitive animal living, if it still survives, in the Altai Mountain region of Mongolia and northern Sinkiang. This area has been the habitat of the Przewalski horse for many thousands of years but man, in encroaching upon the water holes, has made its continued existence extremely precarious. These wild horses were discovered by the Russian explorer Col. N. M. Przewalski in 1881, and eventually Carl Hagenbeck acquired a transport for his zoo in Hamburg. Since then many zoos across the world are breeding the Asiatic Wild Horse. There are now about 200 Przewalski horses in zoos. A stud book is kept in

In Israel, both horses and camels are used by the Bedouin tribes.

A colourful *fiesta* in Seville.

A team of horses taking part in the *Oktoberfest* in Munich.

A gaucho at work on a cattle *estancia* in Entre Rios.

A 'pack trip' cavalcade in Montana.

Small numbers of the ancient Przewalski still exist in the wild state in Mongolia.

Prague, which has the largest herd in captivity.

Further north, in the almost unexplored region of Siberia between the rivers Kolyma Omolon, wild horses are said to exist. The area lies within the same latitude, although thousands of miles further east, as that of the extinct Lofoten ponies. Very little is known about these wild horses, although a few descriptions suggest that they are probably about 12·2 hands high and that their long coats turn white in winter. The horses of the north Siberian Jakuts are probably very similar. Among these people it is not only forbidden to hit a horse but even a bad word said to a horse is regarded as a crime.

Probably the Asiatic Wild Horse, or a variety of it, is the ancestor of all Chinese breeds although in more recent and mediaeval times there were importations from other lands. The horse was native and wild, and was hunted and eaten by Stone Age man before 2000 BC, although the cultural domestic evolution of both man and horse in the Far East is not as ancient as it was in Europe and parts of Asia. There existed in early historical China the legend of a mythical horse-breeder called Ch'eng-ya, meaning 'capable to drive', at a time when horses are thought not yet to have been domesticated.

There are districts in China where horse-breeding is of very ancient origin, particularly that of the Hsia (*c.* 2000 BC), where the invention of the horse-drawn chariot seems to have originated. The Hsia, at least, used white horses with black heads, while their neighbours the Chou, a partially nomadic tribe at the time of their conquest of China who bred vast numbers of horses, preferred yellow horses with red manes.

In the valley of the Hwang Ho, the Shangs, in 1639 BC, developed their feudal institutions on the strength of their horse-drawn war-chariots, and from then on for the next 2500 years the Chinese were a great equestrian people. A number of Shang dynasty two- and four-horse chariot burials have been found; in a Chou tomb twelve chariots with seventy-two horses were found – six horses to a chariot, a double 'terceriga'.

Around 166 BC the northern Hsung-nu or Huns invaded China with 140,000 horsemen. A few years later the western Han emperor Wu-ti heard from his emissary Chang Ch'ien, who was away for twelve years visiting Yüeh-chih tribes (Scythians) of the famous 'blood-sweating' horses of Ferghana. At this time the tribe of the Wu-sun lived in the Ili valley and bred first-class horses. Some of these were acquired for the Chinese emperor.

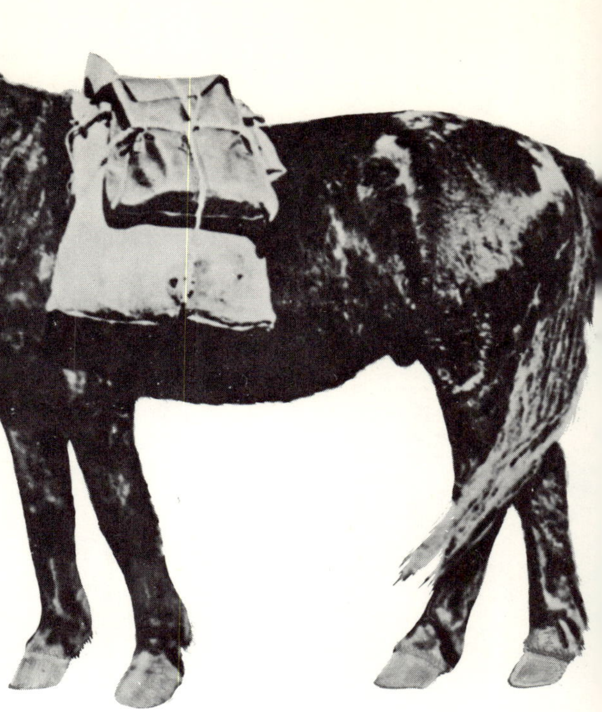

The Mongolian pony. These ponies are still raised in large numbers by the nomadic herdsmen.

The Mongolians are an autonomous people of numerous tribes living in Inner and Outer Mongolia and Sinkiang. Since time immemorial they have been nomadic herdsmen raising vast numbers of horses and cattle. Mongolian ponies are generally nondescript, characterised by long, often flowing, manes and tails, and of a very hardy constitution. The ponies are milked for three months after foaling and the milk is made into *koumiss*. The Sanho is a recognised breed originating from native stock and imported Russian horses. It stands at about 14 hands and is used both for riding and for draught work. Another breed is the heavier Heilung Kiang, bred in the north-west, which evolved through crossing with the Baltic Ardennes, Don, Kabardin and Vladimir Draught Horse.

In north-eastern Inner Mongolia the Hailar and Sanpeitze horses are bred. They owe part of their origin to Orlovs, Anglo-Arabs and Thoroughbreds, all of which were introduced from Siberia in Czarist times. They average 15 hands and are used only for riding.

The Wuchumutsin is an ancient breed of better conformation, standing at around 13 hands. The Chinese Kazakh, however, is far better known and is found in the eastern and north-eastern regions of Sinkiang on the borders of Kazakhstan. These mountain ponies are extremely hardy. In 1841, large herds or droves were exported as far as the plains of India.

The Ili horse is bred in the valley of the Ili, and according to some writers this Ili is of ancient stock and is descended from the Wusun horse of the Han Dynasty, 206 BC to AD 220. It is an active saddle and pack horse. One Ili horse beat 137 other horses in a race over 126 km., taking just over seven and a half hours.

One of the most interesting breeds for colour is the small common-looking Tibetan pony, called Nanfan, which is bred on the grass tableland. They are especially able climbers. Many are white, some yellow dun or brindle. A dun pony was described in 1899 as having a black stripe down the spine with broad black stripes over the shoulders, flanks and legs, and dappled spots over the hindquarters.

The Chinese Sining horse is believed to be a cross between the Nanfan and Mongolian pony. They are considered to be the best ponies of north-west China, being bred and raised on mountain pastures at over 2,000 feet. They are used for pack work between China and Tibet. Other Chinese pony breeds are the Tatung of Tsinghai and the Kjetsui-Sikang pony, and the mountain breeds of south China, all of which are hardy and sure-footed.

Both large and small asses, mules and hinnies are bred in a number of districts in China.

Chapter eleven

International types

THE PALOMINO

The Palomino, 'Golden Horse of the West', is not strictly speaking a breed, and in Britain is not considered as such. The name describes a type of horse or pony; in this case, one noted for its golden coloured coat and white mane and tail.

The origin of this attractive coat colouring is not known, and history does not record when it first appeared. It is believed, however, to be one of the very early colours, mention having been made of it in as early as 800 BC. Palomino-coloured horses appear to have been depicted on the Bayeux tapestry, which was worked in France during the eleventh century.

The British Palomino Society, formed in 1955, states that the ideal coat colour should be that of a 'newly-minted gold coin', but as few animals fit this requirement exactly the Society will accept in its Register a coat colour three shades lighter or darker than a gold coin. The mane and tail, however, must be pure white, not flaxen or light chestnut, and the eyes must be dark, wall or blue eyes not being acceptable. White markings on the face and legs are permissible. Any horse or pony fulfilling the exacting conditions of entry may be registered with the Society, regardless of size or breeding; it is this lack of uniformity which makes the possibility of the Palomino becoming a breed in this country a remote one.

In North America, however, Palominos are recognised as a breed, and two Palomino Associations have been founded in the United States, one in California in 1941 and the other in Texas in 1946.

The purpose of these Associations is to 'provide for the registration, purity of blood and improvement in the breeding' of Palominos, and in 1942 the Association laid down that no horse should be considered for registration if he 'showed coarse, draught horse, Shetland or Paint breeding'. In order to qualify for the breeding registry of Palominos, it is necessary for one parent both of stallions and of mares to be listed in the Palomino Horse Breeding Association. The other parent must be of Arabian, Quarter Horse or Thoroughbred bloodlines.

The body colour qualifications for the Palomino in the United States are similar to those in Britain. However, more flexibility in mane and tail colouring is allowed: up to 15 per cent of hairs may be of a darker hue than the permissible flaxen colour. White leg markings, though, must not reach higher than knee or hock, and, as in Britain, the eyes must both be dark. The American Palomino is similar in conformation to the Quarter Horse (see page 56), and varies in height from between 14 to 16 hands. They are found all over the United States and are used extensively as cow ponies, for rodeos and also, of course, for pleasure riding.

The Palomino colouring tends to darken with age, and it is for this reason that the Register of the British Palomino Society is divided into two sections, one for youngstock, aged one to five years, and the Permanent Register for adult animals. No animal under a year is considered for entry and some of the youngsters who entered the youngstock section and showed ideal colouring as yearlings or two-year-olds may not fulfil their early promise, and will become too dark or even change to dun, cream, chestnut or occasionally grey as they grow older. Manes and tails may also

A Palomino brood mare with her foal at the Kingsettle stud.

deepen, and hence a wide divergence of colour will be noticed among the older youngsters, especially the four- or five-year-olds.

By six years of age, however, the coat colour is considered to be set, and this is, therefore, the age chosen for animals to be inspected for entry into the Permanent Register. It is usual for visiting inspectors to examine the youngstock entries, but Inspection Centres are set up for the older animals, and two inspectors, appointed by the Society, are required to examine the animals for entry into the Permanent Register regardless of whether or not they were entered in the youngstock section. Inspection Centres are frequently held in conjunction with shows during June, July and August, the summer months being chosen as the coat colour becomes darker at this time of year.

In the show ring the judging of Palominos has to be a compromise between those animals who show a perfect, or near-perfect, colouring, and those whose conformation and action is of the highest possible standard. However, colour should not be the absolute criterion. An outstanding animal whose colour is not so good can be placed above one with better colour but poorer conformation, and unless animals have good conformation and movement they are unlikely to head the line-up in any Palomino class, however good their colour. In classes at shows which are affiliated to the British Palomino Society the judges will be those appointed by the Society, and the animals shown must be registered and shown with manes and tails free and not plaited. Among the British native ponies the breeds most frequently showing Palomino colouring are the Welsh Cobs and ponies, and the Connemaras; Palomino Arabs are also, of course, a not uncommon sight.

The breeding of Palominos is a rather chancy business. The crosses most widely used to produce this colour are Palomino cross chestnut, Palomino cross Palomino, chestnut cross albino and Palomino cross albino. A crossing of chestnut and grey has also been known to produce Palomino offspring of a good colour. The most likely crosses to throw a Palomino foal are the first two mentioned, but the results are by no means certain and duns and creams may also make an appearance in the breeding patterns. The Palomino cross Palomino mating is likely to result in 50 per cent Palomino, 25 per cent chestnut and 25 per cent albino, while offspring from a Palomino and chestnut crossing are likely to be 50 per cent Palomino and 50 per cent chestnut.

Palominos, like horses or ponies of any other colour, are used primarily for pleasure riding, and they can be seen as frequently in the showring as the hunting field, driving or jumping. Western riding is increasing in popularity and the Palomino, with its eye-catching colour, is being used more and more in this field of activity with a considerable degree of success.

SPOTTED HORSES

A breed of spotted horses is known to have existed both in Asia and in Europe at least 20,000 years ago. The first evidence of them is to be found in the primitive cave drawings in Lascaux and Pêche-Merle in France. It is in fact through art that our knowledge of the spotted horse can continue to be traced from these early days. Egyptian pictures of about 1400 BC and also much of the early art of Asia depict the spotted horse and clearly demonstrate that he was held in great esteem. We know, too, that the spotted horse was first domesticated somewhere in Asia, about 3,500 years ago, and that from here they spread during the next centuries all over the world.

The greatest horse breeders of ancient times were the Persians, who claim that the spotted horse breed descends from Rakush, the spotted war horse belonging to the hero Rustom, who lived about 400 BC. It is from the Persians that the Chinese obtained their horses – there are spotted horses depicted on vases and wall-hangings and forming part of statues from about 100 BC onwards. One Chinese emperor, Wu Ti, even went to war for twenty-five years with the express intention of obtaining a herd of these coveted horses for himself. During the T'ang Dynasty (AD 618–907) these horses were particularly popular, and were known as the Heavenly Horses.

A spotted horse with leopard markings. The striped hooves are also characteristic.

The more recent history of the spotted horse can be traced in many countries. They have been known in Spain from AD 100, and a fine type of riding horse – some showing spotted markings – was bred there during the following centuries. Spanish horses are thought to have been widely exported quite early on, being sent – among other countries – to Austria, Britain and Mexico. In its early days, around 1760, the Spanish Riding School of Vienna used spotted horses which almost certainly came from Spain, though now of course the School only uses the famous Lipizzaner breed.

From the twelfth century onwards spotted horses are depicted in illuminated manuscripts in Britain, usually carrying saints or nobles. In a manuscript dated about 1397 one of the followers of King Richard II is mounted on a pale chestnut-roan horse with chestnut spots. The first written record dates from about 1685, in the reign of Charles II: a grey horse with red on his quarters was imported into England and named Bloody Buttocks. It is likely that some spotted horses in Britain today are descended from imported Spanish Jennets, a good type of Spanish riding horse with easy paces and an excellent temperament. There is also an old indigenous strain of British spotted horse and a spotted strain of ponies in Wales.

Louis XIV of France is shown in both a tapestry and a painting dated 1672, riding a fine leopard-spotted stallion, so spotted horses must have appeared in France by that date. In seventeenth- and eighteenth-century paintings in Europe spotted horses are frequently to be found as the mounts of kings and nobles, usually performing the classical airs of *haute école*; judging from their conformation, these horses carry some Andalusian blood in them.

In the seventeenth century, too, spotted horses began to be bred in Denmark, and later the famous Knabstrup breed, now sadly virtually extinct in its pure form, was founded. In Sweden and in Holland spotted horses appeared during the eighteenth century, and there is still a spotted strain of the Swedish Gotland pony.

This brief survey will demonstrate the widespread appearance both in time and in place of the spotted horse. In North America the Appaloosa horse has been stabilised through careful breeding into a definite breed bearing particular characteristics and qualities of conformation, markings and ability. This is, however, the only distinct example of a *breed* of spotted horse; the spotted horse otherwise remains a *type*.

In the United States the Appaloosa Horse Club is devoted to the improvement of the American breed. In Britain the British Spotted Horse and Pony Society was founded in 1946, and aims to encourage breeding of spotted strains and eventually to establish a pure breed of well spotted horses and ponies of high quality in other respects. Spotted horses can be of any height, though draught horses and the 'vanner' type are not considered suitable. There are three kinds of marking: leopard – spots of any colour on a white or light-coloured background; blanket – animals with a white rump or back on which spots of any colour occur; and snowflake – white spots on a foundation of any colour. Other characteristics of the spotted horse are a sparse mane and tail, mottled bare skin, white sclera round the eye (as in humans) and vertically-striped hooves.

The showy, distinctive appearance of spotted horses has for hundreds of years made them outstandingly popular for any kind of display work, from carrying heads of state to performing in circuses. They are also both courageous and tractable, qualities which have assured their continuing demand as riding and driving horses and as ponies for children.

DONKEYS AND MULES

The forebears of today's domesticated donkeys are the wild asses of north Africa, which spread to Europe through Greece via Asia Minor. There are still extant today two distinct species of wild ass: the African and the Asiatic.

The history of the ass goes back a very long way. Wall paintings show that the ass was first hunted as meat, in about 6000 BC. Some time between then and 3000 BC it was domesticated. The original purpose of keeping asses is not known, but perhaps – as with horses – it was for food, obtained from the mares' milk and the meat of colt foals.

Early in their history asses were put to work. In Old Kingdom Egypt they were used as an alternative to the ox; before the invention of the wheel or the practice of riding they were tamed and used as pack animals by the nomadic desert-dwellers of the Near East. Donkeys appear at every stage of Hebrew history.

The introduction of the ass into Europe quickly led to its spread throughout the Mediterranean lands where wine was produced. It has played – and continues to play – a very special part wherever vines are cultivated. Its steady surefootedness on steeply terraced hill-sides, its narrowness and delicacy of step which enable it to till the land between the rows of vines, its patient reliability – all these have made the donkey a valuable possession in wine-producing areas.

The wild ass thrives in hot, dry and arid desert conditions. These qualities of survival have stayed with the domesticated donkey which, with the exception of the Tibetan *kiang*, does not thrive in a climate predominantly cold or, above all, wet. The natural climatic limitations of the donkey can be seen from France, where they are quite common in the south and seldom seen in the north of the country.

Following from this, the British donkey is smaller than other types, and is rather less hardy. His coat grows woollier, but even so affords inadequate protection against the cold and damp of the British Isles.

Donkeys in Britain are now mainly kept as pets, though in the nineteenth century they were useful in many modest ways. In southern Europe they continue to work as they have always worked: on the farms, as pack animals and also for riding. They can be found all over Greece – on the mainland and the islands – in Albania, Crete, Cyprus, Spain, Italy and Malta. Spain and Malta both boast leading strains of donkey; the Poitou ass of France is valued as the largest; in the early years of this century Cyprus exported thousands of asses each year.

In the United States donkeys – and mules – were first imported by settlers. Small, poor quality stock was taken over from England, and later good quality Spanish and Maltese stock was imported to improve the situation. On the whole there was more demand for donkeys in the southern than the northern states. This was partly because of climatic conditions – a parallel situation to that in Europe – and partly because the demand for mules was for various reasons greater in the south.

The popularity of mules has dwindled considerably with the advent of generally available motorised transport. Strictly speaking, a mule is the foal of a mare by a male ass (jackass); the foal of a donkey mare by a stallion horse is a jennet. Because mules are almost invariably sterile, to continue to breed them it is necessary to have available a supply of jack-asses. It thus follows that the history of the mule can to a certain extent be traced by that of the ass.

In the ancient world mules were greatly valued for their size. As a result of what is known as 'hybrid vigour' mules often grow to a size greater than that of either of their parents. At a time when horses were no taller than 13 hands, mules were much in demand as the mounts of generals, standard-bearers and others who needed to see and to be seen on the battlefield.

Mules have filled many different roles. Their size and strength makes them valuable as draught and pack animals. When priests were forbidden to ride horses, they rode mules; indeed, in the Middle Ages they were ridden not only by ordinary priests but by Popes, Doges of Venice and royalty. Up to about 1700 Europe was quite sharply divided: in the more northern countries, such as Germany, Britain

A good-sized Cypriot donkey in the mountainous district near Kyrenia.

and the Low Countries, where donkeys do not thrive, horses were mostly to be seen; whereas in the southern lands of Spain, Italy and Portugal, for example, mules were prevalent. In Spain, too, mules were used for carriage-work when carriages became fashionable in the seventeenth century, and no attempt was made to breed carriage horses.

Mules were taken from Spain to the Spanish lands of the Americas, and have been in considerable demand as pack and riding animals. They can carry about twice as great a weight as the indigenous llamas and alpacas originally used for pack-carrying in, for example, Peru. In North America east of the Mississippi they were largely used in the cultivation of cotton and of maize (on which fodder the mule thrives), and in the construction of roads and railways. During the Indian wars the mule-trains provided a vital link between the United States cavalry and its supply bases.

The future of the ass is an insecure one. In Australia and the United States, where donkeys were once in general use as working animals, their value for this purpose has been superseded and they have reverted to a feral state, running in large herds and constituting a considerable nuisance to bushmen and farmers. In the United States, it should be said, there are now several societies devoted to the promotion and protection of the donkey in that country. In the Arab countries they are still in great demand at the moment, but the time will come when their usefulness dwindles, and again there will be a superfluity of them. Perhaps the donkey is only safe in countries where the terrain makes the adoption of motorised equipment impracticable, and others where they are considered as pets. This seems rather unjust, when one considers that the donkey has worked harder in the service of man than has any other animal.

Muleteams such as these, parading at the Cape Show in South Africa, are rarely seen.

PART THREE

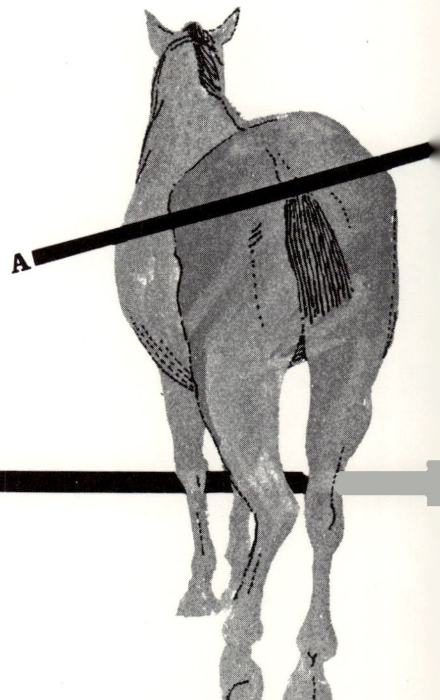

The health of the horse

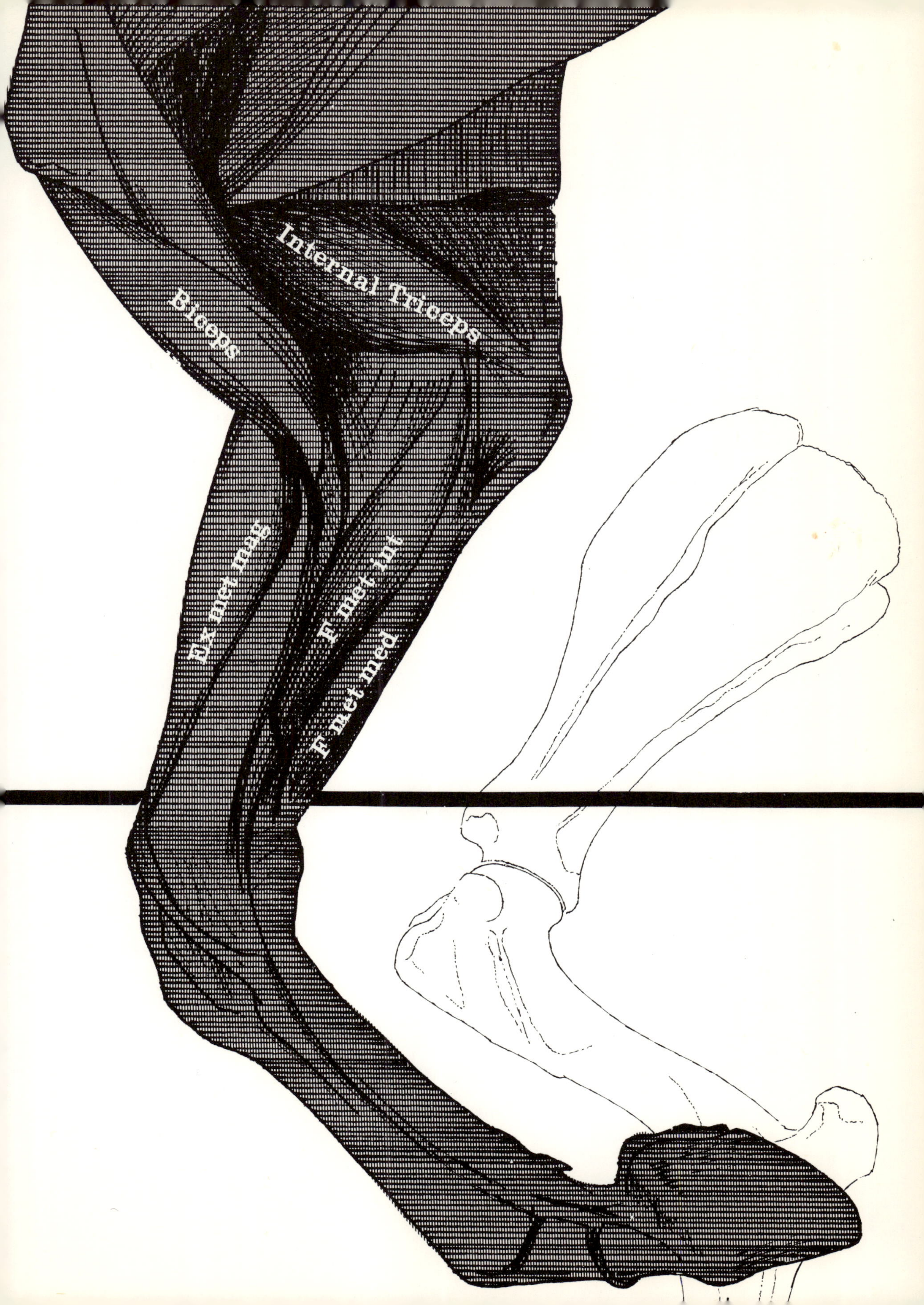
Internal Triceps
Biceps
Ex met mag
F met int
F met med

Chapter one

The physiology of the horse

Physiology is the science which treats of the organs of an animal body and their functions in maintaining health.

The horse's body is composed of a myriad of living cells, each one a separate entity like the bricks of a building; even tissues such as bone and tendon, which to the naked eye appear to be homogeneous, are cellular in nature. Each cell needs to be nourished and to have its waste matter taken away. Every cell and tissue has its own function; the object of the whole organism (the horse) is to feed itself and reproduce.

The framework of the body is made of rigid bones which provide support and facilitate locomotion. The basis of the body is the spinal column, which runs from the head to the tail enclosing the spinal cord–a prolongation of the brain. The nervous system is like a telegraphic system conveying messages; a signal is conveyed from an organ, say a muscle, via a nerve to the spinal cord. It is either dealt with there at the lower level, unconsciously, or it is conveyed up to the brain and dealt with consciously. A message for contraction of the muscle is sent along the appropriate nerve to that muscle. A to-and-fro system like this exists in respect of each structure, be it muscle, blood vessel, liver or urinary bladder.

The ribs, which form the chest, are attached to the spinal column, and the heart and lungs are found within the cavity they form.

The forelimbs are attached to the sides of the chest by muscles, and the hind limbs are attached to the pelvis, which is a bony girdle directly fixed to the spine. The chest is enclosed behind by a sheet of muscle called the diaphragm, which is domed with its concavity facing the abdomen; when this sheet of muscle is flattened by its fibres contracting it creates a suction in the chest. The abdomen lies behind the diaphragm and contains the organs of digestion, the uro-genital system and other soft structures.

Three types of muscle are found in the body: skeletal (which reacts only when stimulated), smooth or involuntary (which is capable of independent activity) and cardiac or heart muscle, also capable of independent activity without outside stimulus.

The skeletal muscles move the bones by contracting, and bend the limbs and other bone-based structures at their joints. When they act the muscles contract and become thicker; they can only pull, they cannot push or exert any action by their lengthening, which is a passive action. Hence every muscle must have its antagonist to produce the opposite action; one muscle bends a joint, its antagonist straightens that joint. The muscles are in a state of 'tonus' or slight tension when they are not contracting. This tonus provides control so that joints do not extend or flex too violently. (Inco-ordination, which occurs in unfit or tired horses, leads to violent movement and consequent damage.) Tonus is responsible for maintaining posture and assists in venous circulation.

Tendons are non-elastic prolongations of muscles, which they attach to bones; tendons are very tough and strong, their function being to transmit the pull of the muscles. Ligaments resemble tendons but are not attached to muscles. They unite bones at joints, holding the bones in place while permitting joint flexion without interfering with movement.

Blood is the fluid by which all cells are nourished. It conveys to the tissues oxygen and all substances necessary for the life and growth of cells, and takes away waste materials, including carbon dioxide. Blood consists of a straw-coloured fluid known as plasma, in which the red blood corpuscles are suspended. These are responsible for the gaseous transport and number about seven million to the cubic millimetre. They present a total surface of some 14,000 square metres (compared with a body surface of the horse of around five square metres). White blood cells are also suspended in the plasma. They are larger than the red corpuscles and their function is largely concerned with inflammation and destroying germs. The fluid part of the blood has the property of clotting when it escapes from the blood vessels; in this way it seals the leak and stops blood loss.

The blood is forced round the body by the action of the heart, a hollow muscular organ with contractions which pump the contained blood into the great arteries. The arteries lead from the heart and continually branch and rebranch, forming ever smaller vessels. The smaller vessels are known as arterioles and the smallest of all, only large enough for one blood corpuscle to pass through at a time, are the capillaries. These minute capillaries are only a fraction of a millimetre in length, and they open and close as required for the local circulation. The walls of the arteries are muscular and can be altered to regulate the flow of blood through them. The walls of the capillaries are so thin that blood cells can pass through them, as can dissolved gases and other substances. This fluid bathes the cells in the inter-cellular spaces; the cells, which also have permeable walls, abstract what they require from the fluid and discharge their waste into it. The capillaries lead into the venules (minute veins) in which the flow is towards the heart. Veins have valves in the form of flaps at intervals on their course, ensuring a one-way flow. The venules unite to form larger and larger veins and finally the venae cavae, which discharge into the right side of the heart.

Next to many veins are the vessels of the lymphatic system, which ramify throughout the body. The thin-walled lymph vessels collect nourishment from the intestines and remove fluid from between the tissue spaces; uniting into larger vessels, they eventually flow into the venous blood stream near the heart. Lymph glands are situated throughout the lymphatic system and 'filter' the lymph as it passes through. Invading germs are held up in these glands and dealt with by the body defences; hence the lymph glands of an area may swell when the part they drain becomes infected.

The heart weighs about nine pounds. It contains four cavities, the left and right auricles and the left and right ventricles; each auricle opens into its corresponding ventricle, and the two halves of the heart are connected outside the organ only. Large veins bringing blood that has circulated round the body discharge into the right auricle. The opening between the right auricle and the ventricle below is guarded by a valve with three flaps, preventing backflow when the ventricle contracts. From the right ventricle the blood leaves through the pulmonary artery, whose orifice is also guarded by a three-flapped valve. The blood from the right ventricle passes to the lungs, where the arteries ramify, the blood in the capillaries comes into close contact with the inspired air and gaseous exchanges take place. The oxygenated blood then returns to the heart through the pulmonary veins, which enter the left auricle. The opening between the two parts of the left heart is guarded by a two-flapped valve called the mitral valve (because of a supposed resemblance to a bishop's mitre). From the left ventricle the blood leaves by the aorta, the great artery of the body. Here again there is a valve. The aorta sends branches to the forelimbs, head and neck and then runs under the spine towards the tail, sending branches to the intestinal organs and the hindquarters. The muscular wall of the left ventricle is much thicker than that of the right ventricle, as the right side of the heart only has to pump blood a short distance to the lungs while the blood from the left side has to be forced round the whole body.

In the muscle of the right auricle is a pacemaker, with fibres that run to all parts of the heart. This initiates a contraction of the heart muscle which spreads throughout the organ, causing a heartbeat. When the heart contracts the blood in it is squeezed out into the arteries, the valves preventing flow in a return direction. The contracting of the heart as a whole begins at the mouths of the great veins, shutting them off and preventing regurgitation. As the heart contracts the auriculo-ventricular valves are forced into contact and stop the blood flowing into the auricles. There is a safety device which allows some blood to flow back into the veins if the heart is labouring in severe exercise: it is seen as a pulsation in the jugular vein in the neck.

The semi-lunar valves guard the openings of the arteries from the ventricles. They open when the pressure in the heart exceeds that in the arteries, and close when the relative pressures change.

After contracting (systole) the muscle of the heart relaxes (diastole) and blood runs in again from the veins.

The output of the heart varies with exercise. Each beat forces into the arteries about 1/15th of the total blood of the body; the total work of the ventricles has been put at about 1/30th h.p.

One of the first charges on nutrition is the work of the heart, which is plentifully supplied with its own blood vessels. The great vessels provide little resistance to the flow of the blood but much pressure is needed to push the blood through the capillaries; these tiny vessels are so small that a piece of muscle the size of a pinhead may contain thousands of parallel capillaries. This fine subdivision causes the blood to present a very large surface to the tissues–a cubic centimetre of blood may be spread over a square metre.

It takes about forty-five seconds for blood from any part of the body to flow to the heart and circulate back again, and during that time a volume of blood equal to that in the whole body will pass through the heart.

The blood in the lung capillaries comes into contact with air and gaseous exchange takes place. Air is drawn into the lungs by muscular expansion of the thorax and is expelled by elastic recoil. In many animals the two sides of the chest are separate, but this is not so in the horse. The lungs communicate, which means that if the chest wall is perforated on one side both lungs will collapse.

In normal respiration, the muscles that act on the ribs by turning them outwards are all that are needed, but violent exercise calls on the muscles that pass from the shoulders and forelimbs and from the head and spine. Thus an exhausted animal will stand with its forelegs spread out and fixed so that these muscles can assist in respiration.

The diaphragm plays an important part in deeper breathing; when its muscle fibres contract the diaphragm flattens and so enlarges the chest cavity. In flattening, the diaphragm pushes against the liver and stomach; these are in turn resisted by the intestines. Such resistance is increased when these organs are full, and the diaphragm movement is hampered.

In the horse the tidal air of quiet breathing is about 3,000 c.c.; in forced breathing this may reach 15,000 c.c. The long soft palate of the horse prevents breathing through the mouth,

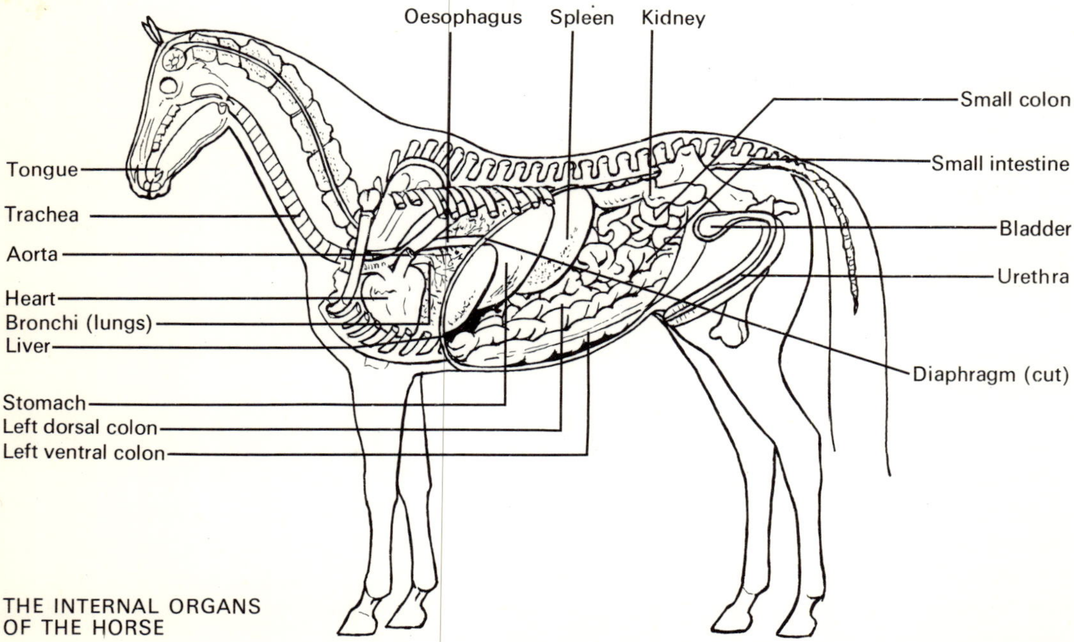

THE INTERNAL ORGANS OF THE HORSE

and air can only reach the lungs by passing through the nostrils and over the turbinate bones. In this way it is drawn over a large expanse of warm and moist mucous membrane before entering the lungs.

Air in the lungs comes in contact with the blood through the delicate covering of the vessels. The blood loses carbon dioxide and takes up oxygen. Venous blood is carried to the lungs by the pulmonary artery, spread out in the capillaries over a large surface, and exposed to the air present in the air cells, which cover an enormous area.

The oxygenated blood returns to the heart and is then pumped round the body, where it comes in contact with the cells. The pressure of oxygen in the cells is always tending to become less than that in the surrounding tissue fluid (lymph), so oxygen passes from the lymph to the cells. The tissue fluid then draws from the plasma within the blood vessels, and the plasma draws from the haemoglobin in the red blood corpuscles.

Blood going to the lungs in the pulmonary artery is a deep dull red. When oxygenated and returning to the heart, it is bright scarlet.

All the substances required for tissues to live must be in the form of proximate principles or be capable of conversion into them. None of these proximate principles can be absent from the diet if the animal is to remain in health. Every tissue element is bathed in lymph, derived from the blood, which carries proximate principles, from which each cell selects what it requires. The selected material from the lymph is made to form part of the protoplasm of the cell. All the operations in the living body ultimately have to do with the storing up or expending of energy, and all the energy expended is supplied by the food and transported by the blood.

Chemical changes are always taking place in living muscle, heat continually being produced and oxygen absorbed. Glucose is constantly taken from the blood, the amount taken being increased during activity. Carbon dioxide is constantly being given off, the amount also increasing during activity.

The abdomen contains the organs of digestion, including the liver and pancreas, both of which have additional functions. The liver is the great chemical laboratory of the body, where most detoxification takes place; it is also a manufactory and store for glycogen (which can be converted into sugar when required), and contains many of the body's defences.

The blood is constantly flowing through the kidneys, where waste material and water are extracted and formed into urine, which passes from the kidneys through the ureters and into the urinary bladder. This is a muscular sac in which urine is stored until a suitable opportunity arises for its evacuation. This is via the urethra through the penis in the male or into the vagina in the female. Both the penis and the vagina are also, of course, part of the reproductive system.

Near the kidneys are the adrenal glands, part of the hormonal system of ductless glands. This system is composed of glands with secretions that enter the bloodstream and so pass to the whole body, producing profound effects upon it. The adrenal glands produce adrenalin, which has been called the hormone of fright, flight and fight, or the emergency hormone, because alarm stimulates its flow and prepares the body for sudden action. Other ductless glands include the pituitary gland (below the brain), the thyroids and parathyroids (in the neck) and the thymus (at the entrance to the chest). The 'islets of langerhans' in the pancreas are ductless glands secreting insulin, which controls the blood sugar level.

The whole of the body is covered with skin, which is divided into two layers, a deeper sensitive one and an outer insensitive one. This latter is constantly being shed as dandruff or scurf. The deeper layer contains nerve endings, blood vessels, sweat and oil glands, and hair bulbs. The sweat glands constantly produce sweat even when the body is at rest, though the secretion then may be so slight as to be invisible. The coat is normally shed in spring and autumn to be replaced by another coat, thicker for the winter and finer in summer. In cold weather small muscles pull the hairs erect, increasing the volume of entrapped air, to give extra warmth; oil glands provide a protective covering for the skin.

The digestive system

The principal constituents of a horse's food are carbohydrates, proteins, fats, water, vitamins, minerals and trace elements. The last four of these can be absorbed through the wall of the bowel into the blood in their original state, but the other three must first be broken down into simpler substances. This breaking down is digestion, which is accomplished by keeping the food at the temperature of a warm bath in the presence of water and enzymes (substances which promote the necessary chemical processes) for several days.

All these processes occur in the alimentary tract, a muscular tube some thirty yards long which runs from the mouth to the anus. It has several large sacculations throughout its length and has a capacity of about fifty gallons. The enzymes are secreted by the glands which line the intestines, and by the pancreas and liver which discharge their secretions into the bowel a few inches behind the stomach.

Food is drawn into the mouth by the lips and tongue and is mixed with saliva, which has a lubricating and transporting function. Three pairs of salivary glands are present in the horse. These are the parotid glands, which lie one under each ear with secretions that pass into the cheeks through ducts opening near the upper molar teeth on each side; the submaxillary glands, which are found between the jaw bones and discharge through ducts on the floor of the mouth under the free part of the tongue; and the sub-mandibular glands under the tongue, which discharge on either side of it. In the course of a day these glands together produce eight or nine gallons of saliva, most of which comes from the parotid glands. Secretion is stimulated by the sight and sound of preparation of food and by its presence in the mouth.

The food is ground between the molar teeth and mixed with saliva. The amount of saliva varies with the dryness of the food; hay takes up about four times its volume of saliva, oats twice their volume and grass half its volume. When sufficiently ground and mixed with saliva the food is swallowed down the oesophagus and enters the stomach (the first of the swellings in the course of the alimentary tract).

For the size of the animal the stomach of the horse is a comparatively small organ, with a capacity of only three or four gallons. It lies high up in the body and is not in contact with the abdominal wall. The lining of the stomach secretes acid, which plays a part in the digestion of certain minerals and also assists the function of another stomach secretion, called pepsin, whose action is to begin to break down proteins.

Proteins are complex substances which are made up of simpler ones known as amino-acids. There are some two dozen such amino-acids utilised by the horse and these combine to form the proteins; as proteins may contain thousands of amino-acid molecules the number of possible combinations is enormous.

Little digestion and no absorption takes place in the stomach and the swallowed food lies in it in layers, being warmed up to body temperature and soaking in the swallowed saliva and gastric juices. Certain foods, such as oats, contain their own enzymes which begin to act in the stomach. These enzymes are destroyed by boiling.

As the horse continues to eat the stomach fills, but allows the first food in to pass out into the small intestine. The rate of flow out of the stomach increases as it fills, and when it is about half full the flow out is approximately equal to that of the food being taken in. When water is taken it passes over the food already in the stomach and goes straight out into the small intestine.

Nothing is absorbed from the food in the stomach – absorption takes place in the intestines. The total surface of the intestines is about ten square yards, but this is increased some ten times by the presence of villi, microscopic projections rather like velvet pile which cover the inside of the bowel.

In the intestine the food comes into contact with more enzymes which facilitate the breakdown of the proteins. These come from the intestinal glands, the liver and pancreas. The secretion of the liver is bile, which intensifies the action of the pancreatic juices and helps convert the fats and oils (the difference between these substances is that fats are solid and oils are liquid at ordinary temperatures) into simpler substances. The juices in the intestine convert the fats into soaps and fatty acids, both of which can pass through the bowel wall into the lymphatics, where they recombine to form fats and oils and are transported by the lymphatic system into the blood.

Carbohydrates occur in the food as starches, sugar and cellulose (the woody part of plants). Starches are converted by digestion into sugars. Most of the sugars cannot pass through the bowel walls and have to be made into the simple sugar glucose before they can be absorbed. Cellulose at this stage is not digested.

It is essential that the intestinal contents should be thoroughly mixed with the enzymes and fluid for digestion to continue, and constant churning is needed to ensure this and to bring the products into contact with the bowel wall so that they can be absorbed. This mixing is brought about by peristalsis or movement of the muscular bowel walls. Several types of peristalisis occur in the horse. In one the tube constricts, and the constriction passes along the gut pushing along some of the food inside. In another movement a length of the bowel is swung like a pendulum, swishing the fluid from one end of the section to the other. Food is passed along the bowel in this way from the stomach to the large bowel. By anti-peristalsis it is also passed in the reverse direction until it has passed in each direction several times. This process continues for several hours.

All this time food is passing out from the stomach into the first part of the intestine. This continues after the horse has stopped feeding though the rate of stomach evacuation slows down as the organ empties. It takes several days' starvation for the stomach to become completely empty; thus with three or more feeds a day there is always some food already in the stomach when a horse is fed.

The small intestine is followed by the large intestines, comprising the caecum (a large diverticulum of some eight gallons' capacity which has stirrup-like muscular bands supporting it), the great colon, with a capacity of sixteen gallons, lying in two large horseshoe curves and about ten inches in diameter, and finally the small colon. The small colon ends in the rectum, whose termination is the muscular ring forming a valve known as the anus.

There is a muscular valve between the small intestine and the caecum, and food is confined to the small intestine until this valve relaxes and permits the material to pass into the caecum. There are no glands discharging here but the enzymes which were added earlier in the tract continue their work. Much nourishment has already been absorbed but much remains, and absorption continues throughout the length of the large intestines. After a time the caecum is partially emptied by the muscular bands contracting and thus shortening the sac, forcing the fluid contents into the great colon.

Digestion continues here for several days, with the contents passing along first in one direction and then the other. As well as the type of digestion which occurs in the small intestine there is a further method involving bacteria which break down the fibrous part of the food, releasing the carbohydrates from the cellulose. This action is accompanied by the production of heat and gas. Some of the gas is absorbed and some passes out as flatus. The churning and bubbling make a noise which can often be heard when standing close to the horse.

Eventually the bowel contents pass into the rectum, where by compression and absorption much of the fluid is removed and the material pressed into balls of dung which still contain about 75 per cent water. These are stored in the rectum until expelled from the body.

The whole digestive process, from food entering the mouth to the residue being expelled as dung, takes three or four days.

The flow of secretions depends upon the sight and taste of food and upon routine. The nature of the secretions depends upon the food to which the horse is accustomed, and strange foods are digested slowly at first because the intestines do not contain the appropriate enzymes. Juices do not flow freely when a horse is alarmed or frightened by strange surroundings or company, and changes of routine also interfere with secretion. Any of these conditions may give rise to indigestion, which leads to abdominal pain (colic). In the common type of colic sections of bowel are thrown into painful spasm. Another form of colic is associated with the production of gas in the intestines, which expand and twist into loops. This may result from injudicious feeding and is exceedingly painful; it can end fatally with a bursting of the bowel. The muscular walls, especially those of the large bowel, may contract so violently that the bowel is telescoped into itself. This condition, which is known as intussception, is also very painful and often fatal.

These conditions should all be avoided without difficulty if due care is taken over feeding: avoid sudden changes of diet, feed regularly in comfortable surroundings, and all should be well.

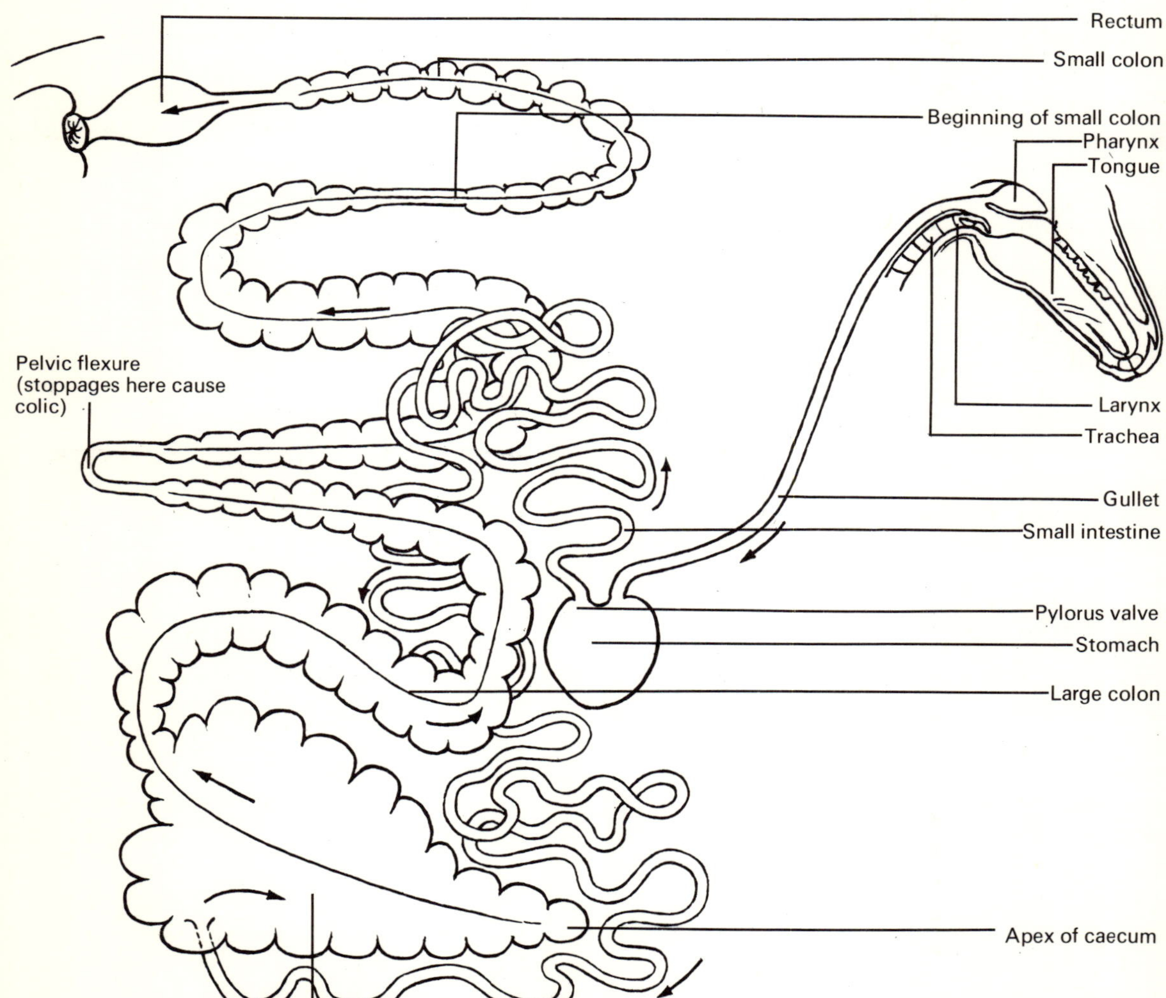

THE DIGESTIVE SYSTEM

Chapter three

The structure of the foot

Whether a horse is standing or in motion, its full weight and its load have to be borne by the feet, often by one foot alone. When a heavy horse, carrying a rider, takes a high jump a tremendous thrust has to be taken by the hoofs. The foot not only has to be enormously strong but it also has to be elastic, durable, and able to absorb shock.

The foot consists of a horny box developed from the skin enclosing the last bones of the limb, which are cushioned by fibro-fatty pads and an abundant blood supply.

The end bone of the leg is the pedal bone (also called the coffin bone or third phalanx) which corresponds to the last bone in the second finger of man–that which is covered by the fingernail. Part of the second pastern bone also lies within the hoof. The navicular bone, lying at the back of these two bones and incorporated in their joint, is the third bone within the foot.

The pedal bone is roughly the shape of the foot and is of pumice-like consistency. When bearing weight the under-surface is not horizontal but at a slight angle to the ground. Blood vessels ramify within the bone, for the foot is filled with blood vessels and an escape route must be provided for the blood when these vessels are squeezed between the bone and the horny hoof. Each wing of the pedal bone has an extension of cartilage (the lateral cartilages) which can be felt above the coronet towards the heels.

The outside of the pedal bone is covered with fleshy material from which horny leaves (sensitive laminae) grow. These are the link between the bone and the wall of the hoof, from which corresponding horny laminae grow and interleave with the sensitive structures. There are about six hundred vertical laminae and each has a hundred or more secondary laminae so that the horny and fleshy leaves are closely and firmly bonded.

The second phalanx articulates with the pedal bone just below the level of the coronet, the joint being called the coffin joint. At the back of this joint and with a bearing on both bones is the navicular bone, which is in the nature of a pulley, over which the flexor tendon runs. This tendon is the deep flexor tendon, whose upper part can be felt in the cannon region as the middle of the three fibrous structures lying behind the bone. It originates in a muscle above the knee and its function is to bend the pedal joint backwards. To pull it forwards is the function of the extensor tendon which runs down the front of the leg and is attached to the upper front of the pedal bone. The superficial flexor tendon inserts into the back of the second pastern bone.

Under the pedal bone lies the fleshy or sensitive sole, which is compressed between that bone and the sole of the foot. It secretes the horny sole through minute papillae. Behind the fleshy sole, towards the heel, is a fibro-fatty pad called the digital cushion, which absorbs much

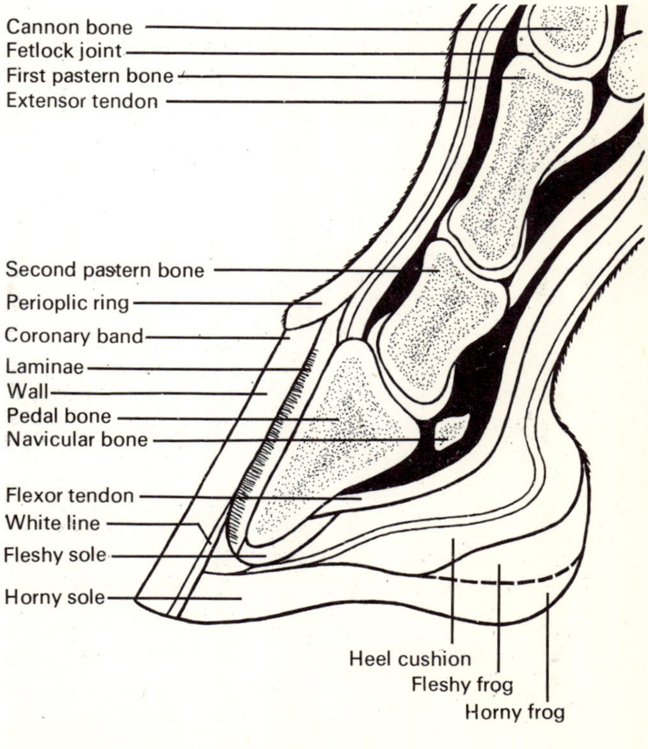

of the concussion when the foot comes to the ground. The digital cushion is composed of elastic and non-elastic fibres enveloped in fat and is compressed between the bones of the foot, the frog and the lateral cartilages. The cartilages provide part of the shock-absorbing mechanism by bending outwards with elastic resistance.

All these sensitive structures are enclosed in the horny hoof. That part which can be seen when the horse is standing is known as the wall and is divided into the toe, the quarters (sides) and the heels, though there are no definite lines of demarcation.

The horn of the wall grows down from the coronary cushion which lies at the top of the hoof and is an extension of the skin of the leg. Microscopically, the horn of the wall has numerous vertical tubes through which nourishment passes down the wall from the coronary band. Horn is a hard, tough, gluey substance in which these hair-like tubes are embedded; it may be as elastic as india-rubber or as hard as a brick, according to its function. The difference is mainly due to the amount of water it contains. Confusion sometimes exists about the terms coronary band (or cushion) and coronet; the band is the structure from which the horn grows, and the coronet is that part of the hoof in which the band lies. The rate of growth is slow, and the horn takes up to a year to grow from the coronary band at the toe to the ground, where it is worn away in the natural state or cut back by the farrier in the case of a shod animal. The rate of horn growth is fairly constant, so at the heels – where the depth of the wall is about half that at the toe – it takes some six months for the heel to grow down. Horn contains no blood vessels or nerves and is insensitive.

The new horn is protected by periople secreted by glands in the perioplic ring just above the coronary band. Periople has been described as varnish; it is in fact a paper-thin covering over the new horn below the coronary band, usually worn away lower down the hoof. It checks evaporation and consequent cracking of the horn below. The periople covers the back of the foot to form the bulbs of the heel. When soaked in water periople turns white.

The walls of the hoof turn inwards at the heels to form the bars on each side of the horny frog. It is as if the cross-section of the hoof were a circle which had been pushed inwards.

It will readily be appreciated that this formation of the wall permits expansion under pressure which could not occur if the hoof were circular. When pressure is taken on the foot the wall expands sideways, and examination of a worn shoe which has not been tightly nailed at the heels will show that the iron has been polished bright by the friction of the moving horn. The wall is thicker at the toe than at the heel, an average specimen will be about half an inch thick at the toe and three-eighths of an inch at the heels.

When the foot is picked up the sole and frog can be seen. The sole lies within the wall and is domed upwards. This design is, of course, stronger than a flat sole would be and acts as a shock absorber by flattening under pressure and exerting a sideways force on the walls. The hind feet have deeper concavities than the forefeet and are oval in shape, unlike the forefeet which are round.

The sole is separated from the wall by the white line, formed of unpigmented, relatively soft horn, which runs around its periphery. This is an important structure from the farrier's point of view, as it is essential that he keeps his nails outside this line to avoid the sensitive structures. Unlike the wall, the sole does not rely upon abrasion to wear it away but flakes off when it reaches a certain thickness – it never requires trimming.

The horny frog lies between the bars. It is of an india-rubber consistency although, like the rest of the hoof covering, it is of horny nature. It transmits weight passed through the fleshy frog to the ground and has a tendency under pressure to expand the heels. It has an anti-slip action when the horse is on hard going and acts like a tyre tread on soft going. Triangular in shape, the frog presents an angle in front called the point of the frog. The depressions on either side are called the lateral lacunae, and the depression dividing the heel of the frog is known as the cleft.

All parts of the under-surface of the hoof are capable of bearing weight, and the structures called upon to do so depend on the nature of the going. In the unshod horse the bottom of the wall and the frog take the pressure on hard going; on soft ground, such as sand, the hoof sinks in so that pressure is also spread over the whole of the sole. Again, according to the nature of the going a braking effect can be made by the wall and the frog (which grips a hard surface and digs into a soft one). In the shod horse weight is taken on the wall and the sole just inside the white line and, with good shoeing, on the frog.

Arteries which supply blood to the foot, accompanied by veins and nerves, run down the back of the leg (between the suspensory ligament and the flexor tendons in the cannon region). Near the coronet these arteries begin to ramify within the foot; there is an extensive plexus of vessels in the hoof (wounds of the foot bleed very freely). Unlike veins in other parts of the body, the foot veins have no valves to stop return flow of blood and the flow through them depends largely on weight being applied and relieved intermittently, thus squeezing out the contained blood into the veins of the legs to pass back to the heart.

In the case of unshod horses the wall is usually worn off level with the adjacent sole, but if the ground is soft it may grow over-long and split or break. On very rough or hard ground the wear may exceed growth and lead to foot soreness.

Feet should not be too small, neither naturally nor—which is much worse—as a result of unskilled treatment. Small feet may take an insufficient grip of the ground and may be structurally weak and unable adequately to absorb concussion. Odd-sized feet should always be regarded with suspicion, as they may indicate long-standing disease. Unduly large feet are also undesirable, as they call for large and heavy shoes which are tiring and, in the long run, damaging to the animal.

Flat feet are usually associated with weakness of the general structure and are most common in animals reared on marshy land. The horn is generally thin and weak and such feet are frequently associated with weak sloping pasterns. Upright feet are deep from coronet to ground and the wall is more than usually vertical. They are associated with upright pasterns, which lead to excess jarring and concussion with evil side effects.

The old saying has it that 'no foot, no horse', and in view of the great strain it has to bear it is not surprising that the foot is a common seat of disease. As the strength and ability of the horse increase it is called upon to do more, and the feet need to be correspondingly more perfect.

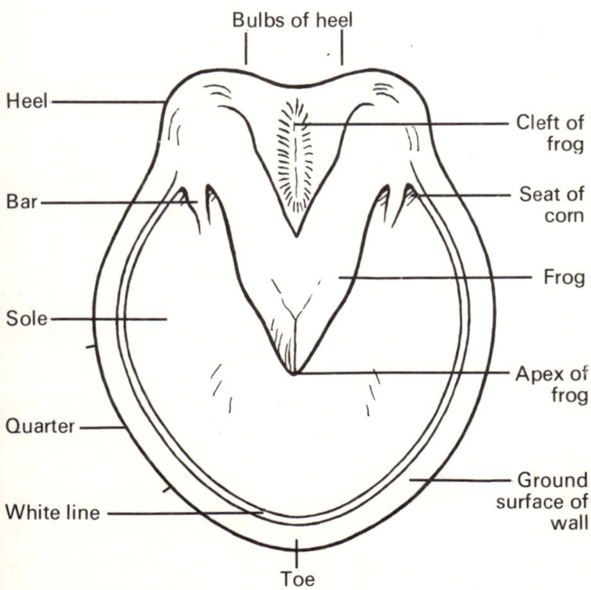

UNDERSIDE OF THE FOOT

A hoof showing the horny laminae on the wall and bars.

Lameness : some causes and treatments

Inherent weakness, which makes tissues less able to stand up to work, may lead to lameness. Some forms of lameness can be prevented merely by taking care; it is significant that cases of foot wounds caused by treading on nails by Army horses dropped abruptly when the soldiers concerned were 'put on a charge'. Sore shins are likely to be found if horses are worked when they are very young; strain leads to curbs; movements which impose great demands on the hocks should not be demanded from horses with weakness in these joints.

Horses are kept for working, not for molly-coddling, but work puts on the limbs stresses that did not have to be borne in the wild. Horses in work jar their feet on hard metalled roads; their legs must bear the added weight of a rider and the strain in stopping, turning and landing from jumps; the natural shock-absorbing mechanism of the foot is partly nullified by iron shoes which increase jar, raise the frog from the ground and interfere with the expansion of the foot, while the nails weaken the wall.

Most lameness in the forelimb occurs in the foot; and in the hind limb the hock is the weak spot. Concussion is the great offender in front and strain behind.

Pus in the foot results from infected material reaching the soft structures in the hoof. This may be grit working its way up, but usually follows a penetrating wound caused by treading on a spike or from a shoe nail being driven into the sensitive structures. A loose shoe which turns on the foot, so that the horse pricks its sole with one of the nails, is a common cause of this type of injury. Penetrating wounds which go through the frog are especially serious, as the injury may involve the coffin joint. Even in those cases where this does not happen it is much more difficult to make an effective drainage hole through the rubbery frog than through the hard horn of the sole.

Sometimes the offending nail can be seen sticking in the foot. With other injuries the site must be found by nipping the foot with pincers to find the sensitive area, or by paring away the sole to reveal the entry hole.

Treatment comprises removing the cause, cutting a hole in the sole to permit escape of pus, tubbing and poulticing. Subsequently a plate of metal wedged under the shoe branches to protect the injured part until it has healed will help to keep out dirt. In some cases antibiotics are needed, especially if the coffin joint has been damaged. There should always be cover against tetanus, as the site and nature of these wounds invite infection with this disease.

Thrush, which can often be avoided, may lead to lameness. It is characterised by an offensive discharge from the cleft of the frog; the cleft sometimes splits up between the bulbs of the heel. The disease is favoured by constant moisture; hence it is brought about by wet and dirty bedding and by neglecting to pick out the feet regularly. Cleanliness and packing the cleft with thick string soaked in tincture of iodine, Stockholm tar or a soluble sulphonamide usually effects a cure.

Navicular disease comes on when a horse is in its prime and almost invariably affects the fore-feet only. It is an inflammation of the navicular bone and surrounding tissues. The bone forms part of the coffin joint and is attached to the deep flexor tendon, by which it is firmly held against the other bones, the firmness increasing as more weight is put on the limb. Changes which occur in the bone include ulceration, rarefaction and bony outgrowths.

Hereditary predisposition, mineral imbalance or improper feeding in youth, bad shoeing and lack of frog pressure, faulty conformation and concussion have all been blamed for the appearance of this disease. Factors increasing pressure on the navicular bone, such as long toes or long sloping pasterns, aggravate the condition, as do short upright pasterns, by increasing concussion.

Pain can be relieved with drugs. The horse's

gait may be improved by shoeing with rolled toes, while leathers will reduce concussion. But these measures are only palliative. At present there is no known cure for the disease and the operation of neurectomy (which is not entirely satisfactory) may be the only way to keep the horse in work.

Laminitis, or fever of the feet, is a disease with many causes. It is an inflammation of the sensitive leaves which unite the horny hoof to the foot. In the acute stage it is exceedingly painful and may turn into the chronic form, resulting in permanent disability. Predisposing causes are heavy-topped animals such as fat ponies, unfitness, hot weather and badly-shaped feet. Exciting causes include overeating (as when a horse gains access to the feed bin, is put in a lush pasture or is given a big feed of strange food), prolonged concussion, and blood poisoning from infectious disease or colic.

Modern antihistaminic drugs and anti-inflammatory agents are used to reduce the reaction to foods, and these, together with forced exercise to drive the blood out of the feet, will often effect a cure if administered promptly; anaesthetising the nerves of the hoofs will enable a very lame horse to take exercise. Cold applications and standing the horse in water help to reduce the inflammation, and a soft laxative diet is indicated.

The chronic form of laminitis may either follow an acute attack or appear spontaneously. Lameness varies but is not always marked. The outlook is always bad but drastic operations on the foot with exaggerated lowering of the heels have sometimes resulted in improvement.

Neglecting to have horses regularly shod leads to several forms of lameness, including corns and strain of the tendons. A corn is a bruise of the sole in the angle between the wall and the bars. A mild corn is dry, but more serious cases become moist or suppurating and may even become complicated by infection spreading to adjacent structures. They are most common in the forefeet, especially at the inner angle of large flat feet. The usual cause is an injury from the heel of the shoe pressing on the seat of corn; when a shoe is left on too long the shoe is carried forward as the horn grows until it is embedded in the seat of corn. Treatment comprises removing the offending shoe and paring away the horn over the corn. If there is infection, drainage must be provided and the horn cut away deeply. Tubbing or poulticing will then control infection. Afterwards a shoe must be fitted that does not press on the affected area.

Ringbone is a serious form of lameness associated with concussion, and may also follow strain from twisting or a blow. Ringbones are bony outgrowths on the pastern bones. They are classified in several ways, popular terms being 'true' and 'false' ringbone. True ringbone affects the pastern joint or the coffin joint, while false ringbone is clear of the joints. If a horse with false ringbone is rested in the early stages of the disease and inflammation reduced with cold applications, recovery is the rule; stubborn cases are sometimes needle-point fired. True ringbone is incurable and the prognosis always grave, although unnerving may enable the horse to continue work.

Although there is a hereditary tendency for splints and they sometimes form spontaneously, strain is the usual cause. In the young horse the splint bones are attached to the cannon bone by fibrous tissue, which changes into bone as the animal grows older. The heads of the splint bones form part of the knee joint and if extra strain is thrown on the inner splint bone head the fibrous tissue joining it to the main bone will be torn; bone laid down during the healing forms a 'splint'. The strain is likely to occur when working a young horse, especially if it is turned sharply or if its conformation is such that the inner side of the knee joint takes a greater share of its weight than normal. Cold applications, rest and pressure bandages are the usual, and satisfactory, methods of treatment. A blemish usually persists, and premature return to work will cause it to be larger than would otherwise be the case.

The treatment of strained tendons still leaves much to be desired. A tendon is said to be strained when some (rarely all) of its fibres are ruptured. This is likely to occur in a young, unfit or tired horse, or when an animal gallops into a sudden change of going. Types of conformation which put the tendons at a disadvantage, such as long sloping pasterns, long toes, low heels or 'tying in' below the knees, all predispose to strain.

First aid consists of cold applications and pressure bandages. Massage is helpful. Modern treatment includes surgical interference by splitting the tendon longitudinally and using a technique whereby a piece of tendon from another part of the body is inserted into the middle of the affected tendon. Various treatments, including X-ray and infra-red therapy, galvanisation, faradisation, darsonvalisation, electro-massage, ultrasonic applications and the 'black box' have all been used, and some startling and extravagant claims made for their efficacy. Firing has its advocates.

Spavin is generally associated with chronic arthritis of the hock and some animals have a hereditary predisposition to it. Defective conformation increases the liability to strain. Rest for two months while consolidation takes place generally results in lameness disappearing; the use of pain-relieving drugs while work continues may result in the final spavin being larger than it otherwise would have been. Refractory cases are treated by needle-point firing, by cutting tendons near the lesion or even by neurectomy.

A strong ligament, the calcaneo-cuboid ligament, lies at the back of the hock, uniting the point of the hock with the back of the cannon bone. The powerful muscles which straighten the hock, for example when a horse jumps or rears, transmit their pull through this bone and put strain on the ligament. If it partly gives way a swelling forms. This is a curb, which is most likely to occur in young horses, especially those with poorly-formed hocks. Resting and cold applications generally limit the damage and the horse is able to return to work, although a blemish persists. It has been reported that lungeing the patient over small jumps after the initial inflammation has subsided hastens recovery.

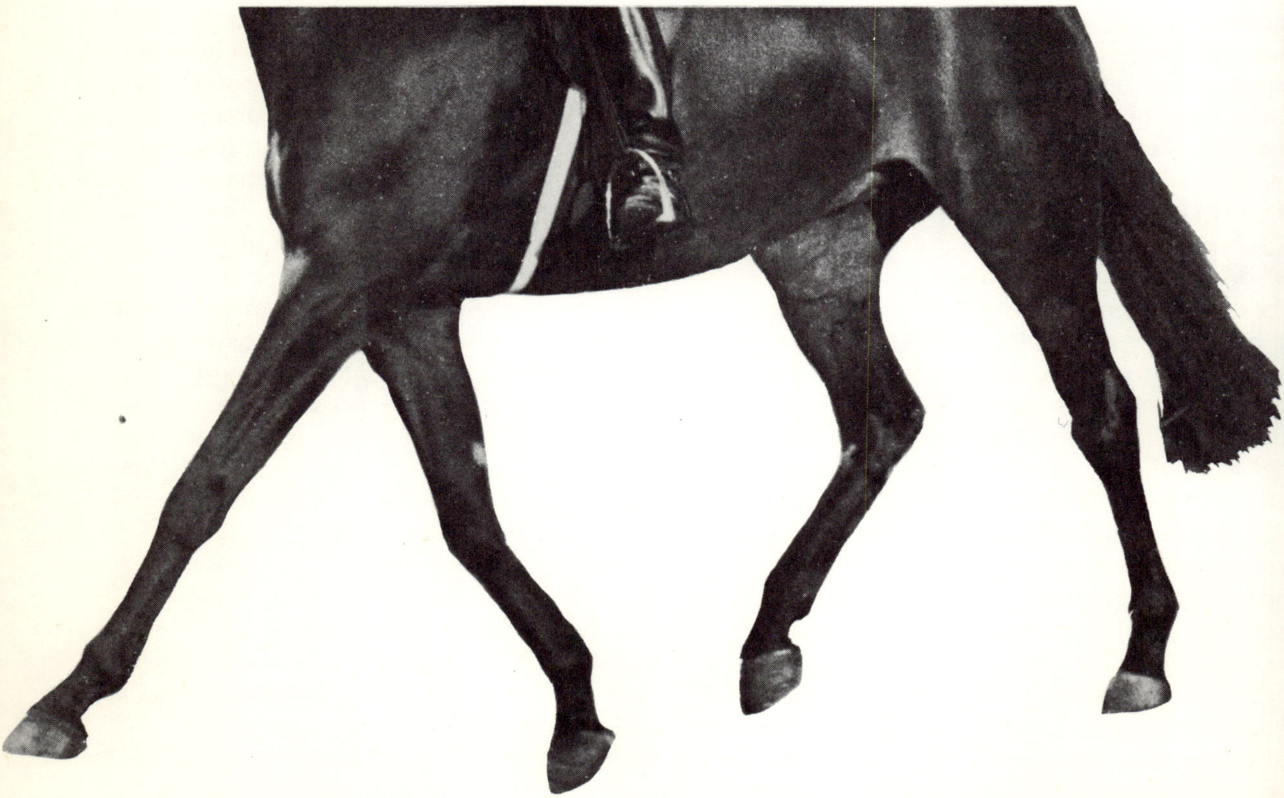

Chapter five

Indications of sickness and health

'The eye of the master maketh the horse grow fat.' Regular observation of horses is essential if they are to be kept in the best of health.

A healthy horse is alert, stands squarely on its legs (resting one hind leg is normal and is not a cause for anxiety, but resting a foreleg indicates disease), its attitude is of awareness of its surroundings, its ears are pricked or moving and it turns its head to see any person who comes near. It is ready for its food at mealtimes and eats its full ration steadily and without quidding. Its droppings are well-formed and moist with no parasites or mucus visible in them, without strong offensive odour, and sufficiently firm to break when dropped to the ground; urine is passed several times a day in amounts of a quart or more and is light yellow and rather thick. The eyes are bright and free from dull spots or blemishes and without discharge. The coat shines, lies flat and is freely movable over the underlying tissues; the dock area is clean and shining and free from worm eggs. The legs are not puffy and the bones, tendons and ligaments are clearly defined. The horn of the hoofs is not cracked or ridged; the frog is large and in contact with the ground and there is no offensive odour or discharge from the cleft; the horse is not lame. Temperature and pulse are normal and breathing is regular and unhurried.

Before examining a horse in detail one should approach quietly and observe from a distance. This is the time when one notices if the animal is resting a foreleg, observes the nature of its breathing, its posture and movements, and looks at the facial expression. Not until one has obtained a general impression of the animal and its condition and surroundings should a closer approach be made.

Temperature

The temperature is taken from the rectum using a clinical thermometer–the stub-ended, half-minute type being best. The mercury should first be shaken down well below the normal temperature and the instrument lubricated with oil, soap or other suitable substance. The tail is quietly raised and the thermometer inserted gently into the rectum; at first the horse may keep the sphincter tightly closed so that there is difficulty in inserting the bulb, but with steady pressure and slight twisting movements a well lubricated thermometer usually slides in easily. The thermometer should be directed to one side so that it is in contact with the lining of the bowel and not just in the faeces–there is a marked difference in the two temperatures. The normal temperature is 100–101°F, but it may rise a degree or so after exercise or in hot weather and is up to half a degree higher in the evening than in the morning. In infectious diseases a rise in temperature is often the first symptom, and during an outbreak regular temperature-taking enables horses in the early stages of infection to be detected, and thus taken out of work and treated without delay. A rise in temperature is part of the phenomenon of inflammation or reaction to infectious disease and helps the animal in its fight against the bacteria or viruses. Although pyrexia (as a high temperature is called) may make the patient uncomfortable it is usually undesirable to reduce the temperature without taking other steps to overcome infection. An exception is heat-stroke, when there is a failure of the heat-regulating mechanism.

Markedly sub-normal temperature is a grave symptom of ill-health.

Pulse and heart

The pulse is a guide to the condition of the circulation. When the heart contracts it drives blood through the arteries–this wave is felt as the pulse. The common sites for observing the pulse in the horse are at the angle of the jaw and inside the forearm; in both these places the artery being felt is under the skin and lies against bone. There is one pulse for each beat of the heart, and the rate and nature of the pulse show how the heart is behaving. The heart beats about forty times a minute but increases with exercise, excitement and when the digestive organs are replete. There is an increase in the rate in the primary stages of fever and when there is inflammation of the abdominal organs. The type of pulse reflects the action of the heart. Thus what is known as a quick pulse, in which the number of beats per minute remains normal but the duration of each pulse is shorter, indicates that there is some heart disease causing the cardiac muscle to contract rapidly. A slow pulse, in which the pulse is longer than normal, may be associated with resistance to the flow of blood from the heart such as occurs in narrowing of the great arteries. Congestion of the lungs leads to a full pulse in which the arteries carry more blood than usual. A hard pulse occurs when the arterial walls are contracted and the vessel becomes less compressible against the underlying bone–this is one of the symptoms of laminitis, in which disease the pulse is also quick, full and bounding.

Sometimes the pulse is intermittent, a beat or beats being missed. The omission may be regular (occurring at, say, every fourth beat) or irregular (occurring at varying intervals). Intermittency may be compatible with apparent health and is more likely to be of consequence when it occurs after exercise. In this condition, knowledge of the individual horse is valuable.

Skin

The condition of the skin at any time is a good indication of a horse's overall state of health. It should be glossy and blooming, not harsh, dull, staring or ragged. Nor should it be hidebound–i.e. tightly adherent to the underlying tissues–but should move freely when the flat of the hand is run down the neck and over the ribs. A hidebound animal may be suffering from malnutrition or a chronic wasting disease such as worm infestation. The condition also occurs in acute illness and when there has been a loss of water from the body, as in diarrhoea.

In acute fever the skin feels hot to the touch; localised heat may indicate inflammation.

Common skin conditions are, firstly, louse infestation, causing rubbing and ragged coat and perhaps louse eggs on the hairs; mange causes stamping and bald patches on the back of the legs, with crust formation; bot eggs can be seen stuck on the leg hairs; ringworm shows as patches of broken hairs with scaly skin; and sweet itch leads to rubbed mane and tail.

Mucous membranes

The unpigmented tissue of these membranes (a 'skin' lining the eyes, mouth, nostrils, vulva, etc.) allows the colour of the blood in the capillaries to show through and thus indicate the condition of the blood. In health the mucous membranes are salmon pink, moist and glistening. If an abnormality is found in one portion of mucous membrane and not in others it points to localised disease. Pallid membranes are paler than normal, indicating a reduced flow of red blood, such as occurs in bleeding or in infestation with blood-sucking parasites like lice. The nature of the pulse may help to ascertain the cause of pallid mucous membranes. An injected membrane (deeper red than normal) indicates inflammation and is an early sign of bacterial or viral infection and raised temperature. When bile pigments are broken down (as in some liver diseases) the yellow pigment they contain circulates in the blood stream and turns the tissues yellow. This is most easily seen in the mucous membranes and is called jaundice. When the blood lacks oxygen, as in diseases of respiration and of the heart, the mucous membranes turn a leaden blue colour and are 'cyanosed'.

Eyes

Discharges from the eyes tend to be watery at first, and in general infection later change to a sticky fluid. When only one eye is affected it is likely to be a localised illness in that eye, but when it occurs in both eyes, systematic changes are probably indicated.

Eyes sunken in their sockets show that fat has been absorbed from the back of the eyeballs, as occurs in chronic wasting disease or in old age.

Nasal discharges

Any discharge should be investigated. A thin watery discharge occurs in the early stages of some general diseases such as strangles and becomes purulent as the disease develops. If the discharge is from one nostril only it may be due to sinusitis from a diseased tooth or infection of the guttural pouch on that side. In health, tears flow from the eyes into the nostrils and watery eyes will lead to a thin nasal discharge. If the tear ducts become blocked then the discharge will run down the cheeks.

Mouth

The lining of the mouth should be moist and salmon pink. An offensive odour may come from diseased teeth or from wounds to the lining of the mouth (often due to sharp teeth); it also arises from alimentary disorders.

Ears

Disease of the ears is uncommon in the horse, but mites in them can lead to head-shaking. Blind horses have abnormally mobile ears: special attention should be paid to the eyes of a horse whose ears are always moving, particularly if it also picks its feet up high.

Lymphatic glands

Situated on the lymphatic ducts of the body, these act as 'sieves' in which invading organisms are held up and dealt with by the body defences. When this happens the glands often swell and become tender, and they may turn into abscesses which burst. This occurs with the glands under the jaws in cases of strangles. Tumour cells are also arrested by lymphatic glands, which then become enlarged. Any enlargement of the lymphatic glands suggests abnormality in the parts draining through them.

Respiratory system

The respiratory rate in a resting horse is about twelve a minute; breathing should be effortless. The rate is increased by excitement and exercise or by high temperature; it may be counted by watching the rise and fall of the flanks or steamy breath in cold weather. Any noise in breathing means that something is wrong. It may be due to laryngeal paralysis, stenosis or tumours. Difficult breathing is shown by dilated nostrils, head stretched out and exaggerated movements of the chest and abdomen; it is due to such things as nostrils or sinuses blocked with discharge, inflammation of the throat, pressure from harness, disease of the lungs or pressure on the diaphragm from the abdominal contents.

If the chest is painful the animal will use its abdominal muscles for breathing and fix its ribs, causing a ridge, known as the 'pleuritic ridge', to appear behind the last rib.

Broken wind leads to a double expiratory movement, in which the flanks collapse inwards as the animal breathes out and then make a further expiratory movement to drive out the air from the lung vesicles, which have lost their elasticity and ability to collapse due to disease.

Coughing is a sudden forced expiratory movement by which undesirable matter is expelled from the breathing apparatus. This may be foreign material, such as particles of dust or hay, or it may be mucus due to inflammation. A cough can also be due to nervous irritability. In infectious diseases such as strangles the cough is harsh, dry and painful at first and then becomes soft, easier and less distressing as the disease progresses and more fluid is present. Harshness is due to the breathing passages being dry; when they are moist the cough becomes softer. A cough may also arise in connection with heart and lung disease as well as being due to local irritation and infection.

Abdomen

When the hind part of the abdomen is less in girth than normal the animal 'runs up' and is said to be 'tucked up'. This occurs in abdominal pain, debilitating diseases and also in exhaustion.

The normal noises made in the bowels are called 'borborygmi' and may sometimes be heard when standing away from the horse. They are sounds of health and can always be heard if the examiner's ear is pressed against the abdomen. The nature of the sounds differs with the area auscultated and only practice will enable the listener to detect the abnormal. Sounds may disappear in impacted colic. In addition to other symptoms, pain in the abdomen (colic) causes the horse to look round at its flanks and to paw the ground. Impaction of the bowel may cause pressure on the urinary bladder and make the horse strain as if to urinate.

Many symptoms of disease are related, and it is important that all the available signs are considered before any decision is made about an illness.

Diseases and ailments

The Skin

The common diseases of the skin are lousiness, ringworm, sweet itch, mud fever and cracked heels.

Lousiness–Horses readily become infested with lice unless well and regularly groomed. Infestation is most common in animals at grass, especially if in poor condition, and thus is most often seen in late winter and early spring, although summer cases are by no means rare.

Affected animals show bare patches from which hair has been rubbed due to the itching caused by the parasites. This may be so intense that the patient does not merely rub out the hair but causes a raw patch. The rubbed places occur on all parts of the body but most often on the quarters and sides of the neck. Most infected animals show apparent pleasure if scratched on the base of the mane and withers.

In Britain two kinds of lice affect horses: biting and sucking; differentiation is unimportant to the horse owner. The species vary in size, from being just visible to the naked eye up to $\frac{1}{16}$ in. long. While the large ones, especially on grey horses, may sometimes be seen from a distance of yards, in many cases prolonged and careful search is necessary before any lice can be found; they are slate grey in colour.

Effective proprietary louse powders and washes are on the market, and the washes are to be preferred. Where practicable, it is best to clip the horse before applying the dressing, as this provides less cover for the vermin and enables the dressing to reach them better. When a long coat is removed it is often found that the infestation is much greater than was at first suspected. The condition is contagious and as the lice lay eggs which stick on the hairs, the clippings should be burned or, if this cannot be done, buried deep in the dung heap.

As affected horses are likely to be in poor condition and may be anaemic, steps should be taken to build them up and to give them iron tonics if needed. A check should also be made to see if red worms are present.

Although lousiness is sometimes glossed over as 'a touch of the frost' it should always be treated seriously.

Ringworm–As the name implies this infestation commonly appears as a circular patch, but in the horse lesions may be of any shape.

The disease is caused by a fungus which grows on the hairs. In a typical case there are areas one to two inches in diameter surrounded by a crusty ring and with a scaly centre from which broken hairs protrude. Sometimes the places itch and make the horse rub raw patches so that sores and scaly areas appear side by side and the ringworm lesions may be overlooked.

Tincture of iodine is an effective and time-honoured treatment still satisfactory; it should be painted on to the affected places daily. In recent years medicaments have been developed which are given to affected animals by mouth, and the later ones are safe, effective and a convenient method of treating the disease, especially when numerous horses are involved or it is not feasible to clip the coat. Another hopeful modern approach to the disease is the development of a vaccine.

In the laboratory several forms of equine ringworm are recognised. They can also infest humans, and care should be taken to avoid contact of human skin with lesions on infected animals.

As ringworm is contagious, precautions should be taken to prevent it spreading from one animal to another by way of grooming kit, saddlery, rugs, and the like, The spores of ringworm are difficult to kill and can persist for years on stable woodwork against which infected animals have rubbed. Flaming with a blowlamp destroys the spores in buildings; obviously great care is needed to prevent fire.

Sweet itch–This malady is not yet fully understood. It affects the long coarse hairs of the mane and tail and is accompanied by irritation which makes the animal rub itself until raw areas appear. The coarse hairs are broken and rubbed out and the underlying skin often thrown into large wrinkles.

Midges have been blamed for spreading the disease, and stabling susceptible animals during the evenings of the sweet itch season (summer and autumn) certainly reduces the frequency of attacks.

Administration of thyroid tablets throughout the month of February seems to relieve the severity of attacks and lessen their number in animals which in previous years have suffered from the disease. Soothing dressings, such as calamine lotion, are of help in treatment and benzyl benzoate is the basis of some popular proprietary treatments. Antihistamines and corticosteroids are also used by veterinary surgeons to control the itch.

Mud fever–Mud fever is a form of dermatitis most frequently seen on the skin of the belly and inside the limbs. It is associated with certain types of soil, and not all horses are susceptible. Prevention consists of washing off the mud if the horse is still wet after work and careful drying with a chamois leather or warm dry bran, avoiding friction; dry mud can be gently brushed off. After cleaning, soothing lotions are applied. Cod liver oil is a good dressing but rather unpleasant to use.

Cracked heels–Cracked heels resemble mud fever of the back of the pasterns. The condition is more often found on white heels than on those with dark hair. The condition can often be prevented by coating the heels with petroleum jelly or lanoline before work. Treatment is as for mud fever, and bandaging hot, oven-dried bran on the washed legs often helps.

Digestive tract

Sharp teeth, colic and worm infestation are the common alimentary ailments.

Sharp teeth–Before food is swallowed it is mixed with saliva and finely ground between the molars or cheek teeth. These teeth move from side to side, and to facilitate their grinding the upper jaw is wider than the lower, with the rows of teeth correspondingly farther apart. The result of this is that as the teeth are ground down the outer borders of the upper molars and the inner (tongue side) borders of the lower teeth are worn less than the rest of the surface and may become so sharp that they cut the tongue and cheeks. The pain from this will make the horse chew its food less before swallowing and it may eat less, too. Failure to masticate properly is shown by the presence of whole oat grains in the dung. Such interference with the chewing may lead to colic and to unthriftiness as nourishment is being lost in the faeces.

The sharp teeth can be felt by holding the horse's tongue and passing one's finger over the tooth margins. Care is needed to avoid being bitten by the horse or cut on the sharp edges of its teeth.

Treatment consists of filing away the sharp edges with a special rasp, and although painless this is often alarming to the horse.

Colic–The term colic is used to indicate abdominal pain and is usually associated with indigestion.

Spasmodic colic is the most common form. Pain comes and goes, with intervals of up to an hour between spasms. The horse looks round at its flanks, paws the ground, lies down and tries to roll, may stand as if trying to stale, and sweats. As the pain becomes more severe the horse assumes an anxious look and may groan. The pulse rate increases, and in severe cases the temperature rises.

Causes of colic include improper feeding, cooked feeds, sharp teeth, worm infestation, large feeds when exhausted, sudden changes of food, and watering after feeding.

Treatment consists of relieving spasm and pain and shifting any irritant from the bowel. In recent years several drugs have been synthesised which are a great step forward in the treatment of colic. Cases which do not respond to first aid will need professional treatment, and as it is important that the veterinary surgeon should know what drugs have been given it is as well that he should be asked to provide the colic drinks for the first aid chest.

Pain also occurs when the bowels are distended with gas (flatulent colic), a condition which may lead to rupture of the bowel. In another form the bowel becomes impacted with a dry mass of food. This causes dull pain and may need treatment for several days before it can be softened and shifted. Contraction of the muscular bands of the bowel wall can lead to telescoping of the gut, with fatal results.

Twist is a form of the ailment in which the mesentery (the membrane by which the bowel is suspended from just below the backbone) becomes twisted on itself, so cutting off the blood supply by pressing on the blood vessels which traverse the mesentery. It is unlikely that rolling leads to twist, as is popularly believed, and if an affected animal gains relief by rolling it seems inhuman to stop it doing so.

In the early stages of colic walking the patient is helpful, though care should be taken not to take it to places–such as public roads–where it would be dangerous or inconvenient for the horse to lie down, as it may be difficult to get it to rise. Nor should it be allowed to roll in a stable, where it might get cast under some fitting such as a manger.

Differentiation of the various forms of colic calls for an internal examination by inserting the arm in the rectum and palpating the bowels. For this, an expert knowledge of the anatomy and situation of the various parts of the alimentary tract is, of course, essential.

Worms–Horses are hosts to three types of worms in the bowel: tapeworms, large white round worms and strongyles (red worms).

Tapeworms seldom cause a problem and in many cases can be dealt with by giving half an ounce of freshly ground areca nut.

Ascarids, as the large white round worms are called, are more alarming in appearance than the other parasites. They are white or yellowish, $\frac{1}{4}$ in. or more in diameter and as much as a foot long. After successful treatment a horse may void a gallon of such parasites. Toluene is a commonly used drug.

Strongyles cause the most trouble and are a frequent cause of unthriftness and even death, especially in young stock. Red worms may be seen in the droppings of infested animals but some worms are so small as to be almost invisible. The worm eggs are numerous in the dung–sometimes many thousands to the ounce–but can only be detected with a microscope. The adult worms live in the bowel, but in the course of their life cycle migrate through the abdomen and may cause irreparable damage to the blood vessels; the worms also cause anaemia and weakness by sucking blood from the gut. The eggs in the dung hatch into larvae, which crawl up grass stems and are eaten by horses, leading to further infection.

Many effective red worm medicines are on the market, but control depends not only upon regular dosing of susceptible animals but upon preventing further infection by dosing horses before turning them out to grass. The collection of droppings from fields, or at least reducing the intensity of infestation by harrowing the dung so that it is quickly dried, is also necessary. When infestation of a pasture has built up over the years it is said to be 'horse sick'.

A worm called the oxyurid lives in the large intestine. The female lays her eggs around the anus, where they can be seen as waxy masses. These worms cause irritation and infested horses rub their tails sore and lose condition. There are effective proprietary remedies against these worms, and carbolic ointment may be used to allay the itching and to move the egg masses.

Systemic diseases

Horses are subject to numerous diseases, and so many viruses have been discovered in recent years that there are as yet no popular names for the diseases which they cause. The invisibility of viruses with orthodox microscopes makes their identification very difficult.

Influenza and some coughs are associated with viruses. Influenza sweeps across the country from time to time, often with years between epidemics. It is highly contagious, and although seldom fatal except in young stock can lead to permanent impairment of affected animals if they are not rested. Fortunately, effective preventive vaccines are now readily available. Affected horses often have a high temperature, they lose their appetites and develop a cough. The disease must run its course, but rest is essential and linctuses or electuaries can be used to relieve the cough.

There are other causes of cough against which there are as yet no effective vaccines. In the treatment of these coughs relief of the symptoms is aimed at, together with suppression of secondary bacterial infection with antibiotics or sulphonamides.

Strangles–This is an infectious and contagious disease mainly of young horses. It is widespread, and as hitherto isolated young animals are especially susceptible, cases are common in young stock which pass through markets or are conveyed in public transport.

Affected animals are off their food, have a high temperature and shiver, and then develop thick nasal discharge and cough. The lymph glands, which lie between the bones of the lower jaw, become swollen and tender. They may swell to the size of cricket balls or bigger. These glands form abscesses full of pus. When the abscesses burst, pus escapes and usually there is a marked improvement in the patient's condition. The pus and all discharges are infectious and can transmit the disease to other horses, either directly or by contamination of the attendant's hands, clothing or other articles which may be affected.

Treatment consists of easing the nasal discharge by steam inhalations, fomenting the abscesses to hasten their 'ripening', warmth and fresh air and careful feeding. In some cases antibiotics and sulphonamides are used, but these drugs should not be administered indiscriminately.

Usually the abscesses are confined to the glands of the head, but in some cases the abscesses appear elsewhere, sometimes with fatal results.

Recovered animals have some immunity for the rest of their lives.

Tetanus–Also called lockjaw, because of the common symptom of being unable to open the mouth, tetanus is a widely-spread disease caused by a germ which occurs in horse dung. The germ will not grow in the presence of oxygen, and the disease is therefore associated with penetrating wounds such as those caused by a horse stepping on a nail. These puncture wounds heal at the surface, leaving the tetanus germs carried in by the nail without air. As horses' feet are usually contaminated with faeces, wounds of the feet are particularly liable to become infected with tetanus, but punctured wounds on any part of the body can provide entry for the disease germs. Large open wounds, because of exposure to air, are much less likely to lead to tetanus.

An affected animal is first seen to move stiffly. Sudden noise will cause the third eyelid to flick across the eye and the tail sticks out stiffly if the horse is moved back. The lack of co-ordination of gait increases hourly, and the patient may be unable to open its mouth to feed. In most untreated cases death follows in days or weeks.

Treatment consists of giving large doses of hyper-immune serum (blood from recovered horses) and such antibiotics as are effective against the tetanus germ. Nursing is important– the horse must be insulated from stimuli which might send it into muscular spasms. The stable should be darkened, nearby horses shifted and noise (even the rattling of buckets) stopped. Liquid nourishment is needed when the jaws are locked.

Immediate protection can be given to a horse at the time of injury by hyper-immune serum, but this protection only lasts for a few weeks. Immunity of much longer duration follows the use of tetanus toxoid, but the immunity is not immediate and this preparation cannot therefore be used in emergency.

Azoturia–Azoturia is one of the 'Monday morning' diseases, ailments which occur when horses are put to work after a period of idleness. It is most often seen in those animals which have been kept on full rations while resting. Symptoms show when the horse is working again: it is in great distress, sweats and trembles and shows muscular stiffness and hardening of the muscles, especially those of the quarters. If urine is passed it is dark brown from the muscle pigment excreted by the kidneys. It is essential that an ailing horse should be immediately rested and taken to a stable by horsebox even if there is only a short distance to go. Professional treatment is required and should be obtained without delay.

Broken wind or pulmonary emphysema–This is a dread disease of the lungs due to rupture of the microscopic alveoli in which the bronchioles terminate. The walls of these alveoli burst and the area of membrane available for gaseous exchange during breathing is decreased. This causes the animal to breathe more rapidly in order to get enough oxygen into its blood, but because the ruptured alveoli have lost their elasticity so an extra effort has to be made by the abdominal muscles to expel the contained air. As a result, the horse makes two movements of its belly when breathing out, and the tense border of the ribs forms a prominent arch which is characteristic of the ailment. As the disease advances more alveoli burst and the distressed breathing, at first only noticeable on exertion, occurs during light work and finally at rest. Coughing of a peculiar 'hollow' type develops and often flatus is passed with each cough. Circulation in the lung tissue is restricted, throwing a burden on the heart, which enlarges to cope with the additional load. The enlargement alters the valve seating in that organ and valvular disease of the heart adds to the horse's distress, because blood intended for the lungs is pumped back into the veins.

There are numerous theories about the cause of broken wind, and probably more than one exciting factor exists. The nature of the food given seems to play a part, and certain hay crops have been associated with a high incidence of the disease, possibly due to an allergy to some fungus on the hay. Horses showing severe symptoms have recovered when moved to other areas – these were probably cases of allergy rather than of true broken wind.

The drugs known as corticosteroids often reduce the symptoms, and careful management may enable lightly affected animals to continue work. Tests should be made to try to find if any food exacerbates the disease in a particular case, and care should be taken to ensure that affected animals are only worked when their stomachs are almost empty. The use of materials other than straw for bedding and the removal of hay from the diet often benefits the patient, as does cod liver oil and the damping of food to keep down irritant dust.

Roaring – Another respiratory disease is roaring or whistling. The names denote the type of noise made when the horse is breathing hard. The disease is associated with paralysis of the nerve leading to the vocal cord. It is nearly always that on the left that is affected, and the symptoms are seen most often in big horses such as heavyweight hunters. The paralysis allows the cord to impinge on the airway in the larynx and the horse cannot get enough air through the narrowed channel, so breathing becomes laboured when exertion is called for. The flaccid vocal cord vibrates in the current of air as the horse breathes in, causing a roaring or whistling sound. The cord is blown out of the way and does not impede the outward flow of air from the lungs, and the noise is thus heard only on inspiration.

The noise, especially in the early stages of the disease, may be intermittent and slight; whistlers may become roarers; cramping of the larynx by making the horse raise and flex its head may increase the disability. The harm in roaring results not from the noise but from the inability to breathe in enough air, causing partial asphyxiation.

Treatment is surgical, either by inserting a tube into the windpipe below the obstruction or by removing the offending vocal cord through an incision into the larynx. There are objections to both these techniques, although in most cases they do enable the patient to continue work. A new technique, in which an elastic prosthesis is attached to the vocal cord to pull it aside, holds promise.

Injuries – Injuries are of course common, and range from scratches to severe lacerations or even to severing a limb when barbed wire becomes wrapped around the leg of a frantically struggling horse. Terrible injuries are sometimes inflicted in motor accidents.

Harsh antiseptics and disinfectants have no place in modern treatment, in which the aim is to remove foreign matter and severely damaged tissue as gently as possible – generally by a fine jet of running water; to stop bleeding by pressure; to bring the edges of severed tissue together by suturing or bandaging; and to control infection with antibiotics or sulphonamides. Sedatives, tranquillisers and pain relievers are nowadays quite readily available and are used as needed to ease the patient and to prevent interference with the wound.

Minor infected wounds are poulticed after cleaning. Wounds of the feet caused by stepping on sharp objects need treatment to promote drainage, and it is even more important than with wounds inflicted elsewhere to ensure that the horse is protected against tetanus.

Saddle sores are treated on general principles, with especial attention to removing the cause. This is done not by putting padding over the sore but by building it up elsewhere so that the saddle no longer touches the injured part.

Despite its efficient natural protection the eye is injured at times. Common causes are hay seeds (especially if the old-fashioned overhead hay racks are used), and injuries from thorns and whip lashes. Great care must be taken in removing foreign bodies and early removal is desirable, as such things as awns become more firmly embedded as time passes. It is wise to have eye ointment in the stable medicine chest and to apply it several times a day after removing a foreign body or in treatment of other forms of injury to the eye.

Finally, weeping of an eye may be due to blockage of the tear ducts, the passages from the eyelids to the inside of the nostrils down which tears normally flow. If they are stopped up by sandy material or discharges from the eye the tears run down the cheeks, often scalding them. Treatment is given by blowing the ducts clear with a catheter, a simple operation to perform, but not one for the layman to attempt.

Ageing by the teeth

Although an opinion about the age of a horse can be formed from some aspects of its general appearance, such as the frizzy tail of a foal or the drawn face of an old animal, it is by examination of the teeth that age is usually determined. The assessment is made by determining which teeth have erupted and by the wear, shape and angle of growth of the front (incisor) teeth. The principles of ageing are simple, but many variations and difficulties arise in practice and even experts often differ.

Horses have two sets of teeth during their lifetime. The first set (known as milk or deciduous teeth) erupt early in life and are later shed; they are followed by the permanent teeth, which have to last a lifetime. The milk teeth are not simply pushed out by the permanent ones but have their roots absorbed and are then shed to make way for the second set.

A horse has twenty-four milk teeth, made up of six upper and six lower front teeth (incisors, nippers or pincers), and three upper and three lower molars (grinders or cheek teeth) on each side. These are replaced by six permanent incisors in each jaw and six permanent upper and lower molars on each side – a total of thirty-six. In addition, four teeth known as tushes erupt, one upper and one lower on each side of the jaw between the incisors and the molars, but nearer the former than the latter. Tushes are normally found only in males, but small ones not infrequently occur in mares.

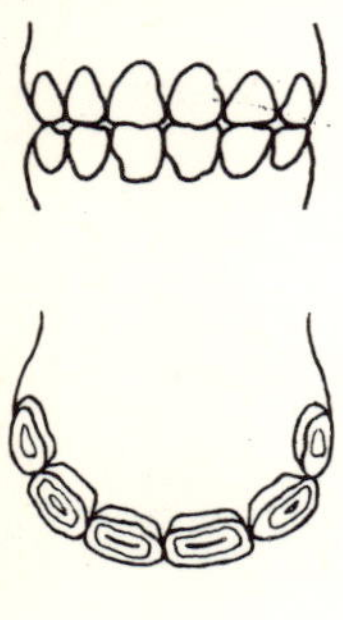

2 years
(all milk teeth)

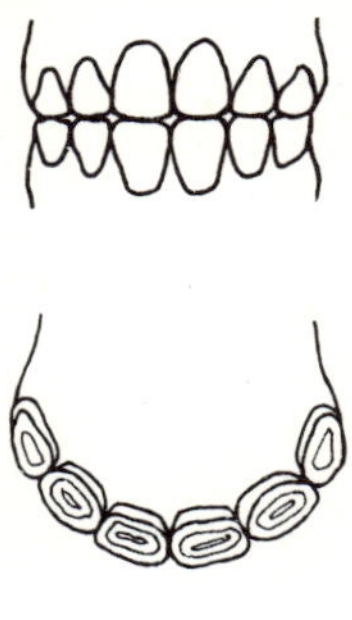

3 years
(first permanent teeth in front)

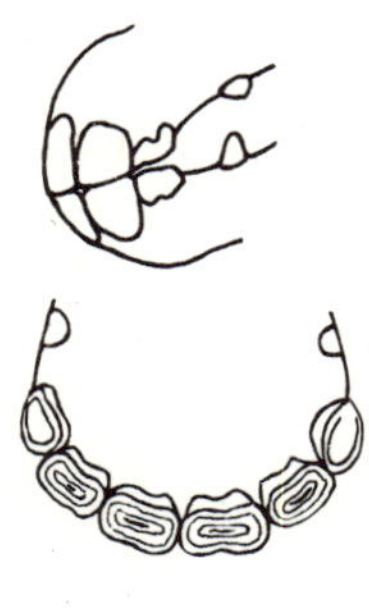

4 years
(last milk teeth in back)

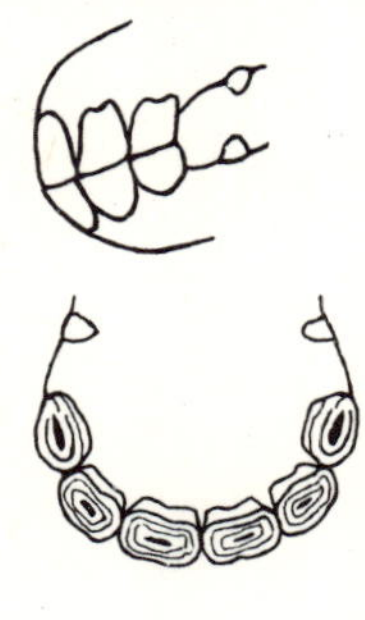

5 years
(all permanent teeth)

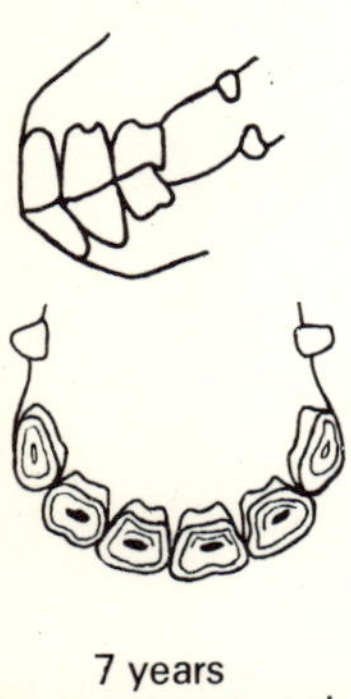

7 years
(hook appears on corner incisor)

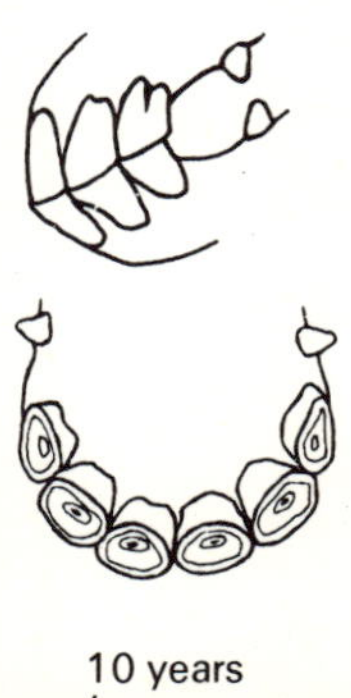

10 years
(Galvayne's groove appears)

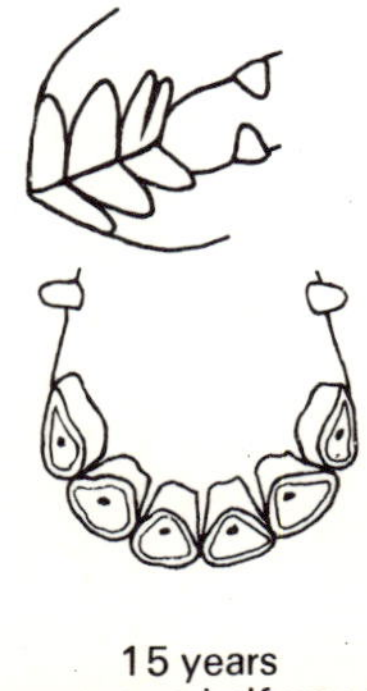

15 years
(groove grows halfway down)

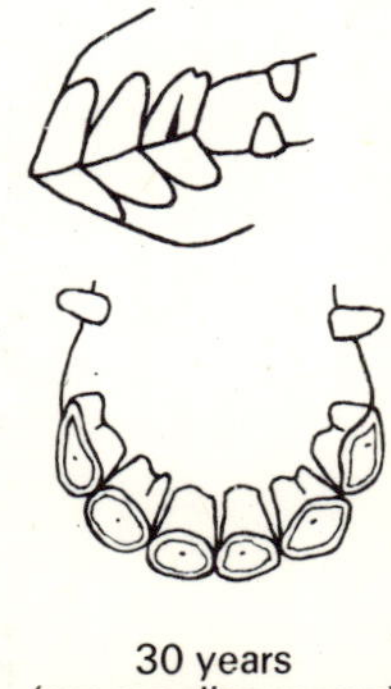

30 years
(groove disappears)

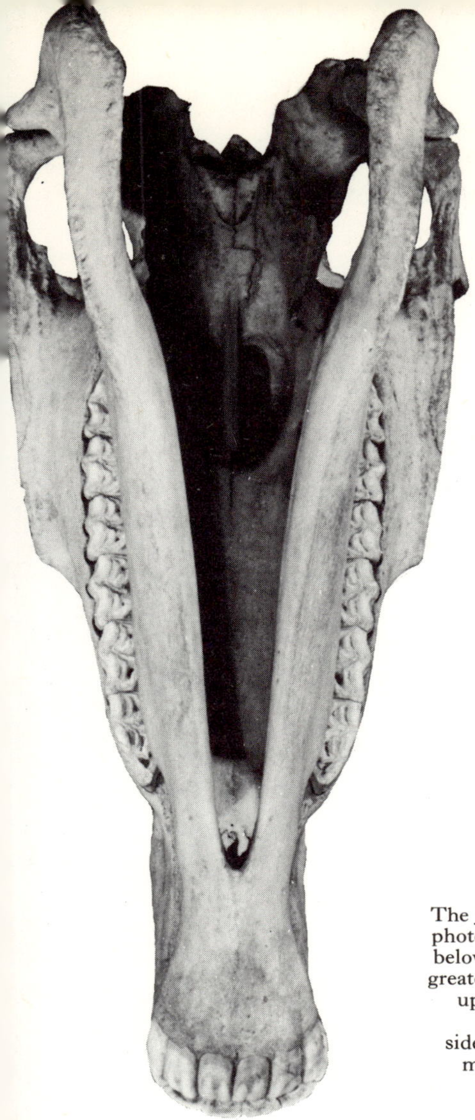

The jaws of a horse photographed from below, showing the greater width of the upper jaw which enables the sideways grinding movement to be made.

Wolf teeth are small vestigial teeth in front of the normal molars.

Up to the age of two years and six months (when the first permanent incisors appear) age is ascertained from the number of teeth in the mouth. By the time it is ten days old a foal has four incisor teeth; these are the two middle pairs, called the centrals. The next pairs (the laterals) appear at four to six weeks, and the corner pairs show at six to ten months. By this time, therefore, there will be a complete incisor set of six teeth in the upper and in the lower jaw.

It is important to distinguish between the temporary and permanent incisors, though this can be difficult. The main differences are that the temporary incisors have a distinct neck, the root being much narrower than the exposed crown, that milk teeth are whiter than permanents, and that there are five vertical ridges and grooves on the outer side of the temporary teeth while in permanents these ridges are more prominent and are two in number on the upper teeth and single on the lower.

The wear of temporary teeth is not such a consistent guide to age as that of the permanent teeth, but signs of wear do appear on the crowns – on the centrals at twelve months, the laterals at eighteen months, and on the corners at two years.

All the temporary cheek teeth have erupted by the time a foal is two weeks old.

The first permanent incisor teeth appear at two and a half years. When permanent and temporary teeth exist side by side differentiation is usually easy because of the difference in size. The laterals appear at three and a half years and the corners a year later at four and a half years, by which time the horse has a 'full mouth', the tushes and permanent molars all having erupted earlier. The tushes of the male come through at four years and dispel doubt about whether the incisors present are of the first or second set.

Although the incisors erupt at approximately the times given, they are not immediately in contact with those in the opposite jaw and it is only about six months after eruption that they start to wear against their complementary teeth.

The eruption of the permanent molar teeth is also a guide to the age of the animal, but the would-be observer may have difficulty in getting even a glimpse of the back teeth of a restless young animal and may well decide that the information to be gleaned from the incisors will suffice his purpose. In any case it is often difficult to differentiate between permanent and temporary cheek teeth, and observation may at best be confined to the number of teeth present.

The small wolf teeth are readily distinguished from the other molars. For practical purposes these may be ignored in ageing, and the remaining cheek teeth numbered 1 to 6 from the front backwards. Numbers, 4, 5 and 6 are not preceded by milk teeth.

The fourth teeth appear at between nine to twelve months, the fifth at two years and the sixth at four years, by which time all the temporary molars will have been succeeded by permanent grinders.

All the permanent teeth are present and in wear by the time a horse is five years old. Thereafter, age is determined by changes in the incisor teeth. It must be understood that these teeth continue to grow throughout the horse's life, pushing out through the gums to compensate for the wearing away of the crowns. Attrition takes place on the biting surfaces (tables) and changes in appearance occur as the tables wear. The rate of wear is fairly uniform, but it can be misleading when animals are reared on sandy soil, which has an abrasive effect, or when, as with Thoroughbreds, they are given hard food early in life.

At five years the corner teeth are just coming into wear, but the enamel which covers the free portion is rolled over and unworn and the teeth have sometimes been likened to seashells in appearance. An unworn permanent incisor has a depression on its surface called the infundibulum or mark, due to an in-folding of the enamel. As the crown is worn down two rings of enamel are seen. It takes about three years' wear to grind the teeth down to the bottom of the in-folding, during which time the mark shows as a dark hollow stained by food.

Thus a five-year-old horse shows the marks on all its incisor teeth but in a six-year-old the marks have gone from the centrals, at seven from the laterals, and at eight from the corners. Although the infundibula have worn away in each case, the sites may still be indicated by irregular rings.

This progress applies only to the lower teeth. Those in the upper jaw have deeper marks which take twice as long to wear away. The upper jaw changes are less consistent, but roughly speaking the marks will have gone from the upper centrals at nine years, the laterals at ten years, and the corners at eleven years.

At seven years it is usual to find that the back corner of the upper corner incisors on each side is not in wear; it forms a projection known as the 'seven year hook'.

The disappearance of the marks is the most accurate indication of the animal's age and it is usual to speak of a horse over eight as 'aged' or 'past mark of mouth', but there are other indications of age after this time and the term aged is also used to mean horses over fifteen years.

At ten years a depression appears in the middle of the tooth on the outer surface of the upper corners. This is 'Galvayne's groove', which by fifteen reaches half way down the tooth, by twenty goes all the way down, by twenty-five has half grown out and by thirty has disappeared. These ages are only approximate, for the groove is not consistent.

The 'dental star' is a yellowish-brown mark which appears on the tables of the incisors at about eight years; it is often not seen on the corners until later. It comes just in front of where the infundibulum was.

The shape of the tables also alters with age. At first they are elliptical, narrow from back to front; by the time the mark has disappeared they are still elliptical but squarer than before. In the aged horse the teeth are triangular with the apices towards the tongue.

When the incisors erupt through the gums they do so more or less at right-angles, and viewed from the side with the head horizontal the incisors form a vertical straight line. As they wear down they project forwards more until, in the very old horse, they are nearly horizontal.

When ageing horses it must be remembered that Thoroughbreds have an arbitrary birthday on 1st January and other horses on 1st May. Teeth erupt earlier in horses bred under intensive systems than in those on free range. The upper teeth usually come through earlier than those in the lower jaw, but being bigger they wear more slowly and look younger.

Chapter eight

Modern veterinary practice

The most important development in veterinary medicine is that recent advances in anaesthesia have enabled great strides to be made in veterinary surgery. Not that the older anaesthetics did not serve their purpose in preventing pain–that could be done by both local and general anaesthesia–but the new drugs and the apparatus for their administration (mainly by tubes which are passed through the mouth and larynx into the windpipe) permit safe anaesthesia for much longer periods. A horse to be operated upon has the surgical site prepared beforehand and is brought into the padded operating theatre. Here it is injected with a short-acting anaesthetic and sinks to the ground. The endo-tracheal tube is passed in the unconscious animal, which is then moved by crane on to the padded operating table whose top lies flush with the floor. The endo-tracheal tube is coupled to the anaesthetic machine and the gaseous anaesthetic takes over as the effect of the injection wears off. The operating table rises to a comfortable height for the surgeon. Surgical drapes and tourniquets are applied unhurriedly, there is time for X-rays to be taken during the course of the operation, the surgeon approaches his task deliberately in the knowledge that all will be well with the anaesthetic even if he has to spend hours on his procedures.

Associated with anaesthesia are drugs known as tranquillisers or ataractics. These lower the animal's reaction to what is going on around it, without impairing its consciousness. This really just means taking away fright. Used in conjunction with local anaesthetics for short operations they produce the condition described as 'I don't care and I don't feel', greatly facilitating interference with an animal which might otherwise resist because of nervousness. Ataractics alone are used as 'chemical restraint' for painless but alarming procedures such as rasping teeth.

A bonus has been the great value of tranquillisers in the treatment of colic. The relaxation they induce often enables colic to be cured by one injection–a big change from former times when colic cases frequently entailed repeat visits spread over several days.

Pain-killing drugs are available which enable horses with chronic painful lameness to work in comfort. In the past such crocks were often worked at night, when their distress would be less obvious to the public, or the affected part was 'unnerved'.

Bone surgery has shown much progress in recent years and improvement in anaesthesia has also been of great assistance in this field. The surgical opening of joints and open surgery on bones are now commonplace, with antibiotics adding to their safety.

The problem of fracture of leg bones is still largely unsolved. There are many reasons for this. Except in the case of stud animals anything less than complete recovery is unsatisfactory (in other surgical fields a persistent unpainful lameness or even amputation may be quite acceptable), the great weight and strength of the patient call for extremely strong splinting, whether internal or external, and the cost of operation and maintenance may render surgical treatment unjustifiable. Nevertheless, many fractures which would have necessitated slaughter thirty years ago can now be successfully treated.

It must, however, be realised that some highly-publicised treatments of fractures have been those in which a splinter of bone was torn off, and not fractures through the entire bone resulting in angulation of the limb. Such splinters are screwed back into position and the bulk of the limb bone, whose integrity has not been interrupted, is able to bear the weight of the animal. This procedure, which entails cutting down on the break, converts a simple fracture into a compound one, with all the risks of infection, and it is here that the advantages of modern anaesthesia and the ability to use aseptic techniques with the added protection of antibiotics are shown. This type of operation does, of course, demand full hospital facilities.

In recent years an improvement in internal fixation has been the adoption of fixation under pressure, known as compression fixation. Among the advantages of this technique are that there is stability as soon as the patient recovers from the anaesthesia, there is little pain, and the rehabilitation period is short. In fractures near joints the method results in little excess bony deposition, mobility is retained and the animal is able to race again. In this technique engines are used to press the broken ends of the bones together under pressure before they are held in place with metal plates and screws.

Not only veterinary surgeons have been involved in the development of this system, for metallurgists have been called upon to solve the problems of producing a metal strong enough for the purpose and possessing the necessary qualities of malleability and freedom from undesirable chemical and electrical actions when buried in the animal tissues.

Abdominal surgery is more general than it used to be, and some horses suffering from colic which before would have died are now being saved. Not all cases are suitable for surgical intervention, and it is not universally available, as the sick animal must be conveyed to hospital for the inevitably costly operative treatment. However, surgical intervention is increasingly being used on valuable animals. Here modern anaesthetics again show their advantages.

The traditional operation for roaring is still performed but a technique using a prosthesis in the form of an elastic strip to replace the paralysed laryngeal muscles holds out promise, for not all horses subjected to the usual laryngeal ventriculectomy make a complete recovery.

Docking has been illegal in Great Britain for two decades, and long tails have not caused the accidents that gloomy prognosticators foretold.

The introduction of phenothiazine as therapy for redworm infestation in the 1940s was a great advance. It is still widely used although there are now other preparations which have certain advantages; compared with the old worm remedies they are safer and much more effective. These other drugs destroy some species of worms which previously had to be treated with drugs which themselves constituted a danger to the patient.

The two groups of drugs known as chemotherapeutics (of which the sulpha drugs are examples) and antibiotics (such as penicillin) have revolutionised the treatment and prevention of infectious diseases. Chemotherapeutic agents are synthesised chemicals which have the power to kill bacteria and other organisms without injuring the cells of the host. They enter the blood stream of the patient and so reach all parts of the body. Piperazine, acting on worms by narcosis, and quinuronium sulphate, destroying single-celled blood parasites, fall into the category of chemotherapeutic agents.

Antibiotics are anti-bacterial substances of biological origin. They are derived from

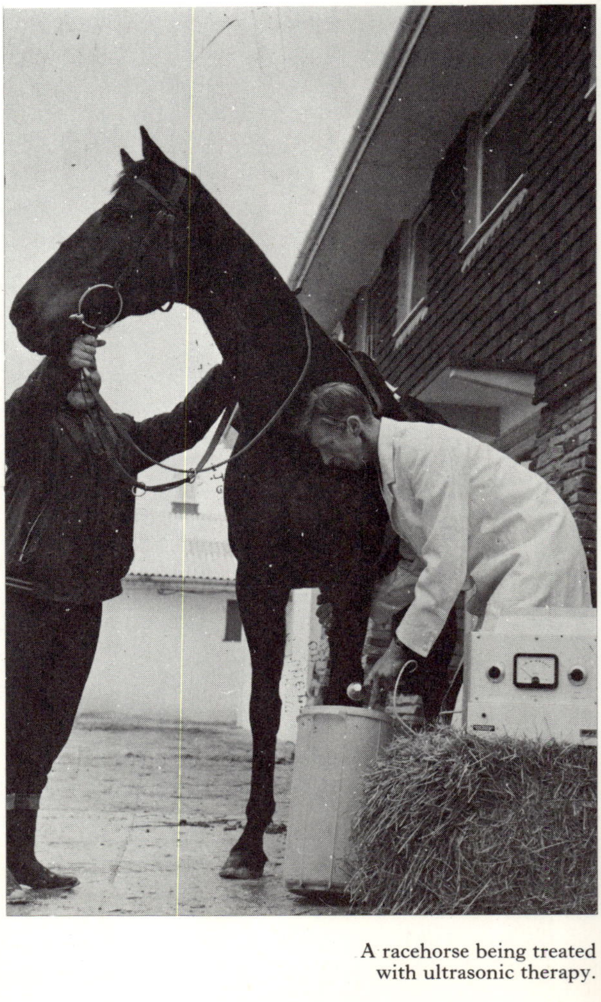

A racehorse being treated with ultrasonic therapy.

moulds, fungi and other natural sources. Penicillin, the first to be used, is still highly effective and the drug of choice for use in its own field; however, it is not effective against all germs and there are other antibiotics which either cover a larger field (have a broader spectrum) or which are effective against a limited number of organisms unaffected by penicillin.

Not all these other preparations have the same safety and lack of toxicity as penicillin. As well as destroying bacteria, the side effects which they produce have to be dealt with and techniques such as the administration of antisera to combat bacterial toxins still have to be used, as have metabolite solutions to overcome the effects of dehydration.

Specifics have not been developed against all infective agents, and as yet there are few effective drugs against disease-producing viruses, although numerous vaccines and antisera have been developed in recent years.

Improvement in laboratory diagnosis has enabled new diseases caused by viruses to be recognised, whereas previously their separate identity was not recognised.

Thanks to modern equipment, much more is known about cardiology, and the ease of permanently recording heart sounds is a great help to the study of diseased conditions. Unfortunately, treatment lags behind diagnosis.

Much has been learned in recent years about the normal blood picture in the horse and this has led to tests to ascertain when a horse reaches its peak of racing fitness as well as helping in diagnosis and treatment.

Although shooting is at least as humane as any other method of destruction, there are aesthetic objections to it. Strong anaesthetics, which can be given intravenously in manageable overdoses, are increasingly used for this unpleasant task; the horse sinks quietly and dies without noise or bloodshed.

A horse under anaesthetic
at the Equine Research Station
near Newmarket.

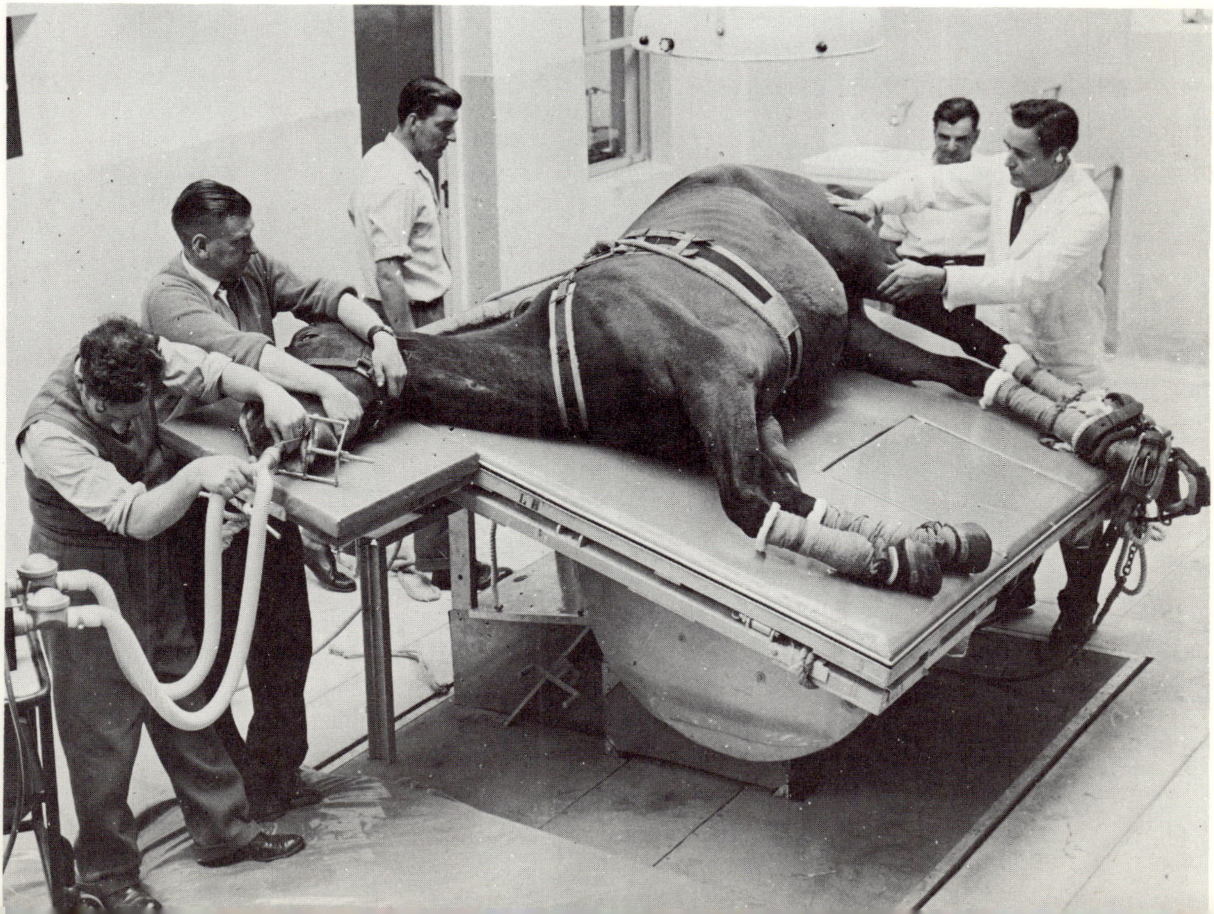

PART FOUR

Care and management

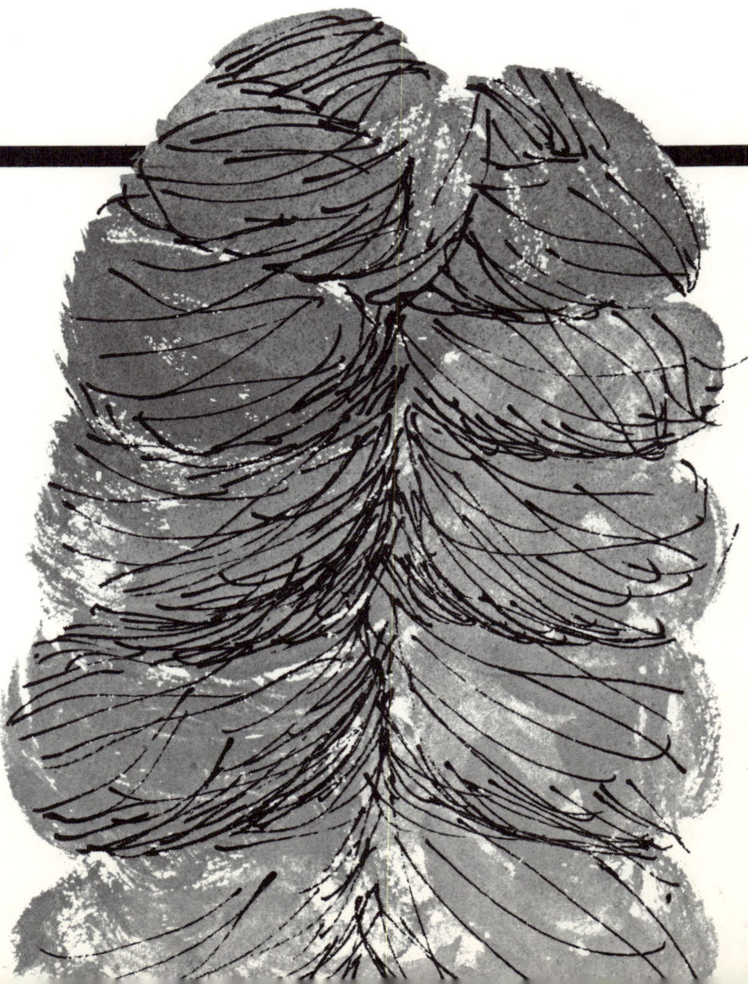

Chapter one

Aspects of management

The first thing to decide when buying a horse is how, and where, you intend to keep it. The most natural environment for a horse, obviously, is to be outside at grass, browsing. However, this does presuppose the availability of a large area of good-quality land; it is not enough to put a horse out in half an acre of poor, over-grazed land and expect it to do well. While horses do adapt themselves quite equably to the more closeted life of the stable, certain character traits need to be borne in mind in this connection, such as the need to avoid boredom and loneliness on the part of the horse.

Perhaps the best solution to the question of which method to use for keeping a horse is the combined system, where the horse is partly stabled and partly put out to grass. This method is particularly useful for horse owners who do not have enough time to give their horses a great deal of exercise every day, and who also wish to save time by eliminating some of the work involved in keeping a horse stabled for the whole twenty-four hours. Perhaps more important than this, though, is the fact that the horse which is allowed out for a few hours a day is closer to its natural way of life. It will be less likely to develop bad habits, to become bored, bad-tempered and over-full of itself than the horse which is cooped up all day except for its exercise period.

Horses and ponies kept at grass need more care and attention than is sometimes appreciated. They need, first of all, enough grass for grazing and of the kinds which they find most palatable. About two to three acres of land per horse should be allowed for, and this should be rested at intervals.

The pasture should be given its own share of attention. The land must drain adequately; on sandy, chalky or gravelly soils this will happen naturally, but where the soil is of clay drainage must be assisted by the provision of ditches or even of piped drainage. It is also a good idea to have the soil analysed, especially if the pasture is old, so that deficiencies can be made up. Lime, phosphates and potash are all needed and can be added to the soil if necessary. The pasture should contain a variety of leafy grasses, including a good proportion of clovers, and be as free of weeds as possible. If the quality of the pasture is really poor, it is advisable to re-seed it.

The land will benefit by being fertilised, too. Liming need take place only once every few years, and the same applies to the application of phosphate fertilisers. Pasture from which a crop of hay is to be taken should be treated with stable manure in the autumn; a nitrogen treatment given in February promotes good growth of grass in the spring.

The maintenance of good pasture is also helped by harrowing and rolling. The former helps to aerate the soil and to spread droppings which, if left in heaps, sour the ground. Rolling will firm the topsoil, which becomes loosened by frosts and heavy rain, and thus helps to provide the right conditions for good plant growth. Because horses are selective feeders, it is a good idea to have the pasture grazed by cattle from time to time, as the latter will eat the longer grasses–horses prefer shorter growths–and the grass level will thus be kept even.

Apart from the quality of the grazing land itself, there are other important considerations for horses kept at grass. They must have a proper water supply, which is best provided by piping it to a trough in the field. The trough should be cleaned regularly so that the water remains fresh. Water for horses should never be allowed to stagnate.

There should always be adequate shelter for horses at grass. Ideally, this should be provided by trees and thick hedges, giving protection from prevailing winds and heavy rain in winter and from sun and flies in the summer. Failing this, a shelter must be erected, sited so that it will give maximum protection.

Before a horse is turned out to grass the field should be carefully inspected to see that it is safe. The fencing should be strong and solid. Hedging is ideal, but where fencing has to be put up posts and rails – the best, though most expensive, method – or heavy gauge wire strung tightly between posts should be used. Chicken mesh wire and barbed wire should not be used. Any potentially dangerous objects, such as stakes, large stones, old rusty nails, discarded tins, and so on, should be removed, as should all poisonous plants and shrubs.

A horse out at grass cannot be left entirely on its own. It should be visited several times a day even when no extra feeds are being given, the water supply checked and the horse itself inspected. A horse ignored for several months will be very much less friendly and co-operative than one which continues to be handled.

With the exception of a horse not being given any work, and out at grass in the late spring and early summer when the grass is at its best – being both abundant and nutritious – almost all horses will need some extra feeding. This is particularly true of those animals, generally ponies, which are kept at grass but which are expected to do a fair amount of work at the same time. Ponies, in particular, tend to become over-fat if left out all day to feed, and it is generally a good idea to bring them in for a few hours each day in order to avoid this. They can be fed a small quantity of hard food during this time. Ponies with a lot of native blood in them will require rather less care and attention than the Thoroughbred blood horse, which needs more preparation and conditioning if it is to perform satisfactorily.

The mental state of a horse is important to its well-being, and this should particularly be borne in mind when stabling is being considered. Loose boxes are preferable to stalls as a form of stabling, for a horse that is tied up all day with only a wall to look at will inevitably become bored. It is not always necessary to have stables built from scratch; existing buildings can often be satisfactorily converted as long as some of the basic requirements can be fulfilled.

Loose boxes should be generous in size, light and airy. As horses are happier in company, boxes should if possible be grouped together so that horses can see each other, and also be situated so that the occupants can survey their surroundings and watch the various activities of the household – equine or human – going on during the day.

There are three essential elements of proper stabling: drainage, insulation and ventilation.

If existing buildings are converted into stables the draining factor can present a problem, and it is sometimes a good idea to have the stable area re-floored. At one time stable floors were drained by having a channel running through them. This is now not common practice; it is better to have the floor very slightly sloped towards the door, and to have a drainage channel running outside the door of each stable. Many stables have no constructed drainage system at all, and this does not matter as long as the horse's bedding is scrupulously removed and replaced. Specially constructed floors – with or without drainage channels – are usually made of concrete, but in many parts of the world floors of hard-packed earth are to be found. Less satisfactory from the drainage point of view, they nevertheless provide better warmth and insulation and are less likely to cause injury.

Insulation is important to the well-being of a stabled horse. Stabling should be so constructed that it keeps out the cold in winter and the heat in summer, and draughts should be avoided. Wooden stables should be lined in some way to achieve this, and the roofing should be completely waterproof and draught free.

At the same time, plenty of ventilation is necessary. If the top half of the stable door is left open, adequate ventilation will automatically be provided, in addition to allowing the horse to look out and take an interest in what is going on. If the box has a window, it should be placed in the same wall as the door in order to avoid cross-draughts, and should open upwards and inwards so the current of air goes over, rather than on to the horse.

Safety is an important part of stable management. The box should not contain any projections – such as old nails – on which the horse can injure itself. Windows should be covered by a metal grille, and electric light switches fixed to the outside of the stable wall. The light itself should be situated well out of the horse's reach. The horse should have plenty of bedding, not only for warmth and comfort but also for protection. Certain fittings are necessary, but these too should be constructed with the safety element in mind. Mangers, for example, should be built into the corner of the box and set at breast height. Mangers need to be cleaned regularly, and for this reason it is just as satisfactory to use heavy feed tins which can be removed after the horse has eaten.

Water must be freely available to the horse at all times, and the simplest way to provide it is by using buckets. It is possible to equip stables with self-filling water bowls, though this is really only economically practicable in a large stable.

Many old stables have hayracks, generally fitted high up on the wall. These should not be used, as horses feeding from them get seeds and dust in their eyes. Hay should rather be fed from nets, which should be tied to a ring in the wall at a height that prevents a pawing horse from getting its foot caught in the net. It is also much easier to calculate how much hay is being given to a horse if it is fed by this method.

Mangers, buckets and the stable itself should all be cleaned out so that they remain fresh. The treatment of bedding depends to a certain extent on what kind of bedding is used. Wheat straw is the most satisfactory form of bedding, though oat straw (which some horses tend to eat) and barley straw are also available. Bracken, sawdust, shavings and peat moss are also used. Opinion varies, but it is usual to clean out straw beds daily, and to pick up droppings and wet straw at intervals during the day. When the bed is remade, straw should be well banked up round the sides of the box for warmth, comfort and protection, and the floor covered in a generous layer. The deep-litter method, where straw is put down to cover earlier layers and the stable cleaned out only when the bed becomes inconveniently high, provides extra warmth and is a great labour-savour, but it does have other disadvantages.

Good stable management relies to some extent on the right tools and equipment. An adequate number of buckets, feed tins, haynets, etc., are an obvious necessity, as are proper stable tools. A barrow, two-pronged pitchfork, four-tine fork, stable shovel, broom and skep are all essential. A piece of sacking in which to carry straw to the stable is also useful and prevents wastage.

The storage of food is also important. Hard foods should be kept in damp-free, mouse-proof containers; hay should if possible be kept indoors (the stable loft is ideal), but if this is not possible it should be stored off the ground and covered. The cover should, however, be removed as often as possible to allow air to reach the hay.

As well as the grooming kit, there are other items of equipment that contribute to the comfort of the horse. A stabled horse should be clothed in a light linen sheet before it is clipped. Once clipped, it is necessary to provide heavier rugs. The essential clothing consists of a top rug, made of jute or canvas and fully lined with wool, and a large woollen blanket. An open-mesh sheet is also extremely useful, providing extra warmth by insulation in cold weather and keeping the body cool when the weather is hot. It also helps to prevent horses from breaking out into a sweat when travelling, after exercise or through nervousness. When one of these rugs is used in conjunction with the top rug, sufficient warmth will be provided in all but really cold

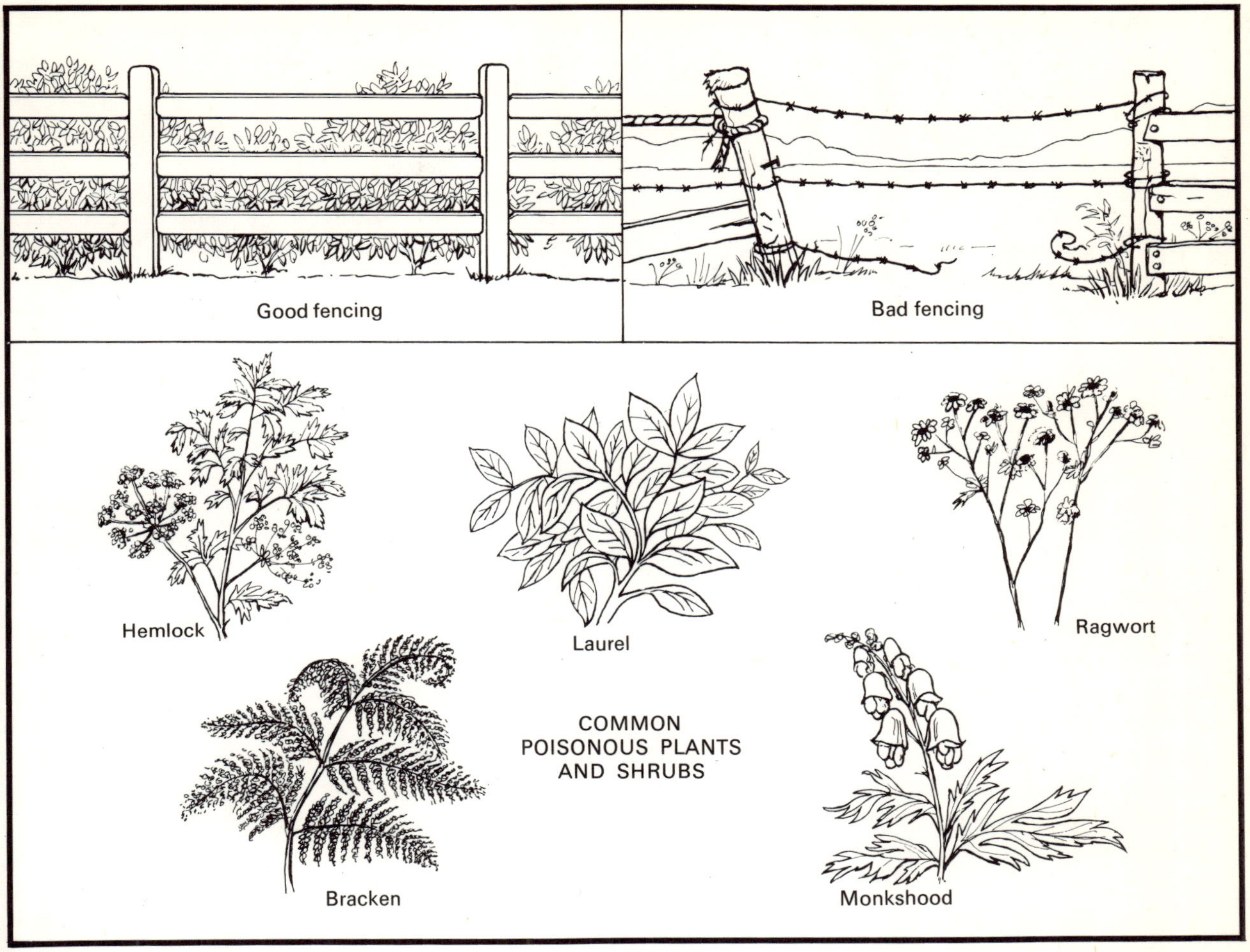

weather, when the blanket can be added. A day rug, made of wool and bound in a contrasting colour, is not strictly necessary but is rather an attractive addition to the horse's wardrobe. A horse kept on the combined system, particularly if it has been partially clipped, will need a really well-fitting, top-quality New Zealand rug for use when it is turned out.

Rugs are kept in place by rollers or surcingles. Rollers can be made of leather or of webbing, and care must be taken to see that they are properly stuffed on either side of the spine or a sore back will result. Rollers should not be fastened too tightly. Surcingles are made of web and are sewn to the canvas of the top rug. Here again, the padding is important to ensure that no pressure is applied to the backbone. Rugs and rollers or surcingles should be fitted with as much care as other items of saddlery.

Stable and exercise bandages should also be available. The former, which are intended to give extra warmth, cover the leg from just below the knee to below the fetlock joint; the latter give support and protection from knocks, etc., during work, and leave the fetlock joint free. Gamgee tissue should be wrapped round the leg before the exercise bandage is put on.

Boots, pads and other protective items which prevent the horse from damaging itself or being damaged, should be bought as they are needed. For horses with a tendency to over-reach, to brush or to knock themselves, these can be invaluable, but they should not be used indiscriminately.

Chapter two

Foods and feeding

No horse, or for that matter pony or donkey, can remain healthy and give of its best unless fed correctly. Every horse must be treated as an individual as far as its feeding is concerned, and it is essential to study each horse's temperament, its likes and dislikes, and its ability to digest certain types of food; for what will suit one horse could well disagree with another.

When feeding horses it is essential to choose only the best quality food. Poor quality food, besides having little feeding value, can be extremely harmful if it is mouldy or musty, as food in this condition will cause serious upsets to the digestive system. Dusty food can also prove harmful to the wind and lungs and cause respiratory troubles.

The nutrients required to keep a horse in good health are not unlike those required by humans: water; carbohydrates; fats; proteins; vitamins and minerals – all of which will, according to how they are fed, determine the horse's performance, health and appearance.

Clean, fresh water must always be available to a horse at all times, and in a clean container – either a bucket or a trough. Horses are fastidious creatures, and they will not drink stale or dirty water; before long they will start to lose condition if no fresh supply is provided. If water is not constantly available, the horse must be offered a fresh drink at least three to four times a day and before every feed. Never give water to a horse after feeding or when it comes in very hot from work – colic can follow if one does. If a horse has a fresh bucket in its box, then it may take a short drink during or after a feed. This is all right; it is only if a horse consumes a whole bucketful after having been fed that there may be trouble.

Stable buckets should be washed out every day, and refilled four to five times during the day – the last time being on one's final evening visit to the stables. Field troughs need to be scrubbed out about once a week if they are large – both automatically-filled ones and those filled by a hose; small troughs and buckets need daily cleaning, and must also be refilled frequently.

Owing to the formation of the horse's stomach it prefers to eat small amounts of food at frequent intervals. The horse at grass, living on its natural food, will eat most of the time. The stabled horse should be allowed to follow this pattern of nature as closely as possible. It is far better to feed three or four times a day than only once or twice. It is also necessary to provide some form of bulk feed to assist the horse's digestion.

In summer, when the grass is at its best, a horse can live on grass alone if it is of good quality, sweet and fresh. Grass by itself will not, however, provide the necessary proteins and oils for horses carrying out strenuous work, or for those feeding young or for growing youngsters except where the grass is exceptionally good. In an ideal year May, June and July will see enough good grass to support a horse, but come August the goodness starts to go and for many extra feeding in the form of either concentrates or hay will be necessary.

A horse fed on grass alone will be in soft condition, and to become fit and hard will require – as well as adequate excercise – either oats or some other form of concentrate in addition to its hay or grass. All horses should have some fresh grass every day during the summer, being either turned out to graze, led out on a long rein to pick the grass along the roadsides or provided with cut grass in the stable. If cut grass is fed, care should be taken to ensure that it is not allowed to heat and start fermenting, which can be dangerous. Lawn mowings are not suitable fodder, even when fresh, as they can ball up inside the horse and cause a blockage.

Grass land needs attention to keep it right, and it is essential to 'top' paddocks frequently to about four inches to ensure a sweet growth, and to collect droppings to prevent the infestation of the paddock with worms, which cause serious harm to horses grazing there. Harrowing is the alternative to picking up droppings and with large paddocks is more feasible; small paddocks must have the droppings picked up once or twice a day or the horses will never thrive, however well fed.

Hay, the means by which we provide bulk to help the horse's digestion in winter and for those stabled all the year round, is made from grass either cut from permanent pasture – meadow hay – or cut from specially seeded pasture – seed hay. The former can be upland hay, which is of good quality, or lowland, which is generally poor and unsuitable for horses. Seed hay can be a mixture of good grasses and clovers or vetches, and also sainfoin which is a member of the trefoil family. Good hay, cut when the sap is still running and well saved (dried and baled or ricked when really dry), should be sweet-smelling, crisp to touch and free from all forms of must, mould or dust. It is best fed in nets or racks, provided that these are set low so that seeds cannot fall into the horse's eyes, or from the ground, though this is wasteful. It is always advisable to shake up the hay first to make sure that no bad patches are sandwiched in the parcel of hay from the bale and to remove loose seeds and dust.

A big horse will need between 14 lb. and 16 lb. of hay a day in winter, or, within reason, as much as it can eat. Smaller horses normally require 12 lb. to 14 lb., depending on how much concentrate they are being fed as well. Chaff or 'choppy' is merely hay that has been passed through a special machine to cut it up small; when in this form it is mixed with the short feed to form bulk and make the horse chew its food properly. The protein content of hay varies from 5 per cent to as much as 15 per cent in top-quality hay. The vitamin and mineral content varies, too, but is only present to any valuable extent in hay under two years old and cut young.

During the winter months all horses and ponies running out at grass, as well as those stabled, will need to be fed. Hay will provide much of the necessary extra feeding, but youngstock, mares in foal and older horses and ponies will also require concentrates to create the body warmth needed to keep them healthy.

We now come to the energy foods. These are needed to provide the body heat that goes either to keep a horse warm or to enable it to become fit enough to do strenuous work without detriment to its health. The principle energy foods are those containing carbohydrates, which include sugars, starches and cellulose. Oats, barley and maize contain about 60 per cent sugar and starch, while cellulose is found in the fibrous parts of the oat and barley hulls (the outer case covering the grain) as well as in bran, which is the outer part of wheat. For this reason oats, barley and maize are classed as concentrates – one needs to feed only a relatively small quantity as compared with bulk foods to provide energy.

Fats are necessary in the horse's diet to enable it to absorb certain fat-soluble vitamins, but in their natural state fats are not easily digested by the horse so a means of supplying them has to be found. Linseed, with its very high oil content, is extremely useful. Soaked overnight, then brought to the boil and simmered for a day, it is mixed with either a bran mash or into the evening feed. It is excellent both for putting on flesh and for maintaining it throughout the winter, and for sick horses, as it is easily digested in its cooked form. It also gives a good gloss to the coat, and it is valued for its high protein content.

Protein is the all-important feeding factor in producing fit, healthy horses. It is essential to build muscle, its prime use being in the formation and repairing of tissue, besides being largely responsible for sound and healthy bone, blood, skin, hair and hooves. Insufficient protein leads to weak frames and lack of bone; and hearts incapable of withstanding the rigours of training – as hearts are, after all, only lumps of muscle that act as pumps for the circulation of blood round the body. Those horses needing extra protein are the young, the very old, breeding stock and those in work.

In order to provide the required protein for our horses we can feed either cereal foods – oats, barley, maize and linseed; or the modern compound feeds – horse or cattle nuts (pencils, cubes or cake). The former is the traditional method of feeding, the cereal being mixed with bran, choppy and any extras that might be required, while the latter is the modern method, and is fed on its own. Both require hay in addition to provide bulk. Alternatively, a mixture of cereals and nuts can be fed to form the concentrate ration.

Nuts vary in their protein content and also in their oil and fibre content – the higher the protein, the lower the fibre (composed mainly of dried grass meal, which provides protein as well as fibre). Every bag carries a label giving the percentages of these elements. Horse and pony nuts normally carry 10 per cent protein and anything from 15 per cent to 17·5 per cent fibre, whereas the 14 per cent protein nuts carry only 8 per cent or 9 per cent or less of fibre. The former are excellent for children's ponies and for horses in light work; the latter are essential for those in hard work like hunting, jumping, eventing and racing. Those carrying are higher percentage still are supplementary nuts to be used as a booster. The cattle-rearing nut, carrying 16 per cent protein and a high oil content, comes into its own as a booster nut for horses in hard work. There is also a 12 per cent protein nut, which is good for ponies in hard work and for horses needing more than the standard 10 per cent nut but who are – probably because of temperament – unable to accept the 14 per cent nut.

There are also nuts made for a special purpose. Stud nuts are specially designed for breeding stock and youngsters, for they carry the right ratio of calcium and phosphate, besides other vitamins and minerals; transit nuts should be fed to horses travelling long distances or ones that are confined to the stable through injury, as they are very low in protein.

All nuts have one disadvantage: though easy to feed it is impossible to give variety to the feed. This applies in particular to the 'complete' nut, a nut containing even the hay ration. It can be useful in some cases and for a limited period of time, but it can lead to boredom and stable vices if fed for too long.

One disadvantage of nuts fed on their own lies in their nut form. Should the horse require medicine there is no base to mix it in, and it is very difficult to adjust each horse's intake of, for example, vitamins or minerals as the nuts are fully vitaminised and mineralised. Pound for pound with oats the high protein nuts need feeding at a slightly lower rate, whereas the low protein (10 per cent) nuts need feeding at one-third more for the same value.

Oats – still the best of the cereal foods – can be fed whole, bruised, cracked or crushed, and though they are heating and are therefore inclined to make some horses hot up, they are excellent for those in hard work. Their protein content varies, which is why a high protein nut is useful to boost the ration. Work and high protein foods must match – more work, more food; and the rider, too, must be capable of coping with a fit horse. Oats should be clean, sweet-smelling and rattle in the hand.

Barley is growing in popularity and if cooked – soaked overnight and then cooked till it forms a jelly – is useful for putting on flesh and tempting shy feeders. It can also be fed crushed. Barley is fattening, though, and should be fed with care.

Maize – corn in the United States – is fed flaked and is an excellent source of protein. These cereals, together with linseed and other sources of protein and oils, form the bases of nuts. Beans, which are fed cracked, are a source of protein but should only be fed to horses in very hard work as they make those in normal work far too excitable. Molassine meal – a good source of sugar – gives feeds a pleasant smell and taste, horses usually love it and it is very good for them. Bran is rather hard to find, but is very necessary for mashes and to aid digestion; $\frac{1}{2}$ lb. in each feed is enough, but it must be damped. For a mash, 2–3 lb. should be mixed with boiling water till well moist but not running wet, and then left to cool. It is useful for sick horses and for others once a week as a laxative. Sugar-beet pulp is often used to form bulk. It is usually supplied mixed with molasses, thus also providing sugar, but it must be well soaked overnight and should always be fed fresh.

Modern farming methods mean that much of the natural vitamin and mineral content is lost before we ever buy our feeding stuffs, and supplements are needed to make up the deficiency. There are supplements with a limited vitamin and mineral content and those with a comprehensive one. The latter are best. Some are based on honey, others on cod liver oil or a mixture of cod liver oil and linseed. Seaweed is another source of vitamins and minerals favoured by some. Whatever we choose to feed in the way of supplements we must stick to one brand – more than one and an imbalance will result. No vitamin or mineral supplement should be necessary if a full ration of nuts is fed; extra salt, however, should be supplied to all horses, however their food ration is made up, and can be given either in the form of a lick or in the feed at the rate of an ounce a day.

Roots are another source of vitamins for the horse, and carrots – sliced lengthways – are an excellent addition to the winter diet of stabled horses or those kept out; so, too, are mangels.

Whatever one feeds, some rules for feeding must be followed: water before feeding; feed at regular intervals and little and often; feed plenty of bulk food; do not work a horse until at least an hour after a feed; feed according to the work demanded; do not make sudden changes to the diet; make sure all food is of the best quality, and wash out mangers, feed tins and buckets once a day; never guess the amount of a feed – weigh food carefully to ensure that the horse is getting its due ration, and mix the feed carefully. It is also important to see that the horse's teeth are in good order and do not require filing; bad teeth will lead to a loss of appetite. Lastly, the horse must be kept free from parasites – worms and lice – as to feed unwanted visitors is a waste of money and no horse can do well if he is acting as host to them.

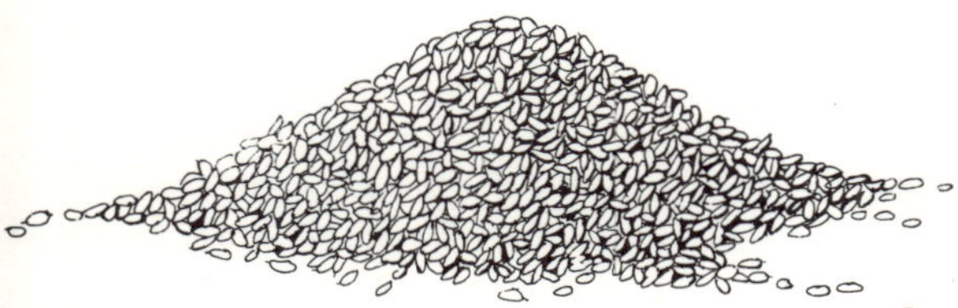

Oats

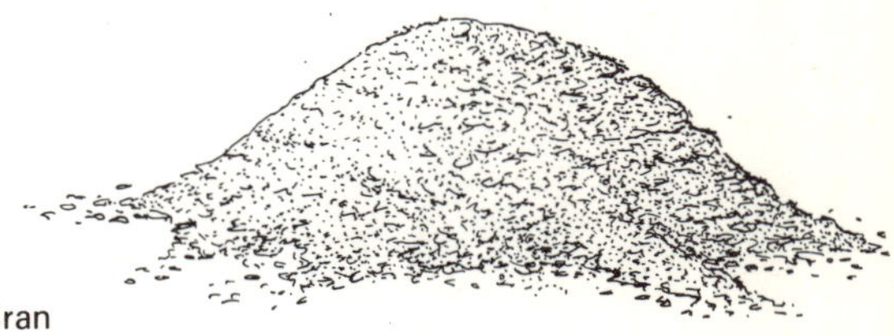

Bran

Additional nutrients

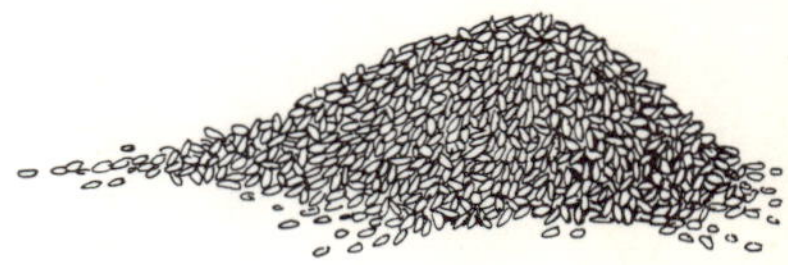

Linseed

Barley

Flaked maize

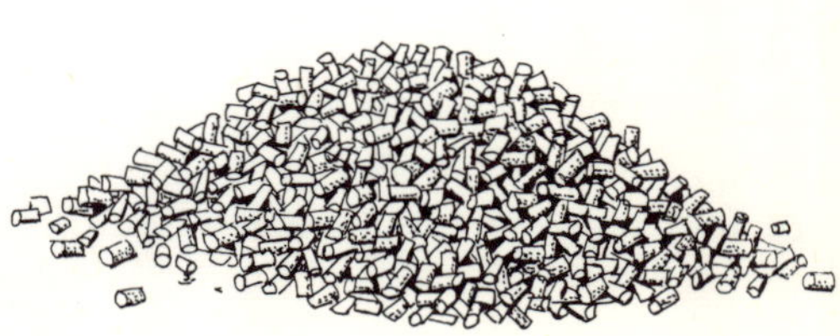

Nuts

Chapter three

Conditioning the horse

In order to get the best out of a horse, and for both horse and rider to enjoy the activities they perform, it is necessary for the horse to be properly conditioned. There are many different levels of fitness in a horse. For example, a racehorse needs to reach a peak of fitness different from that required by the hunter; this again is different from the level suitable for a hack which will only be used for gentle rides at slow paces. It is obvious, though, that no horse can perform effectively, efficiently and safely if it is not fit enough.

The aspect of safety is an important one. A horse that has not been properly conditioned is more likely to suffer from strains and other ailments than one which has undergone a careful training and conditioning programme; a horse with inadequately developed muscles will be less well balanced and weaker than one in which the muscles have slowly been built up by suitable exercise. As a result, where the fit horse will be able to make an extra effort when required–as, perhaps, when jumping a big or tricky fence out hunting–the unfit horse will not be able to do so, and is very much more likely to get into trouble.

The conditioning of a horse divides into three parts: exercise, feeding and grooming. These parts are closely related and must to a certain extent be considered together. A horse that has just come up from grass, and is therefore fat, with soft muscles out of tone, cannot be successfully conditioned just by a change of diet from grass to hard foods and by strenuous grooming. The conditioning process is one of gradual progression, where adjustments to diet, the amount of exercise taken and the grooming given are all an integral part of the process.

There are two other important general points to be borne in mind. The first is that the condition of rider and horse should be advanced side by side–a well-developed, fit horse will generally prove to be a considerable handful to a rider who is himself out of condition. The second point is the physical state of the horse before a conditioning programme is embarked on: a horse that is very fat will need to be given an initial period of very gentle exercise before it is able to cope with more strenuous work; equally, a horse is out of condition if it is too thin, and it must be allowed to put on flesh before being given any substantial amount of work to do. However it is fed, a thin horse will only go on getting thinner if it is asked to do a lot of work.

The exercise given to a horse being conditioned should vary in order to keep the horse interested, to build up its muscles and develop them for the different activities it is going to perform. The amount of exercise should be built up gradually over a period of many weeks. Work at slow paces in the initial stages should include lungeing, loose schooling, work in the *manège*, road work and hacking over different kinds of terrain. Walking and trotting are the two most beneficial paces; the horse should first be worked in the loose school and on the lunge so that its muscles can begin to develop before it is asked to carry the extra weight of the rider. These exercises should be continued even when ridden work has begun. Slow road work will harden the legs, and the horse should be both walked and trotted at this stage. Slow trotting uphill is excellent for the improvement of wind; both uphill and downhill work improve the balance and muscular development of the horse. Balance is also improved by riding the horse over uneven ground.

Schooling should not be entirely restricted to the school. The horse should be prepared to perform some of its schooling lessons while out on exercise, for as well as emphasising the lessons themselves it will keep the horse interested. Short periods of collection, variations of pace, even, perhaps, a few strides of half-pass on a quiet road–all are good for improving the horse's concentration and obedience.

It is difficult to generalise about a training programme for developing a fit horse, but there are certain basic rules to apply. For the first week or two the horse should be given about an hour's exercise a day, the work period being divided between lungeing and slow hacking at the walk and trot. After this initial period the time should be increased to, say, an hour and a half, during which time the horse will be lunged for about fifteen minutes, have about thirty minutes at school work on the flat and again spend the remainder of the time hacking at slow paces. By this time the muscles should be developed enough for the speed to be increased, so that short periods at the canter and some school work over cavalletti are within its capabilities. Towards the end of the second month the overall exercise time can be increased to two hours, and should include at least an hour and a half hacking, with some steady cantering and an occasional short gallop.

In order to remain fit once a reasonable level of fitness has been reached, as it should have been by nine to ten weeks of work, the horse must obviously continue to be given an adequate amount of exercise. It is not enough to expect a horse to hunt one or two days a week and on the other days just to give it a short hour's work. At the same time, a horse that is asked to perform really strenuously–hunting is again a good example–at regular intervals should be allowed a certain amount of rest as well, or its physical condition will begin to deteriorate.

Feeding and grooming should be considered with as much care as the exercise programme, and in conjunction with it. The right food will build up the horse's body and supply it with the material it needs for making muscle, as well as producing the necessary energy. It is perhaps useful to emphasise here that while fat is reduced and muscle improved during the conditioning process, fat is not turned into muscle. It is important to bear this in mind when considering a horse's diet: the excess fat will not itself be turned into muscle by training and exercise, it can only be eliminated while muscle is at the same time built up by the gradual increase in work and development of the diet.

Because horses vary in temperament and their physical ability to use food, each horse will need to be treated individually as far as diet is concerned. A horse that is a 'good doer', one whose body utilises the energy foods to maximum effect, will need less of the body-building, energy-supplying foods than one which is prone to lose condition easily. An excitable horse will need to be conditioned on a diet containing a lower proportion of oats than that suitable for a horse that is naturally lazy or calm and does not hot up.

The amount and kind of food given will again vary depending on what work is to be expected of the horse. For a reasonably well-bred, 16-hand horse with a sensible temperament, the average will range between a diet of, say, 4 lb. of oats, 2 lb. nuts, 2 lb. bran, 1 lb carrots, and 18 lb. hay, fed at intervals during the day in three feeds, for a horse that has just begun its conditioning programme, to that of a fit horse being given a considerable amount of work, which might receive about 10 lb. oats, 3 lb. nuts, 2 lb. bran, 1 lb. carrots and 11 lb. of hay a day. It will be seen from these examples that the amount of hard food (e.g. oats) is increased in proportion to a reduction in the feeding of bulk foods such as hay. Most horses also benefit from being given one bran mash a week and by their diet being altered from time to time. Horses vary in their tastes, but additional foods such as apples usually make a welcome change in diet, and it is also a good idea to allow even a really fit horse to graze for a few minutes each day.

The feeding of ponies is a rather different matter. Many of them either live out at grass all the year round, or are only partly stabled. A high-quality show pony will have different requirements from a hardy native animal used to fending for itself. Oats should be fed to ponies with circumspection, particularly if the rider is not an expert, but there are plenty of other foods available which will help to keep a pony in good working condition. Ponies should not be allowed to become over-fat on very lush grass, as apart from getting very out of condition their susceptibility to contract laminitis will be increased. During the winter a pony at grass should be fed an average of 10 lb. of hay a day, and at least 3 lb. of nuts and 1 lb. of bran should also be given. A pony being given a lot of work should have some bruised barley added to its diet as well.

Grooming, and the general care of the horse, are both important factors. Grooming will develop and tone muscles, and also cleans the body so that it can work most effectively. Grooming removes the natural grease from the horse's coat, and a stabled horse, which does not have to withstand long periods out in the rain and other natural elements, can therefore be groomed to improve its condition because it is protected by being indoors. It is, however, more likely to catch a chill. The stabled horse should not be brought home sweating but should rather be walked the last mile home in order to be given a chance to cool off, though in wet weather it should be kept warm by being brought back at a brisk trot and then thoroughly rubbed down and dried when it reaches home.

The skin is cleaned and stimulated by being groomed; this is not, however, the only benefit of it. Massage, hand rubbing and wisping also play their part in conditioning a horse. A horse or pony out at grass–and it is possible to produce a reasonably fit animal when it is not stabled–should not be over-groomed, because of the loss of the natural protection its coat gives. The stabled horse, however, should first be thoroughly groomed after morning exercise, and then be given wisping and massage later in the day.

The various items used in grooming each has a different function to perform. The dandy brush is most useful for cleaning muddy horses, though because it is stiff and hard it should be used with caution on ticklish horses and on the tender parts of the body. It should not be used to excess on horses out at grass, except for removing mud from the outer surface of the coat.

The body brush is soft-bristled and is used both on the body and on the legs and for brushing the mane and tail, the hairs of which would split were the dandy brush to be used. The bony parts of the horse should be brushed gently with the body brush only, particularly the head. Horses should always be treated considerately during grooming: do not bang them, particularly in the loin area where damage to the kidneys can result from rough treatment. Patience should be exercised with a horse that is head shy–it will almost certainly mean that the horse has been roughly handled in the past. Other standard items of grooming equipment include the water brush, mane comb, hoof pick and stable rubber. Rubber curry combs are replacing the old metal combs, and are generally much more satisfactory, and 'ready-made' hay wisps are now also available.

Most horses enjoy being groomed and quickly learn to co-operate, to move over when they are asked, to pick up their feet for them to be cleaned out and to put up with their more ticklish parts being touched. When a horse is wisped for the first time, the wisping action should be relatively gentle, though the amount of energy applied can be increased with time.

Wisping is a very important part of conditioning. It develops and hardens muscles and improves the horse's circulation. The traditional wisp is made out of a rope of hay, though a chamois leather bag stuffed with hay is just as satisfactory. Wisps should be damped before use, and should only be used on the muscular parts of the horse's body: the quarters, shoulders and neck. They should not be used on the loins, head, belly or legs.

There are other aspects of conditioning linked to grooming. The heavy coat provided by nature to all horses in an environment where winter conditions are more severe than those in summer is a hindrance to peak conditioning. A heavy coat will make a horse taking a lot of exercise sweat to such an extent that it will begin to lose condition. It is for this reason that horses are clipped in winter, the warmth of the heavy winter coat being replaced by rugs.

There are various methods of clipping, the clip used on any particular horse depending on its type, the work it is to be expected to do and the conditions in which it is kept. A full clip involves removing the entire coat, while the hunter clip removes the coat from the body, leaving a saddle patch on the back, but not from the legs. A very fine-skinned horse will probably benefit from the extra protection given to its skin against galls or scalding and against cuts, cracked heels and chills from the legs. A horse with a very heavy coat, however, is probably better off if it is given the full clip.

The blanket clip, suitable for Thoroughbreds with fine coats, leaves the area of the body approximately covered by a sheet, as well as the legs, unclipped, while the trace clip removes hair just from the belly, the flanks and the sides and underneath of the neck. This last is useful for ponies kept at grass which are nevertheless expected to perform energetically during the winter, and for harness horses.

When clipping a horse it is advisable to have an assistant. The horse should be clean, the blades of the clipper sharp, and particular care should be taken to accustom the horse to the noise of the machine and to keep the animal calm so that it does not fidget. Neither mane nor tail should be clipped; if the legs are to be left unclipped it is usually a good idea to trim the heels using scissors and a comb.

The horse's appearance is generally much improved if its mane and tail are kept thinned, though in horses with very fine hairs, such as Arabs, they are not trimmed. Both manes and tails should be thinned a little at a time or the horse's skin will become sore and it will probably object to the next thinning session. Between sessions, the tail should be damped with a water brush and a tail bandage applied to encourage the hairs to lie flat and to give the tail a good shape. Tails can either be left as a switch or can be banged–cut off straight at the bottom, at a level about a hand's breadth below the point of the hock when the tail is carried at its natural level. A thinned mane should be about five inches in length. Once mane and tail have been thinned it is worth keeping them in good shape by thinning them just a little at regular intervals.

Manes are frequently plaited for special occasions, particularly for show events, and for displays are also sometimes braided with ribbon. The number of plaits will vary on the length of the horse's neck and the thickness of the mane, though it is customary to have an even number of plaits along the length of the neck and an extra one on the forelock. Manes can be plaited either by sewing or by using rubber bands; the former method undoubtedly looks neater, but the latter is quicker and easier. Unpulled tails are also sometimes plaited; this is a specialised skill and one best acquired by watching an expert at work.

The most important point to remember where conditioning any horse or pony is concerned is that forethought–thinking about what you are going to want your horse to do, planning to work out the best method of arriving at a suitable level of conditioning, and then applying your knowledge consistently and carefully, will repay in the long run all the extra effort that has been made.

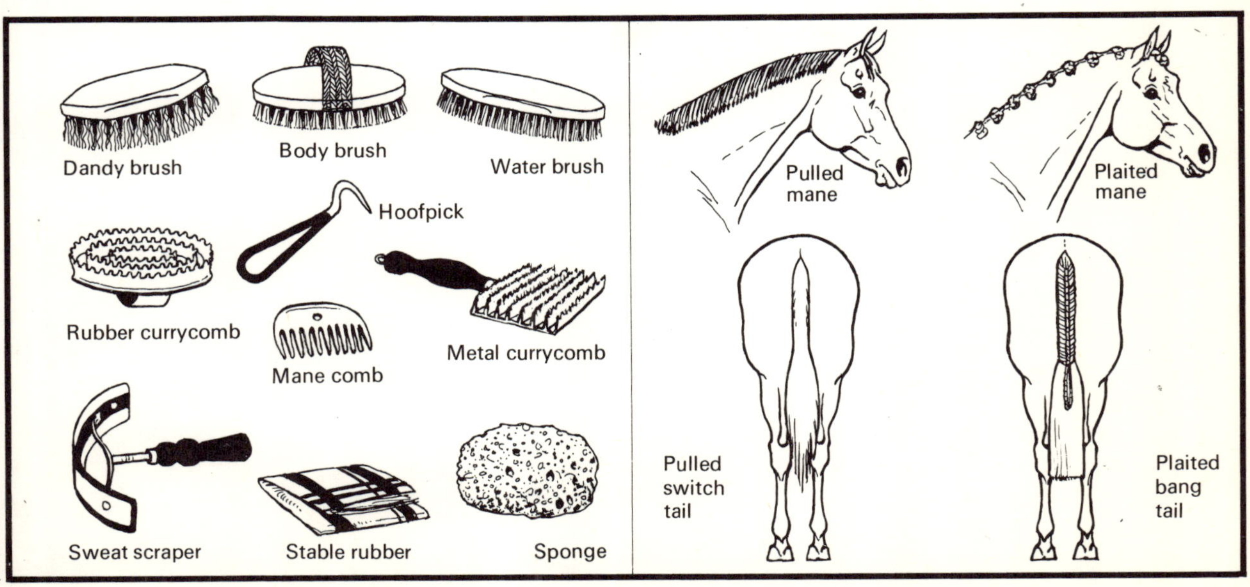

Chapter four

Bitting and saddlery

There are an enormous number of different patterns of saddles and bridles on the market today. Firstly, some understanding of the construction of a saddle is necessary to the appreciation of the difference between a good and bad saddle, or of one well suited to any particular purpose.

The inner framework of the saddle is its tree, traditionally made of beechwood. The tree can vary in shape and size. The head and gullet are strengthened with steel, and a steel plate on the underside of the tree acts as a reinforcement. Laminated woods are now used, and experiments have also been made with other materials. So far only the use of fibreglass for very light racing saddles, a variety popular in Australia, has proved satisfactory. Modern spring tree saddles incorporate two light pieces of steel set on the underside of the tree, giving it greater resilience. Trees vary in length; the spring trees give a greater dip in the seat and need not, therefore, be so long as the rigid variety.

The stirrup bars are fitted onto the tree and are an important part of its structure. Good quality saddles always have forged rather than cast bars. The setting up of the seat is also important to the overall construction of the saddle. Pre-strained webs are stretched along the length of the tree, a piece of canvas is stretched over them, tightly stretched serge is then stitched over this to form the seat shape and the space between it and the canvas stuffed lightly with wool. The pigskin seat, with the skirts covering the bars welted into it, is finally stretched on.

The saddle flaps and panel are also attached to the tree. The panel is an important item, for it affects both the comfort of the horse and the correct position of the rider in the saddle. There are four different basic shapes of panel: full, short, Saumur and Continental patterns. The full panel, the oldest type, prevents close contact between horse and rider, and for this reason the introduction of the short panel was an advantage. As trends in equitation emphasised a more forward seat, the Saumur and Continental panels evolved. The former, cut much further forward, is narrower in the waist for extra contact with the horse, and has an extension to support the knee; the latter is similar in some respects, but is also a much fined down version of the old full panel.

Panels are either made of felt and covered with leather, or are stuffed with wool covered by leather, serge or linen. A leather covering is much the most satisfactory, and well worth the initial extra cost.

The construction and the fitting of saddles are closely linked. Ideally, a saddle should be made specially for a particular horse, though this is not always possible. It is also best to use a saddle only on one horse, for with time it will mould itself to the shape of a horse's back. Again, though, it is not always possible to restrict one's use of a saddle in this way.

Careful fitting of a saddle is important if

A famous firm of saddlers: the Giddens factory.

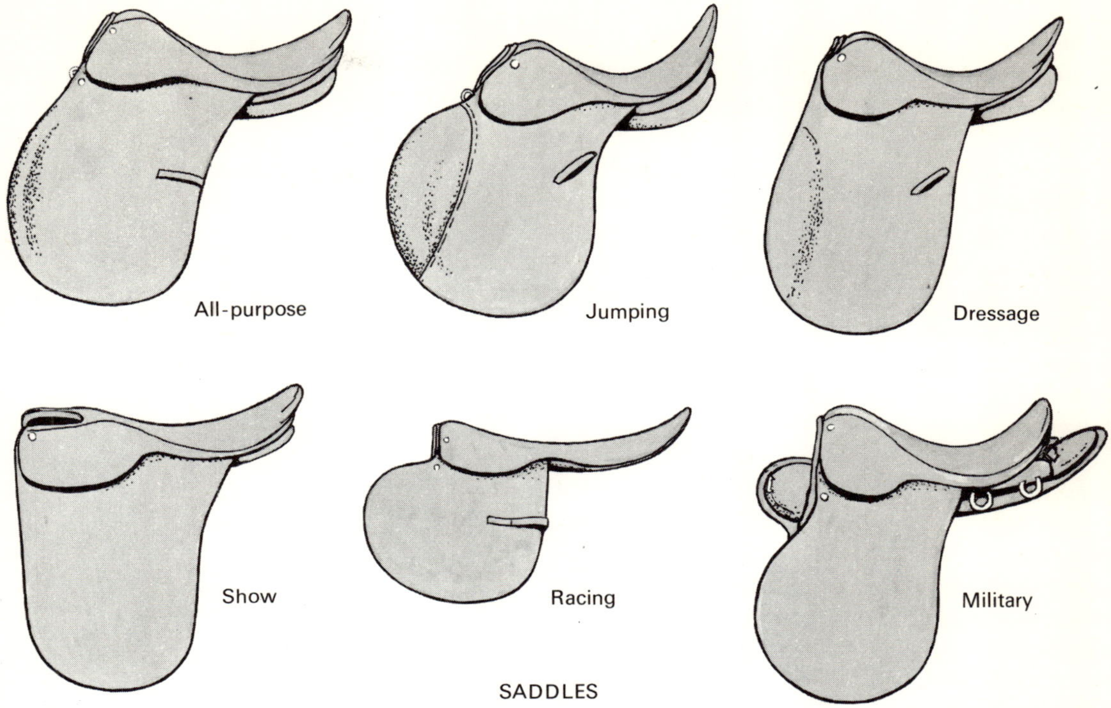

SADDLES

soreness to the horse's back is to be avoided; adjustments to a saddle which does not fit are seldom satisfactory. Trees are made in different widths, and great care should be taken to see that the tree fits the horse's back properly at the outset. The length of the saddle, too, should be suited to the length of the horse's back.

The saddle should clear the withers, and both the length and breadth of the backbone. The saddle should be constructed so that the panel bears evenly on the back, thus ensuring an even distribution of the rider's weight. It is important to maintain the regularity of shape and the resilience of the panel.

Most horses change shape according to their condition, and with use over a long period a saddle may alter in shape too. It is essential to keep an eye on the fit of a saddle and to make sure that these criteria are always fulfilled.

The modern saddle, generally known as the spring tree saddle, has become universally popular in recent years and is now used for most riding activities. Its construction encourages the rider to sit in the correct position, allows maximum contact with the horse and is more comfortable, both for horse and rider, than many of the old-fashioned rigid tree varieties.

The saddle has a long, resilient tree which, because of the 'springs' incorporated in it, gives it considerable flexibility and allows the use of the seat as an aid, transmitting the seat pressure to the horse. The tree is shaped to conform with the shape of the horse's back, thus bringing the rider into close contact with it, and also giving him as narrow a grip as the shape of the horse itself permits. This advantage is maintained by other aspects of the saddle's construction which further eliminate bulk under the thigh, such as the recessed stirrup bars and narrowed panel waist. The saddle carries a strong forward roll on the panel which supports the rider's thigh. This, and the sloped back head, allow the stirrup bars to be positioned further forward than used to be possible; the resulting hang of the leathers helps the rider to hold his lower leg correctly. Another advantage of this saddle is that the girth straps hang relatively far back, so that if a girth of the right length is used the buckles are not felt.

There are two varieties of this saddle, one for general use and one modified for show-jumping. The latter is cut slightly differently to allow for a more forward seat and shorter leathers. Besides these, however, and other saddles different in details of design but produced with the same aims in mind, there are many other kinds of saddle still to be found.

The English hunting saddle, though still widely used, is perhaps the antithesis of all that the modern saddle tries to encourage, and it is probable that its popularity will decline. In some respects it resembles, with its flat seat and straight-cut flap, the show saddle. The show saddle is designed to display to advantage the horse's conformation, in particular its shoulder. It is cut with a completely straight, i.e. vertical, flap and also has a flat seat. This design makes it difficult for the rider to sit correctly, and as both horses and ponies being shown are seen by the judges stripped of their saddles, the value of the show saddle is small.

A good dressage saddle, on the other hand, does have advantages. At first these were just adaptations of the straight-flapped show saddle, but with the increase in popularity of dressage since the war and a clearer understanding of what it tries to achieve, valuable modifications have been made. As in the jumping saddle, the tree is deep, but it is shorter and is straight rather than being sloped back. The stirrup bars are thus positioned further to the rear; this, and the relatively straight cut of the flap, demand an increased length of leather. The position of the leg permits much less grip, but does free the lower leg, placing it so that a finer degree of control and application of the aids can be obtained.

One of the advantages of the modern all-purpose saddle is that it encourages the rider to sit over the centre of the horse's balance. This is also true of the dressage saddle: the balance of a dressage horse moves increasingly to the rear with the necessity for collection and lightness in the forehand, and with a correctly designed saddle the rider is able to assist rather than to interfere with this shift in balance.

Racing saddles are made with some very

SNAFFLES

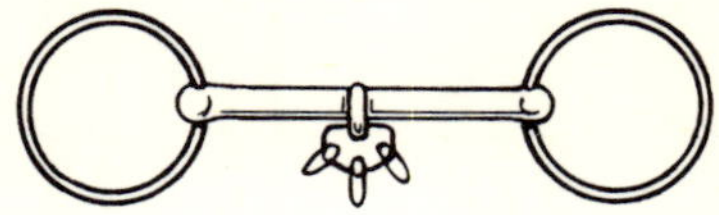

Breaking bit with keys

Loose ring mullen mouth snaffle

Loose ring German mouth snaffle

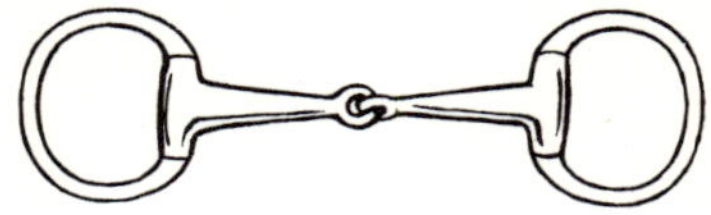

Eggbutt snaffle

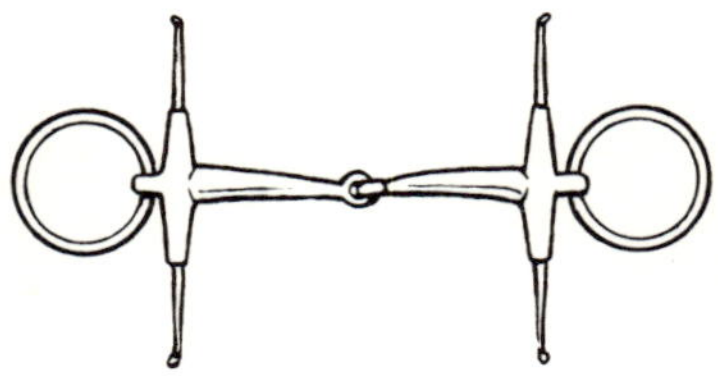

Fulmer cheek snaffle

PELHAMS

Kimblewick

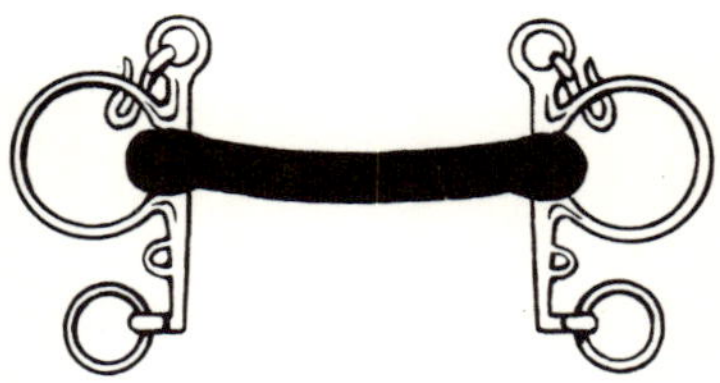

Mullen mouth pelham

WEYMOUTHS

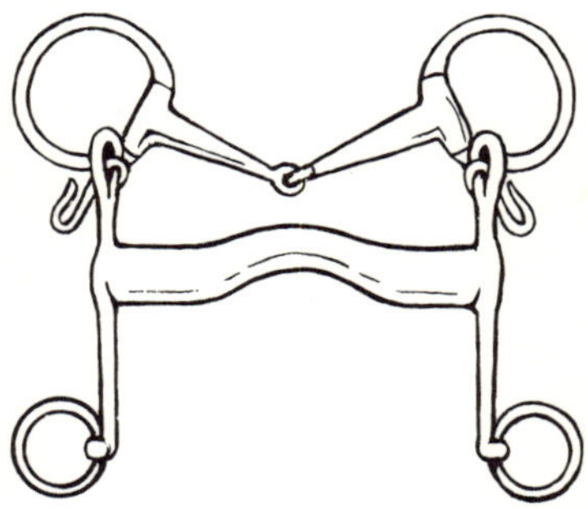

Dressage Weymouth bit and eggbutt bradoon

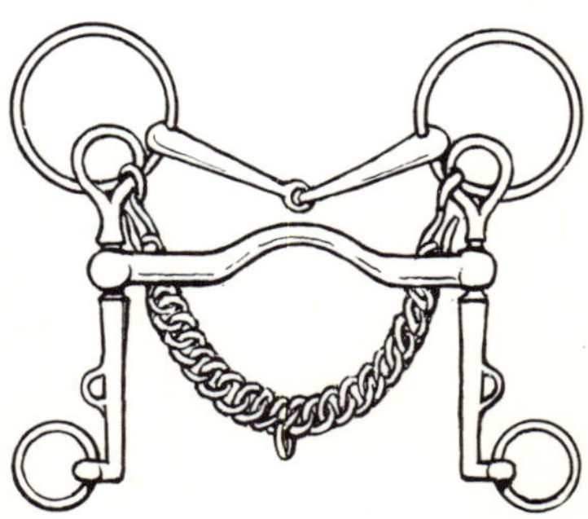

Loose cheek Weymouth bit and bradoon with curb chain

different criteria in mind. The first is weight. A very light flat racing saddle may weigh only ounces; a weighted steeplechase saddle may have to total some thirty pounds, though when a lot of extra weight has to be carried it is advisable to use a weight cloth. The very light saddles do not use stirrup bars (the leathers are looped through a slot round the tree) and the panels are covered in silk or nylon. The heavier saddles generally have serge-stuffed panels, and stirrup bars can of course be used. Recently, Australian styles of saddle have been copied elsewhere, using leather for the panel covering and with very lightweight trees. The trees in all racing saddles are relatively fragile, and these saddles therefore have only a short life. Weight is saved in the size, shape and substance of the flap and in the thickness of the panel, which is reduced to a minimum. Trainers use a heavier, stouter version, known as the race exercise saddle, for exercising purposes.

Various special saddles are made for use by children, beginning with a cheap pad of felt, sometimes the only solution when a compromise must be found between a very small child and a very fat pony. The smaller version of the all-purpose spring tree saddle is much the best and helps children to develop a good seat. It is possible to find saddles that incorporate some of the important advantages of the spring tree saddle, such as a good deep seat and generously forward-cut flaps, but which do not cost quite so much.

Girths are made in several different materials and to a number of different designs. Leather girths, except perhaps with horses that are very soft and out of condition, are much the most satisfactory. There are three kinds: Balding, Atherstone and three-fold girths. These should all be kept clean and supple. Girths are also available made of webbing, sometimes with elastic inserted, of elastic, lampwick, tubular webbing and nylon cord.

Breastplates are useful with horses whose conformation makes saddles slip to the rear. They are widely used in the racing world, and are also valuable in very hilly hunting country.

Stirrup leathers should always be of good quality, and as most people ride slightly unevenly, with more weight on one side than the other, the leathers should be alternated. Stirrup irons should be made from stainless steel or a good metal mixture such as Eglantine. They should always be too big for the foot rather than too small. There are various different designs available, such as the plain iron, Kournakoff, bent top, peacock safety iron, and so on.

It is worth considering in some detail the objects of the bit. It is an extension of the hand, and acts as a regulator for the energy or impulsion created in the quarters of the horse by the rider's seat, back and legs. It is not, or should not be, a means by which a head carriage is imposed on the horse. The origin of the head carriage (and, indeed, of direction) is in the quarters.

As the hind legs become further engaged under the horse's body, so the head is held higher. The bit is only a means of indicating the required position of the head. Should the bit be used to force a particular head position before the horse is physically able to assume it he will react by evading or resisting its action.

The lower jaw, and particularly the bars of the mouth, are the areas most closely associated with the action of the bit, though it also affects other parts of the head. In addition to the bars, the corners of the lips, the tongue, the curb groove, the poll, the nose and the roof of the mouth are all subject to pressure according to the type of bit used.

In the selection of a bit examination of the mouth is essential. The shape of the bars and jawbones, the height of the palate, the size of the tongue and the depth of the declivity in which it lies are all factors that should influence the choice of bit. The bars may be more or less thick and heavily covered; on the other hand, they may be sharp, vee-shaped and lightly covered. The first disposes the mouth towards being less sensitive, the latter towards greater sensitivity. It should be remembered, however, that there is no such thing as a hard-mouthed horse until man has made it so. If the tongue appears to be too large and to overlap the bars, certain problems will arise in the fitting of a curb bit, which will be prevented from acting on the bars. The shape of the jaws will also restrict the types of mouthpiece which will be effective. Careful fitting of bits and bridles is just as important as the correct fitting of a saddle. Not only will the action of the bit be affected by a badly fitting bridle, but chafing and discomfort to the horse will result, and encourage further evasive habits. The width of the bit used and its position in the mouth are both particularly important.

There are innumerable kinds of bits, which is perhaps why excessive emphasis is sometimes still placed on 'finding the key' to a horse's mouth. A bit will assist–or hinder–training and performance; no bit can replace training or achieve what training should achieve.

There are five groups of bits or bitting arrangements: the snaffle, the Weymouth, the Pelham, the gag and the bitless bridle. The basic construction of the bridle applies to all these groups, though the details of design vary according to the purpose for which any particular bridle is designed. A bridle consists of the headpiece and throatlatch, cheekpieces, browband, cavesson noseband and reins. With curb bits, curb chains of various kinds and lipstraps are also used.

An all-purpose bridle or one used for hunting, for example, will be made of plain, fairly substantial leather, while those used for showing will be much lighter and finer, the noseband frequently adorned with extra stitching. The bit and reins can either be stitched on or fastened with studs, billets or buckles. A sewn bridle is both the neatest in appearance and in some ways the most safe; other types of fastening are easier to dismantle and to clean.

The snaffle is the simplest kind of bit, basically consisting of one mouthpiece which acts on the corners of the mouth to produce an upward effect, raising the head. The mildest snaffle is a rubber mullen mouth, or half moon, one. Eggbutt snaffles, with a jointed mouthpiece and fixed rings, are perhaps the most popular. Dee cheek and eggbutt snaffles both developed from the cheek snaffle, one of the oldest kinds. The Fulmer (or Australian loose ring cheek) snaffle also used to be popular, though there are fewer of these bits in use today. Snaffles with loose rings allow a greater play of the bit in the mouth; fixed ring bits cannot pinch the corners of the mouth and the horse cannot evade their action by sliding them through the mouth.

There are many other varieties of snaffle available, such as eggbutts with slots in the rings; German snaffles, which have thick–and therefore comfortable–mouthpieces; the Dick Christian, which has an additional link in the centre of the mouthpiece; and the French bradoon, with a 'spatula' central link.

There are also stronger snaffle bits, such as the twisted snaffle, those with rollers set round or across the mouthpiece, Scorrier or Cornish bits with four rings instead of the usual two, Y-mouth or W-mouth and spring-mouth snaffles.

The Weymouth or double bridle is the most advanced form of bitting. It consists of the bradoon, or snaffle, and the curb bit. It encourages the correct position of the head by the combination of a raising of the head caused by the snaffle and a lowering and nearly-vertical positioning of it produced by the action of the curb. In a horse that has been properly trained in a snaffle, the double bridle will help to achieve a head carriage which gives most effective control.

There are also several varieties of double bridle, though the number of these has been reduced with the increasing popularity of the snaffle. The best known are the slide cheek and fixed cheek Weymouths, the dressage Weymouth and the Banbury.

The Pelham bit attempts to combine in one mouthpiece the action of the snaffle and the double bridle. Though this is, of course, not really possible, many horses do seem to go well in Pelhams. There is a confusing variety of them, the best known being the mullen mouth, arch mouth, Hartwell, Hanoverian and Scamperdale. The Kimblewick, adapted from a Spanish jumping bit, has rapidly become popular and is now frequently seen.

The gag bridle uses a bit similar to the snaffle, but a cheekpiece is passed through the top and bottom of the bit rings and attached to the reins, thus accentuating the upward head-raising action of the ordinary snaffle. The gag is another form of bridle that should only be used by experts, for it tends to stiffen the head carriage. It is best used with an extra pair of reins, attached to the bit rings in the normal way, which can be used when the horse is carrying itself correctly, the gag reins being reserved as a corrective aid.

Bitless bridles are less common than the others. They achieve control by acting on the horse's nose rather than its mouth, and are most useful with horses that for some reason cannot be ridden in a normal bridle. In its simplest form it applies pressure to the nose, the curb groove and, to a lesser extent, the poll. It is also useful when training a horse to jump, as the possibility of the horse being pulled in the mouth, causing pain and a loss of confidence, is avoided.

The choice of reins is largely a matter of personal preference. Plain leather reins are the simplest, and should be nice and narrow. They do, however, slip in wet weather or when covered in sweat, and rubber-covered reins or plaited leather ones are a satisfactory alternative.

Nosebands and martingales are designed to help the action of the bit and to counteract evasive measures of the horse. Properly used, these extras can be very useful; wrongly applied, they can do considerable damage.

The most popular noseband is the drop noseband, which has two important functions. It alters the limited action of the snaffle, making flexion of the lower jaw and poll possible. Its second function is to close the mouth, thus preventing the horse from crossing its jaws, opening its mouth or getting its tongue over the bit, all common evasions.

It is extremely important that a noseband should be correctly fitted. It should lie about 2½ to 3 in. above the nostrils, with the rear strap, which passes under the bit, fitting into the curb groove. It should be fastened snugly but should never be too tight.

There are other nosebands which act on the same principle, that of applying pressure to the nose. The Flash noseband combines the cavesson and drop noseband in one, and can therefore be used in conjunction with a standing martingale. Grakle nosebands have two straps, one fastening above and one below the bit. Pressure is exerted on one point only, where the straps cross over in front, and it is perhaps stronger than the ordinary drop noseband as a result. The Kineton or Puckle noseband is the most severe, consisting of two metal loops with an adjustable connecting nosestrap. It does not close the mouth, but any pressure on the bit is transmitted to the nose.

All martingales are used in order to lower the horse's head and increase the rider's control. The most common is the standing martingale, a single strap which attaches to the noseband at one end, and–like other martingales–to the girth at the other. It is kept in place by a neckstrap. It increases control by exerting pressure on the nose. Standing martingales should be fitted with care, as if they are adjusted too tightly they hamper the horse when jumping.

Running martingales divide in two where they pass through the neckstrap, each end being attached to the reins by a ring. They thus act on the mouth rather than the nose. The action of a running martingale is severe if it is tightly adjusted, and this amount of extra control should only be used by experienced riders with sympathetic hands. If used with a double bridle, the martingale rings should be attached to the curb rein. Leather or rubber stops should always be put onto reins when a running martingale is in use, to prevent the rings from sliding forward.

Various other martingales are also available. The bib, which has an action like that of the running martingale, has a piece of leather between the two branches as a safety precaution. Irish rings, also known as Irish martingales, are not really martingales at all: they only affect the direction of rein pull. The combined martingale acts as both a standing and a running martingale. The pulley martingale resembles the running martingale but consists of a cord, with the rings at either end, that passes freely through a pulley on the main strap and allows the horse greater lateral freedom. The Market Harborough martingale, also known as both the German rein and the English rein, consists of two straps which are attached to the reins at one end and pass through the bit to be attached to the martingale body at the other. The straps are slack while the horse holds its head correctly, but exert a downward pull on the bit as soon as the head is thrown upwards. Though it is the subject of considerable controversy, the Market Harborough martingale has its advantages, one being that it does not restrict the horse's ability to extend while jumping.

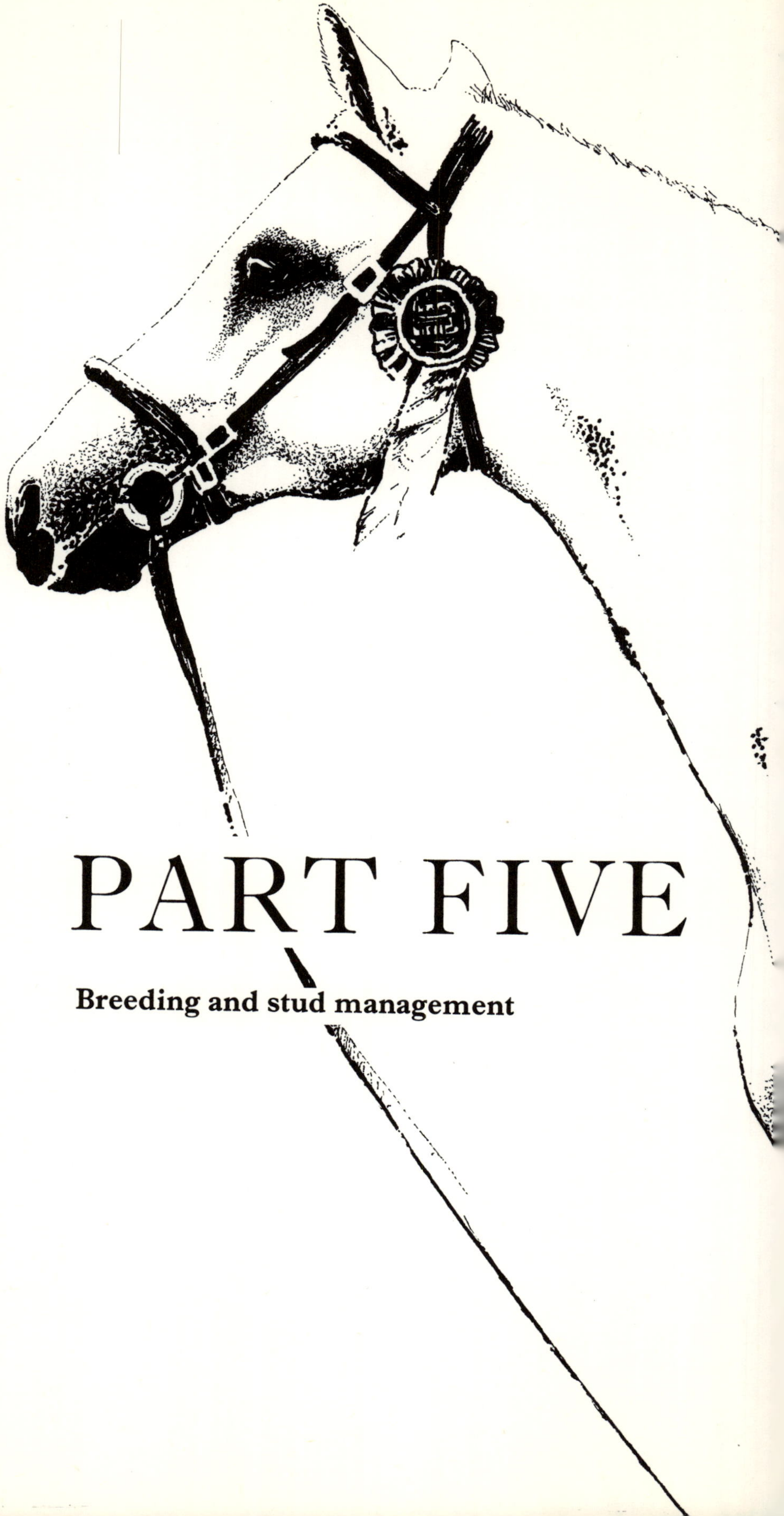

PART FIVE

Breeding and stud management

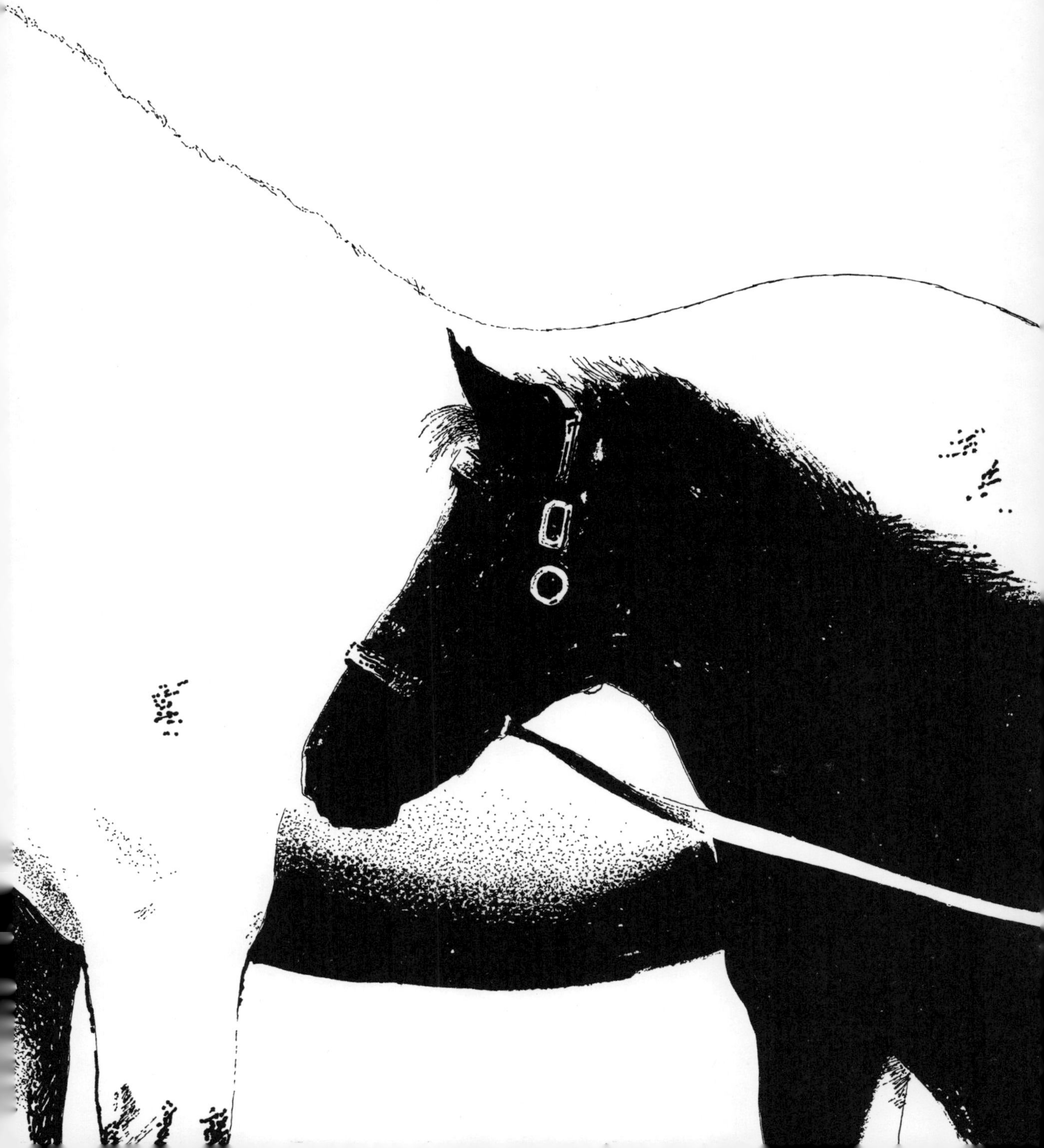

Chapter one

The reproductive system

The genital organs of the mare consist of the vulva (which can be seen below the rectum under the tail) which leads into the vagina, a soft-walled passage whose sides are in contact with one another. The vagina is about eight inches long and runs into the body of the uterus, with a structure known as the cervix at the junction. The body of the uterus is also soft-walled and is about ten inches long in the quiescent state, though it is capable of expanding to three or four feet to contain the foetus. The two horns of the uterus run forward from the body, each ending in a Fallopian tube, a thin-walled, corkscrewing tube some eight inches long when straightened. Close to the end of each Fallopian tube lies an ovary. The ovaries are positioned behind the kidneys and close to the spine.

The function of the ovaries is to produce eggs (ova) and to manufacture secretions (hormones) which control the cyclic changes that occur in the reproductive organs. Other hormones which control the ovaries are produced by the pituitary body, a small gland lying at the base of the brain.

Puberty in the mare occurs at fifteen to twenty-four months of age or later. It may even not occur until the mare is four years old, and this is the cause of much early infertility. Only exceptionally well grown two-year-olds should be bred; it is better to mate as three-year-olds.

Each spring the pituitary hormone stimulates the ovaries to produce the eggs which lie within their substance. The ova are surrounded by cells to form follicles which make their way to the surface of the ovary and then burst, discharging the contained eggs into the adjacent Fallopian tube. The hormones also cause changes to take place in the genital tract and in the mare's disposition. This condition is referred to as 'heat' or being in oestrus.

The period of heat varies from a few weeks to a few days, the tendency being for long heats to appear early in the year and to become gradually shorter until they cease in late summer or autumn. It is only when a mare is on heat that she will accept the stallion. The ovum is shed during the last forty-eight hours of heat and can live within the Fallopian tube for four to six hours. For conception to occur it must contact a sperm during this time.

The production of the hormones which induce oestrus in early spring is dependent upon several factors, the most important of which is the action of daylight upon the mare.

After each heat period the mare returns to normal for about three weeks before she again comes into season. This cycle continues through spring and summer into the autumn.

Mares are most likely to conceive if mated at the third or fourth heat of the year, and foals born of such conception have the lowest death rate. This is attributed to the nourishment from the good spring grass at the time of service and birth.

There are numerous signs to indicate when a mare comes into heat but they are not always all present. Typically, the mare seeks the company of other mares more than usual, swishes her tail and protrudes the clitoris (the small, sensitive, rod-like organ lying within the lips of the vulva), she passes urine frequently and in small quantities, and mucus appears at the lips of the vagina. While experts can also ascertain the phase of her cycle by inserting a hand and arm into the rectum and feeling the ovaries, the most certain way of finding out whether she is ready to be mated is by trying her with a stallion.

This is known as 'teasing', and precautions have to be taken for the safety of the stallion as if the mare is not ready to receive him she may attack him savagely with heels and teeth. The usual practice at studs is for teasing to take place over a padded gate or partition with the mare being brought up on one side and the stallion on the other. Often a stallion of little value is used for this purpose as the risk of injury is considerable. If the mare is ready for service she will adopt the mating posture and hold her tail to one side; if the advance of the stallion is premature she will be restless, bare her teeth and attempt to bite or kick him. When the mares are running together in a field it is the practice in some studs to lead a stallion round the outside of the herd and catch and tease those mares which leave the others and follow him.

Even if the mare appears to be ready there is still a possibility of the stallion being kicked, and it is common practice for hobbles to be worn over the mare's hocks attached to ropes which pass to a neck collar. This is not done to force a service on her but merely as a precaution for the sake of the sire. The optimum time for mating is two to five days before the end of oestrus and before ovulation occurs.

After the stallion has mounted and served the mare his semen will be deposited in the vagina and uterus. It is sucked upwards into the horns of the uterus and eventually reaches the Fallopian tubes. In one or other of these an egg and a sperm cell meet and fuse. The sperms can live for up to thirty hours after emission into the mare. The fertilised ovum descends into the body of the uterus, taking four to six days for the journey. It is nourished there by 'uterine milk' secreted by the uterus under the influence of hormones.

After conception hormones prevent further heats occurring, and failure to come into season again indicates that the mare has 'held', though this is not an infallible guide. The union between the fertilised ovum and the uterus of the mare is a weak one and it is not until about two months after service that they become firmly united. There is a period at about six weeks, when the growing foetus is some two inches long, when the bond is particularly weak and abortion is common at this time. Following such an abortion, provided it is not too late in the year, the mare will come on heat again.

The average gestation period of the mare is 336 days, but it may be as little as 320 days and some foals are carried for over a year. Most mares come back into oestrus seven to ten days after foaling; this is known as the 'foaling heat'. The practice of having mares covered during this period of heat, to which great importance used to be attached, is now decreasing.

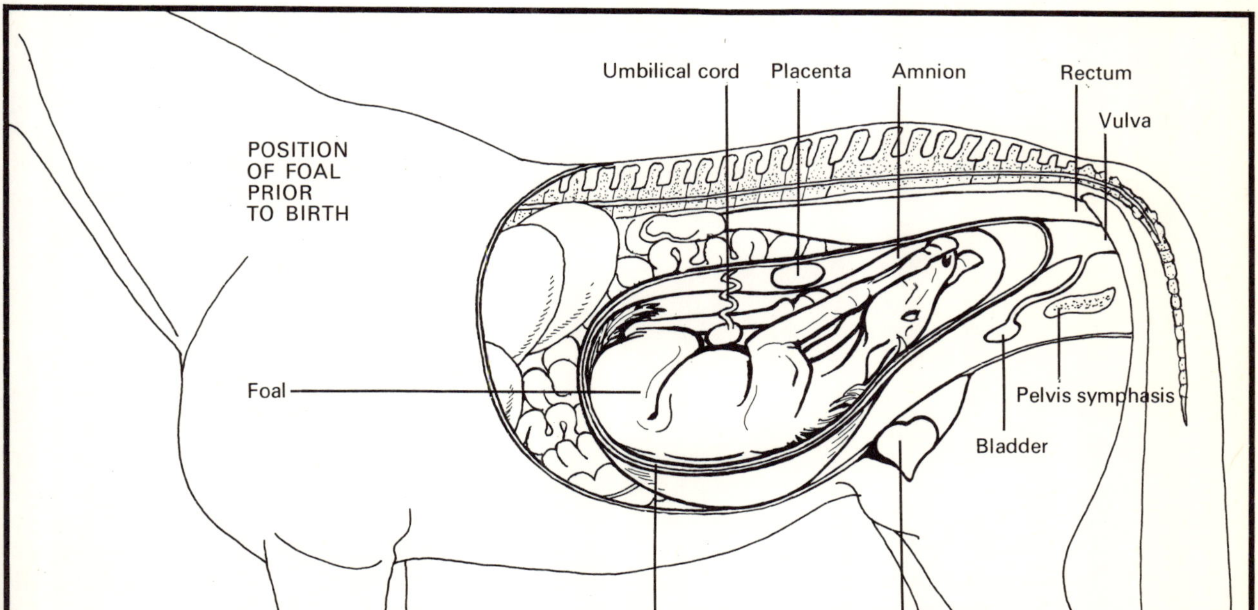

Chapter two

Stallion selection and management

In order to be considered as a prospective stallion, a colt must be an outstanding example of his breed or type. But not only must the future sire be an exceptional individual: he should also be the product of bloodlines which offer some guarantee that he will transmit many of his most impressive qualities to his offspring.

The key to a stallion's success as a sire is prepotency, and this is why you must look carefully at the prospective stallion's family tree as well as his own characteristics when selecting a horse to stand at stud. Due to a chance, fortuitous combination of genetic factors, a horse bred from undistinguished ancestors can sometimes be a truly outstanding individual and yet, because of the lack of firm support of good blood in his background, be a disaster as a sire.

No matter how well-made and well-bred a stallion may be, there is, of course, no guarantee at all that he will make the grade as a stallion. However, an outstanding horse that is backed up by good breeding is always a much safer bet than a similar individual who has suddenly come from nowhere.

The matter of prepotency, the ability of a horse to 'stamp' his stock with his own type and to pass on to them his own qualities, is the reason that selection of a young horse for stud is always something of a gamble, even though he may be a winner of the Derby or a show-ring champion.

In the case of an older horse which has already sired several crops of progeny, selection is a much easier matter, since you can make a reasonable assessment of his probable future performance at stud by looking at what he has already produced. This is also when you see how a horse's genetic inheritance, his 'blood', can influence his results at stud to such a marked degree. For example, a horse with a handsome head and impeccable hind legs may have sired a large percentage of animals with plain heads and 'curby' hocks from mares without these defects. Such a stallion is obviously carrying recessive genes for these undesirable qualities as an inheritance from one or more of his ancestors.

This is why anyone who embarks on the responsible task of selecting a stallion should not only be a fine judge of a horse's make, shape and action, but must also have a very considerable knowledge of the bloodlines of the breed or type to which the prospective sire belongs.

Now let us turn from the horse's bloodlines and look at him as an individual. In the case of the Thoroughbred racehorse, selection of the individual will be easy enough: he will be a colt that has performed exceptionally well on the Turf. 'Mate the best to the best' has always been the basic maxim of the Thoroughbred breeder, and how well it has worked!

Choosing the best in other breeds, however, may not be nearly so simple, since what constitutes an outstanding individual may often be a matter of personal and very subjective opinion. Nevertheless, there are some basic guidelines to follow: try to remain as objective as possible (which is not always so easy when one has bred a fairly successful colt oneself!), and assess the colt on the basis of temperament, conformation, action, constitution, soundness and performance. If you set rigorous standards, and the colt measures up well against all of them, take another searching look at his pedigree, and, if it really does look good, it is worth your trying him at stud.

But before acquiring the colt or, if he already belongs to you, advertising his services, you will need to satisfy yourself on several more points. Is his sexual urge, his libido, normal? And is the quality of his semen such that he is likely to be fertile? A veterinary examination will ease your mind on the latter important question. Once the entire has passed all the demanding tests made to select those few individuals worthy of a chance to perpetuate their characteristics, management will play a fundamental role in his success at stud.

It is essential that the entire should be kept as healthy, fit and contented as possible. This latter point is most important, since a contented horse will be much easier to work with and will make the most of his environment, therefore giving himself the best chances of high fertility, one of the two aims of the stallion owner (the other being, of course, outstanding progeny).

In order to keep a stallion as happy and as level-headed as possible, try to treat him as much as possible like any other horse as you can, within the limits imposed by the fact that he *is* a mature, serving entire. In the ideal set-up, the stallion should run with at least one or two mares that are in foal to him. However, because of their high economic value, many horses are not allowed to do this, and so it is important that special attention be given to their natural and highly-developed need for other equine company. A stallion should always be able to look out of his box at another animal, and should be allowed to see other horses from his paddock or yard, too, unless he is a very silly and excitable type.

The old-fashioned habit of shutting a stallion up like some dangerous wild animal in a dark box or high, close-boarded stallion yard so that he could never see another horse except at covering time was quite mistaken, and was, in fact, an almost certain short-cut to ruining his temper.

In handling the stallion, you will need to establish a basis of trust and respect, and back it up with consistently sympathetic treatment. And you must never show any fear of him! This is essential, since a stallion, who has super-subtle powers of perception, soon knows whether a person is afraid of him or not. If you are, then he may well try to take advantage of you, and will certainly offer neither trust nor respect.

A stallion must be really fit for the covering season, since his state of fitness will have a direct bearing on his fertility. The fitter a horse is, the higher ought his fertility to be. Stallions living out all the time in a fairly large expanse of pasture probably will not need any special conditioning for covering, but the stabled stallion certainly will. The first step you can take towards ensuring that the stabled horse is fit for covering is that of not letting him slacken off entirely to become flabby and fat after the end of the preceding breeding season. To keep him fairly fit is not difficult: just let him run in a paddock of not less than five acres for most of the day, either with one or more mares in foal to him, or by himself. If you cannot make arrangements to turn the stallion into a big paddock for long periods each day, you will have to keep him in reasonable shape by walking, lungeing or riding him.

Ten weeks to two months before the covering season is due to begin you should start work to get your stallion really fit for covering. There are quite a number of ways of doing this. Riding is certainly most fun, and is probably the most efficient way, too, especially when you can turn the horse out to graze for at least two hours every day as well, since exercise under saddle can then be restricted to more energetic work at faster paces. A stallion should always be mounted by a confident, experienced and talented rider, since entires *do* require considerably more equestrian tact than other horses. They should never, of course, be ridden out with other serving stallions or with mares in season. It is always a good idea to give the horse ten or fifteen minutes on the lunge before mounting, as this will allow him time to settle down.

Lungeing is another good way of providing a stallion with the exercise necessary to make him hard fit for the breeding season. But any lungeing programme must always be coupled with letting the horse graze free for a minimum of two hours every day, wherever possible. And for lungeing to be truly effective, both handler and horse must know the procedure thoroughly. The stallion must work calmly and steadily at the end of the long line at all times, walking, trotting and cantering as and when the handler commands.

Walking the stallion for miles every day is the traditional British method of getting a stallion into shape for the season. It is a very effective method, too, although the most time-consuming and laborious. Start with half an hour a day, and increase the time gradually until the horse is being walked on energetically for up to an hour and a half every day. Try to turn him into a paddock for two hours before being walked, as he will then settle to striding along quietly in the rhythmic walk that exercises every muscle of his body so completely, and will not be jogging along over-full of energy and looking for every excuse to play up and let off steam. It will be a big help in the stallion's conditioning process if you can walk him up and down many long, gradual gradients. And, wherever possible, stay off public roads.

Feeding complements exercise, of course, and you will need to increase the stallion's ration as you step up his exercise. Slowly transform it into a high-protein diet, which will help in muscling him up and also in the very important matter of semen production.

Given the great diversity in make and shape of stallions, it is obviously out of the question to lay down any absolute rules about feeding. Suffice it to say that the stallion feeder will need much knowledge of and experience in the feeding of horses, and a shrewd eye for condition, since so much of the health, well-being and fertility of the horse will depend on what goes into his manger. Watering is important, too, and the stallion should have fresh, clean water constantly available. Feed and water containers must be kept scrupulously clean at all times.

Grooming is a must for the stabled stallion (entires running out all the time are never groomed since they need all the natural protection against bad weather that their ungroomed coats can give them). The stallion should always be groomed after he has been exercised; he must be properly strapped and not just brushed over lightly. Grooming is an important part of bringing a horse to his peak of fitness and also of maintaining him there. It should always be begun when you start to get the horse fit, and continued for the duration of the covering season.

A stallion's loose box should be large, with high ceilings, and should have plenty of natural light and adequate ventilation without draughts. If it opens into a paddock so that the door can be hooked back for much of the day to allow the horse to go in and out as he pleases, you will have the ideal, but obviously rather luxurious, arrangement for housing a stabled stallion. Next best is an adjoining runway or yard which will provide him with plenty of scope for exercise, although this should always be combined with his daily period in the paddock.

On a smaller stud, the stallion's box can with advantage be sited in the general yard, where the presence and the coming and going of the other horses will keep him interested and contented. At large establishments, where there is much activity involving many strange mares, it is best to have the stallions' boxes away on their own, but constructed so that the entires can look out of them at each other companionably. And remember that the stallion box or boxes should always be situated close to where covering takes place.

You must pay close attention to worming, vaccination (tetanus immunisation is essential), and the care of feet and teeth. Take care not to let a stallion's hooves wear so much that they become sore, since pain in his feet can even put a horse right off covering; but if a stallion is being exercised only on soft going, then make sure to trim his feet every month to six weeks. And see to his teeth at least once every six months, since tooth trouble can ruin his digestion or his temper or both.

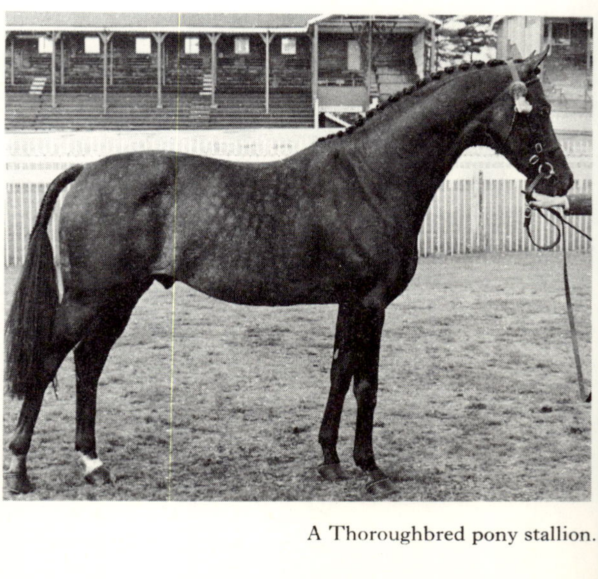

A Thoroughbred pony stallion.

Festooned with the ribbons of the champion, the Arabian stallion, Delos, is led round the arena at the Sydney Royal Easter Show.

The Connemara Pony Show.

H.M. the Queen drives down the course at the start of Royal Ascot.

Hard-pressed, the leaders round the turn at Cologne.

Nijinsky, whose pedigree chart is reproduced below, seen here winning the 2000 guineas at Newmarket.

Nijinsky bay colt 1967	Northern Dancer bay 1961	Nearactic brown 1954	Nearco	Pharos	Phalaris
					Scapa Flow
				Nogara	Havresac II
					Catnip
			Lady Angela	Hyperion	Gainsborough
					Selene
				Sister Sarah	Abbot's Trace
					Sarita
		Natalma bay 1957	Native Dancer	Polynesian	Unbreakable
					Black Polly
				Geisha	Discovery
					Miyako
			Almahmoud	Mahmoud	Blenheim
					Mah Mahal
				Arbitrator	Peace Chance
					Mother Goose
	Flaming Page bay 1959	Bull Page bay 1947	Bull Lea	Bull Dog	Teddy
					Plucky Liege
				Rose Leaves	Ballot
					Colonial
			Our Page	Blue Larkspur	Black Servant
					Blossom Time
				Occult	Dis Donc
					Bonnie Witch
		Flaring Top ch. 1947	Menow	Pharamond	Phalaris
					Selene
				Alcibiades	Supremus
					Regal Roman
			Flaming Top	Omaha	Gallant Fox
					Flambino
				Firetop	Man o' War
					Summit

Chapter three

The brood mare

The better a mare is as an individual, and the better bred she is, the more superior a brood mare should she make, always assuming that she possesses the desirable breeding conformation of sufficient depth and width, and normal genital organs and mammary glands. For these reasons fairly high standards should be set when a mare is selected for sending to stud, bearing in mind that an inferior mare is going to cost every bit as much to keep as a well-made, well-bred one, and that a poor foal will be at least as expensive to raise as a good one.

Nevertheless, this counsel of excellence certainly need not preclude the turning of an old family favourite, now retired from her normal working life, into a brood mare. However, do not expect great results from her as a producer unless she is a fairly well-made individual with a reasonably distinguished pedigree, no matter how good the stallion you send her to may be. And, regardless of how strong a hold she may have on your heartstrings, do not decide to breed from her if she has any important hereditary defects. If you do, you will almost certainly be heading straight for disappointment. Such inheritable defects include sickle and cow hocks, feet that turn in or out too much, pasterns that are too upright and those that are too sloped, a ewe neck, sway back and parrot mouth.

Fairly rigorous standards should be applied to any mare from whom you contemplate breeding. She should measure up to reasonable requirements, in particular of temperament, conformation and soundness of constitution; and also of action and performance. This applies even in the case of a Thoroughbred racing filly who, although she will be selected for stud primarily on the basis of her performance on the racecourse, will stand a much better chance of becoming a top producer if her temperament, conformation and constitution are also good.

In the case of a mare from whom pleasure horses are to be bred, temperament is of the utmost concern. Not only will a bad-tempered mare tend to pass her testy temperament on to her offspring as an inherited quality, but she will reinforce any tendency towards nervousness and irascibility by the example she sets her foal after it has been born. Anyone who has had charge of a large number of brood mares and foals on a stud will tell you how quickly foals begin to imitate their dams, and how much more trusting and friendly the foals of good-tempered mares usually are compared to those with ill-tempered dams.

Conformation is of vital importance, too, since a brood mare must be wide enough and deep enough to carry the foal comfortably, to allow it to grow without restriction and to foal down as easily as possible. Herring-gutted, slab-sided, narrow-hipped mares can cause endless trouble, and the only filly of faulty conformation I would consider breeding from would be one who had been an exceptional performer on the racetrack.

Correct conformation of the genital organs and normal development of the ovaries is essential as well, since a filly with an infantile genital tract will cause only headaches. This is where a preliminary veterinary examination can be a great help. In some cases it proves essential, since a mare with an infantile genital tract and/or underdeveloped ovaries will have a normal vagina and will therefore look perfectly all right to the layman. Another important thing to check up on is the conformation of the mare's anus and vulva. They should lie in the same vertical plane, because if the anus is very 'indented' excreta is liable to pass from it onto and into the vulva, and thence to the mare's genital tract, with the consequent danger of infection and threat to fertility. Keep an eye open, too, for a vulva that is very large, loose and floppy; this is another undesirable formation, for it will make the mare prone to 'wind-sucking', and liable not to get in foal as a result.

However, the problem of a too-large, too-loose vulva can be overcome by having it stitched by a veterinary surgeon. The mare's udder should be normally formed as well and, in the case of a mare who has already had a foal, should be seen to be well developed, with reasonably large, properly shaped teats.

The constitution of the brood mare is also of vital concern: she should be a 'good doer', thereby increasing the chances of her taking good care of the developing foetus, providing it with all the nourishment it needs as long as she is herself fed appropriately.

Mares should not be bred from before they are three years old, and I really prefer them to be four. They should be 'let down' for at least four months before going to the stallion, living a quiet, natural life after the bodily and nervous stress of racecourse or show ring.

Thoroughbred brood mares are sent to the stallion stud early in the year and foal down eleven months later. This time of year is far from ideal either for breeding or for foaling, and other mares should ideally not have their foals before April at the earliest, when full advantage can be taken of the better weather and the spring grass. They should therefore be sent to the stallion in, say, March of the previous year.

At the stallion stud a close watch will be kept on the course of the mare's sexual cycle with the aid of frequent 'teasing' by a male horse. She will be covered once or twice by the stallion during one three- to five-day oestrus period, and if she has conceived to him she will not come back into season again three weeks later. After the mare has missed two oestrus periods, i.e. when six weeks have elapsed since her last service without her showing any signs of coming back into season, a diagnosis of pregnancy can be made by a veterinary surgeon by means of a rectal examination. Some owners suspect that this manual inspection of the mare can cause abortion; however, it is standard practice on the leading Thoroughbred studs.

Once a mare is back home again, certified in foal, she will not be a lot of trouble if she does not have a foal at foot, since she can then live out day and night until cold weather sets in towards the end of the autumn. If she has a foal at foot, however, management will vary depending on whether or not she is a Thoroughbred.

Profile of a Thoroughbred mare and foal owned by H.M. the Queen. Notice the foal's light, well-fitting slip.

It is essential that brood mares at grass should have only quiet company. Other brood mares are ideal; mares should be kept apart from young stock and geldings, whose attentions can cause a mare to 'turn'.

All mares, except those of native pony breeds or cross-breds containing a good proportion of tough native pony blood, should be brought into boxes to sleep once the wet, cold weather begins at the end of the autumn or early in winter. During the last third of their eleven-month pregnancies, when the foetus makes its

A winning hunter brood mare.

greatest burst of development, they should be fed increasing amounts, particularly of protein. Right up to foaling the mare should be kept in good condition, though she should never be allowed to get too fat as this tends to make foaling more difficult. Even during the later stages of pregnancy it is still most important for a mare to have her normal daily exercise.

Keep an eye on the mare's udder during the few weeks before she is due to foal. During this time she should start to 'make a bag': her mammary glands should begin to increase in size and continue to develop until she foals. If your mare is making little or no bag, she will need your help to ensure that she will have enough milk for her foal when it arrives. Correct her diet: she will benefit by being fed fresh grass if any is available, sliced carrots, parsnips and apples, and by the daily addition of dried milk powder to her feed for three or four weeks prior to her foaling. Cubes or meal of dried grass, clover or lucerne will also help, and should continue to be given for up to three weeks after foaling in the case of constitutionally poor milkers.

In the month to three weeks immediately prior to foaling it is also important progressively to cut down the quantity of hay fed to the mare, so that just before the foal is due to arrive she is receiving about two-thirds of her original ration. This is so that the mare's insides will not be packed with bulky fibrous matter when foaling begins. Watch her droppings, too, and if they start to become dry and hard correct matters with bran mashes.

During the last ten days or so before the foal is due is pays to handle the mare's udder gently every day, which gets her accustomed to the sensation of the foal's first search for milk. It is essential to do this with first-foaling mares, who otherwise may object violently to the foal's first sucking. As many first-foaling mares will also object to your handling of their udders, make sure there is someone reliable at the mare's head before you start.

For the mare that will be foaling indoors (many native pony and cross-bred mares are allowed to foal down in the field, where there is less risk of infection than in a box) it is essential to prepare the foaling box properly. It should be thoroughly cleaned out and disinfected, then bedded down with plenty of clean straw which should be banked up both round the walls of the box and against the door to eliminate floor-level draughts and to protect the rather clumsy young foal. Most mares foal down at night, so the foaling box will need adequate artificial lighting.

Mares and foals at grass
well-fenced, quality pasture.

Take care of the final preparations for foaling well ahead of time. Have a good vet on call in case he is needed, and get together the things you will require, which are: two clean buckets for warm water, and old bucket or other container for the afterbirth, cotton wool or gamgee tissue, antiseptic powder or lotion for the foal's navel stump, several rough and fairly absorbent towels, a bottle of liquid paraffin, and soap and a nailbrush for your hands and arms. You must also have a source of hot water at hand.

As the time for foaling approaches, the mare's udder will get bigger and firmer, and then, some six hours to two days before foaling is due to begin, most mares will 'show wax'. Clear, thick, serum-like material will issue from each teat canal and will harden into little beads of a substance that looks like wax. Not all mares wax up, however; on the other hand, a few will do so a number of days before they actually give birth. Another sign of impending birth in most mares is 'slackening'. This is the relaxation of the ligaments of the pelvis to make the passage of the foal easier; it shows most obviously by the appearance of a groove on either side of the root of the tail. When either or both of these symptoms are shown a fairly close watch should be kept on the mare day and night.

Soon before foaling starts the mare will begin to get restless, and will wander round her box uneasily. She will probably turn to look at her flank with a worried expression, kick up sharply with a hind foot from time to time, switch her tail about, and will probably soon start to sweat. When she shows these signs of beginning labour, half-fill your buckets with warm water and mix the antiseptic into them. Then scrub your hands and arms to the elbows. Put several large swabs of cotton wool or gamgee into the other bucket, and keep a close watch on proceedings.

If the birth is taking place normally, the greyish water bag will appear balloon-like between the lips of the vulva and will break, and you will soon see the foal's forelegs, one in advance of the other, appear inside the whitish membrane of the amnion. It is now time for you to check on the position of the foal.

Go into the box and swab around the mare's vulva and anus, inside her buttocks and under her dock. Then lubricate your hand with liquid paraffin, slide it along the top of the foal's forelegs into the mare's vagina and feel for the foal's muzzle. If it is in the correct position for a normal birth your hand will go in only as far as the wrist before you will feel it. If you cannot find the muzzle, the foal's head is turned back and must be brought round before delivery can take place. You will need the veterinary surgeon urgently. While you are waiting for him, try to get the mare back on her feet and walk her slowly round the box. This will probably stop her straining, and will help to make things no worse before professional assistance arrives.

The most important thing to realise is that any presentation other than that of the foal's forelegs followed closely by the head in a 'diving' position means that the vet should be called immediately. The great majority of mares, however, do foal down without any complications at all.

The mare will strain hardest when forcing the bulkiest part of the foal, the chest and shoulders, through her pelvic arch. This is where progress will probably slow down considerably, but once the foal's chest and shoulders are through the little animal, still enveloped completely or almost completely in the amniotic membrane, will come sliding out on to the straw, with only its hind legs, from just above the hocks, still in the mare.

Cup your hand under the foal's eye to prevent it from being injured by straw, and strip the membrane from over its nostrils and head, making sure there is none in its mouth. But do not do anything else at this stage, since much of its blood is still circulating in the placenta, and to break the umbilical cord now would mean that the foal would lose much of its full quota of blood.

Soon the foal will start to struggle and will kick its hind legs free from the mare; the umbilical cord will break off. Dress the stump of the cord with antiseptic powder or lotion, taking the utmost care not to touch it with your hands. Strip the rest of the membrane from the foal and drag the foal round to where the mare can lick it. Then tie up the afterbirth securely with twine so that it hangs at about the level of the mare's hocks when she gets to her feet. If she does not lick the foal to dry it off, towel it over briskly. Then leave the box and watch from outside.

It is best to leave the foal by itself to struggle to its feet, find the udder and learn to suck. However, a first-foaling mare may have to be restrained until she gets used to the idea of being sucked by her offspring. A normal foal will take anything from fifteen to ninety minutes to get to its feet and take its first suck of the essential colostrum, the 'first milk', which is loaded with vitamins and proteins as well as antibodies to combat infection.

Once the foal is on its feet and sucking, there are two things to watch for: that the mare does not retain the afterbirth, and that the foal does not have a 'stoppage' of meconium. If the mare has not expelled the afterbirth within four hours after foaling, you must call the vet. And if the foal does not have a bowel movement within six or eight hours after it has been born it will need an enema of warm soapy water. If the foal still has not passed any meconium by the time it is

eighteen hours old, you will need the vet's assistance.

Thoroughbred foals and their dams can go out during the day in good weather, but should sleep inside at night until the foal is weaned at four to five months. These foals will be taught to lead early in life, and will start to nibble hard food from their dams' mangers when they are about two weeks old. Other foals may start to sleep out with their dams soon after they have been foaled, especially if the weather is fine and mild. All foals, whether or not they are Thoroughbreds, should where possible be wormed regularly, as it is worms that do the most damage to youngsters.

When it is time to wean the foal, separate mare and foal completely so that they can neither see nor hear each other, and cut down on the mare's concentrates to decrease her production of milk. Do not milk out her udder unless it is very full and causing her discomfort, when a small amount may be withdrawn, since milking her will only cause more milk to be secreted. If you do not touch her udder, the pressure which will build up will stop further production of milk in the natural way.

Chapter four

Youngstock management

The very first thing to bear in mind about youngstock is that they are just that – babies and youngsters. This may seem an obvious statement to make, but it is surprising how many people seem to forget this fundamental fact, apparently expecting from the youngstock the behaviour of adult horses, and making hardly any allowances for the fact that in many cases they are the equine equivalents of human toddlers and school beginners.

You do not expect a baby or a toddler to understand you as an adult does or to behave like one, and no more should you ask the foal, weanling or even yearling to comprehend and to react just like the grown horse. You must exercise great patience and much skill when handling and training youngstock, never asking too much at any one time. You will need to progress logically and slowly in the youngster's education, confirming him in good habits step by step, and you must be firm enough to hold his unquestioning respect, yet not so harsh that you upset him. In this regard, the handling of Thoroughbred youngstock, particularly well-fed, well-grown colts, is particularly demanding, and calls for considerable talent and experience.

A well-grown foal and its dam at the Royal Stud at Hampton Court.

Since the methods of managing young Thoroughbreds and those of raising other young horses differ so markedly in many ways, I will deal first with Thoroughbred youngstock, and then with all others.

Thoroughbred foals should wear a little headcollar during the day from the time that they are one or two days old. It should be taken off at night, so that there is no chance of the foal hooking up a foot in it during the hours of darkness in the box. To put the headcollar on the foal for the first days, you should have a helper catch up the mare and hold her alongside one wall of the loose box, keeping her head turned so that she can see her foal all the time. With the aid of another helper, approach the foal from the nearside, and secure it by putting your left arm around its chest and grasping its tail at the root with your other hand. The assistant can then fit the headcollar.

During the first few days that the foal spends inside, a start can be made on teaching it to lead. Have a person on either side of the foal, and pass the end of a rope or webbing lead through the headcollar so that each end is held by one of the handlers. Pass a rolled sack behind the buttocks, with its ends held by each person in his other hand to stop the foal from pulling back. Then lead the mare around the box several times, with the foal following behind. Repeat this drill for a few days, and when mare and foal are led outside for the first time, probably for a short spell of exercise in a covered yard, it can be led along behind its dam in the same fashion. Then the foal can graduate to being led by one person, and finally to being led with its dam by only one handler, who will usually hold the mare with his left hand and the foal with his right.

This initial work with the foal will usually be done by skilled professionals on the stallion stud where the foal will have been born. When your Thoroughbred foal arrives home, it will almost certainly have been taught to wear a headcollar and to be led alongside its dam – but, a public stud in the breeding season being such a hectic place, it will know precious little else. Now it will be up to you to begin to lay the foundations of a good-tempered, well-mannered horse by handling the foal thoroughly, teaching it in the process to stand quietly for the farrier and to be a good patient for the veterinary surgeon.

There are some basic rules to bear in mind. Do not make the handling sessions too long; fifteen minutes a day is sufficient. Proceed slowly and quietly, always exercising much patience, so that you do not frighten the foal. And if it does object to any phase of the handling, do not stop the lesson before the foal has learned it, as otherwise it will only be all the more unco-operative next time. Each handling procedure must be repeated daily for several days until the foal accepts it completely. Start off by handling the legs, then get the foal used to having its feet picked up and having their undersides rapped in imitation of the farrier. Handle all parts of the head, too, as well as the stifle, the groin and under the tail, which can be ticklish spots. And, finally, train the youngster to accept a rectal thermometer, since it is certainly no time to begin thermometer training when a highly-strung, well-developed foal is ill!

You must keep a close check on the foal's feet and limbs from the moment it returns home, with the forefeet being particularly important. Do the feet turn in or out and how much? If they do, then it is imperative to get a skilful farrier to begin on corrective trimming, which can do much towards setting a foal's feet right at this stage, while its bones are still soft and growing. This is particularly important in the case of youngstock bred for sale; foal and yearling buyers pay an enormous amount of attention to feet, legs and action, and want them to deviate as little as possible from the straight.

During the first vital stage while the foal is on the mare, you will have to pay a great deal of attention to their nutrition. They should go out on to good, vigorously-growing pasture every day, and be fed enough of the best quality oats, bran and hay in their box to satisfy their considerable daily needs for energy and protein, minerals and vitamins. A milking mare's energy needs alone may be more than twice her requirements for simple body maintenance, and she will require nearly three times as much protein, two to three times as much calcium and phosphorous, and nearly five times as much vitamin A.

In practical terms, this means that you should feed her from 12 to 15 lb. of oats, 15 to 20 lb. of hay and 1 lb. of bran a day, depending on her size and condition, and also on the amount and quality of pasture you have available. 15 to 20 lb. of hay should be sufficient for most mares, but if your mare wants more, then feed it to her, since you cannot overfeed hay, and lactating mares should get as much as they want. This is not so with oats, of course, and 15 lb. daily, split into three feeds, would be the upper limit. The mare should have rock salt in her manger, and you should mix some well-boiled linseed and its tea into one of her feeds every few days.

The foal should be given about 1 lb. of oats a day for every month of age, and a sprinkling of bran with every feed. It will have free access to its dam's hay. By the time it is ready for weaning at about five months of age, it should be clearing up about 5 lb. of oats, 1 lb. of bran and 7 lb. of hay a day.

Almost as important as good feeding is effective worm control. You should talk this over with your vet and work out a plan of action. Personally, I think that tubing a foal at eight, twelve and sixteen weeks is the ideal start to worm control, but due to circumstances such a system may not be possible in practice.

When the foal is about five months old, it should be weaned off the mare, unless it is very backward or has had a major setback, when weaning should be delayed for another month.

By this time the normal foal will be eating up well and will already be feeling fairly independent of its dam, so weaning should not usually be much of a shock. Colts and fillies are separated at this time.

With the foal away from its dam and her milk supply, you must use all your skills to keep it growing on well, blending feed and water, pasture and exercise and time inside, so that the weanling takes every advantage of its hereditary capacity for growth. Yet you must exercise keen judgement and not 'push' too much either, particularly with regard to feeding, since unbalanced feeding or over-feeding can cause serious limb troubles.

Towards the end of the year the weanlings should be bitted, since it will be much easier to control them with bits in their mouths if they need veterinary treatment, or if you wish to stand them up for visitors or a photographer.

And so into their yearling year. Once spring arrives, the yearlings should be left out for a long time every day, since, along with good feeding, plenty of exercise is essential to their continued physical development.

If the yearlings are being sent to the sales, work should be started to get them ready for that great event some five weeks before the auctions are scheduled. The preparation of yearlings for sale requires much skill, judgement and experience, and a great deal of hard work coupled with meticulous attention to detail.

At first the yearlings are turned out in the morning as usual, but are caught up again in the early afternoon, when bits are buckled on to their headcollars and they are walked in hand for some fifteen minutes. The amount of time spent walking them is gradually increased as the days pass so that at the end of their preparation the yearlings are being walked for up to an hour. You should train them to walk out just as well as they can, so that with time they will step out boldly and walk easily with the handlers beside their shoulders and not dragging them along. This takes time, and has to be done gradually and carefully, but it makes a big difference to the overall impression.

After the walking session is completed, the yearlings should be taken into their boxes and

This winning two-year-old still has plenty of room for growth.

brushed over. This grooming should be done fairly lightly during the first few weeks of the preparation, but should be slowly intensified until, starting about two weeks or ten days before the sales, the yearlings are being thoroughly strapped over after their exercise.

Three to four days before the sales, the farrier should put light plates on the youngsters' forefeet. Then they cannot be turned out together any more because of the risk of hurting one another with shod hoofs. Each animal can be turned out alone for an hour or two, and you can walk them all together in the afternoon. Or perhaps you will find it better, depending on the temperament of the yearlings, to restrict them to walking once or twice a day in hand during those last few days before they go to the sales.

The early life of non-Thoroughbred horses follows a rather different pattern. Many of these foals, which will be born during much better weather than will Thoroughbreds, will begin to live out with their dams five or six days after they have been foaled. Many native pony foals will have been dropped outside, anyway, and may not see the inside of a box until shortly before they are weaned, if then, but they generally thrive all the same.

Naturally, foals that are living out all the time with their dams will not generally receive much or any handling, especially on a bigger stud, although if you can take time to catch up the mares and bring them and their foals into the boxes daily for a handling and leading session, so much the better. These foals will not learn to eat hard feed or hay either, since they will live from their mothers' milk and fresh green grass alone. Their management, in fact, is simplicity itself up until the time they are weaned, unless a foal is injured in some way or falls sick and needs to be brought in for treatment.

It is not necessary to wean these foals quite as early as Thoroughbred youngsters, but if a foal's dam is pregnant again, it should be weaned by the time it is six months old, since continued suckling tends to weaken the mare and make her anaemic. If the mare is not in foal, then the youngster can be left on her until it is seven or eight months of age. An exception, however, would be the very 'colty' colt; in some very forward cases these will start to nip and tease their dams and other in-foal mares, and even playfully try to mount them, by the time they are four to five months old. This precociously masculine behaviour can cause a mare to 'turn', and the colt foal should therefore be weaned just as soon as these tendencies are noted.

Many foals will have been living out without supplementary feeding, and often will not have been handled except during the first few days after birth, and it will be essential to train them to eat grain and hay before you can wean them. For a week or ten days before weaning is to take place you should bring them and their dams into loose boxes, and the mares should be fed with a small amount of hard feed, with some feed in separate receptacles for the foals. The youngsters will soon start to imitate their dams and to nibble feed from their mangers, but after a few days, when they have developed a taste for the feed, they may begin to gobble their dams' rations more than the mares like. Then, if chased away, the foals can go to their own separate feed containers.

Mares and foals should be left in for several hours a day in order to get the youngsters used to eating hay. This time inside will also give the opportunity to get semi-deep litter laid down in the boxes. With this system, only the droppings are taken out each day; the wet straw is left and covered by a thick layer of fresh bedding. By the time the foal is weaned, the box will have a good depth of bedding, the bottom of which will not be easily displaced as the foal rushes wildly around just after it is separated from its dam. This means that the chances of it getting down on to the hard floor and hurting itself are greatly reduced.

To wean, lead the mare into the box with the foal following on at her heels as usual. Then, as the foal goes up smartly to the feed at the far end of the box, take the mare back out of the door – and out of sight and earshot as quickly as possible. A half-wild foal which has never been handled will probably have to remain in the box for anything up to a week, during which period you can gradually get it used to your presence, and eventually stroke and handle it. Finally, you will be able to ease a headcollar on to it. When removing droppings from the weaned foal's box during this time, do not use a pitchfork, but simply lift the straw under the droppings and tip them quietly into a small skep.

When the foal is used to the headcollar, and will accept the stroking of its body without fuss, you can start on leading lessons in the box, remembering to ask for only a little progress at a time. Then, once the introductory lessons have been learned well, you can lead it outside.

As soon as the foal is over the first shock of weaning, it should receive its first worm dose, which should be followed up at monthly intervals for the next eighteenth months. And once the weanling will accept handling of its legs and allow you to pick up its feet, you can start to do something about its hoofs, too.

Ideally, the weanlings should sleep every night in loose boxes, or at least a covered yard, from the time they are weaned until the next spring. The daily handling they will receive during this time will lay a firm basis for their behaviour in later life, and the shelter and good food inside during the cold weather will be of the utmost value. It is a good idea to bit them when they are about nine months old.

Once spring has arrived, the youngsters can live out day and night. You must keep a close eye on them, however, catching them up and handling them frequently, and worming them and trimming their feet regularly. It is best for them to run in the largest paddocks you have, with shelter from colder winds earlier in the year and from the heat later on. Very hilly fields are not suitable except for native ponies, since youngstock can damage their limbs racing down steep slopes.

Once the nutriment starts to go out of the grass about the middle of autumn the youngsters will need supplementary feeding, and if they can spend the nights of their second winter in loose boxes or a covered yard so much the better. However, a large shed open along the front, enclosed on the other three sides, and with its back to the prevailing winds, will be an adequate compromise if you do not have the facilities or labour available to house the yearlings in boxes at night. A plentiful supply of hay can be placed in a rack in such a shed, which should be kept well bedded down.

It is equally true of both Thoroughbred foals and those of other breeds, that the time and trouble taken over their early handling and welfare will be richly rewarded when their formal training begins.

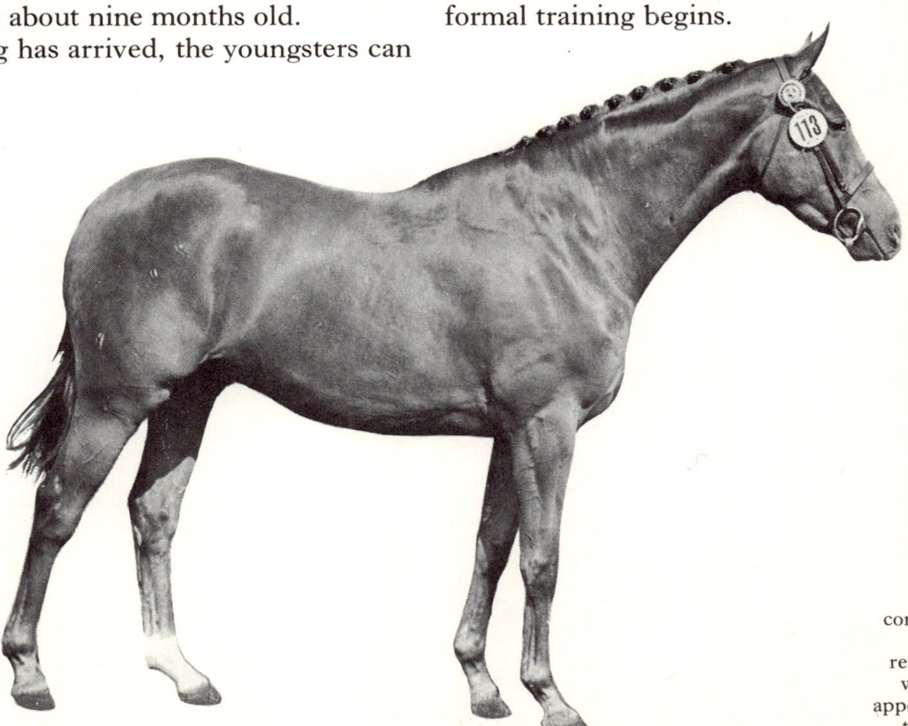

Careful conditioning and training are rewarded by the well-developed appearance of this three-year-old.

Breeding by government control

In a number of countries of the world the state plays a leading role in horse-breeding, setting standards of excellence, helping breeders to achieve them and, in some cases, ensuring that they are met with ruthless rigour. All horse-breeding in communist countries is, obviously, managed by the state; in some non-communist countries government control is also exercised.

Three excellent and differing examples of the part played in horse-breeding by the state are provided by Austria, France and Spain.

In Austria the government's grip on the breeding of the Haflinger is total – and is said to have wrought an enormous improvement in this breed of small, hardy and attractive animals in the short space of only twenty years. In the South Tyrol district of the country Haflinger stallions have been registered for nearly a century, and organised breeding, which is controlled by an Austrian government department, has been under way for some fifty years.

Individual breeders are only permitted to own mares, not stallions. All colt foals are inspected by government officials at weaning, and from an annual colt crop of 300 to 400, only about a score are passed. These foals are taken to the government stud, where they will be raised as possible stallions. All the other colt foals are sold for meat.

A herd of Akhal-Teké mares.

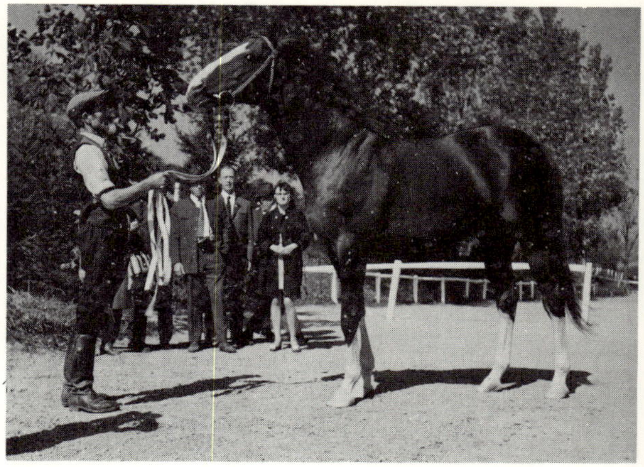

A stallion of the famous Babolna state farm in Hungary.

When the selected colts are three years old, they must undergo another rigorous inspection. Those that pass this examination are inscribed in the Haflinger stud book, and will stand at the government stud or will be boarded out with selected breeders for the covering season. They are made available to the owners of mares for a small fee.

Haflinger mares must undergo close scrutiny by the government officials, too, before they can be entered in the stud book and begin to function as brood mares. They are first checked as weanlings, then again as yearlings. If they pass the latter inspection, they are branded. Finally, also as three-year-olds, they come under the gimlet eyes of the state inspectors once again: their height, girth, bone measurement, colour, conformation, temperament and action are all observed and evaluated.

In France, government control of breeding is also strict, but not nearly as rigid as in Austria, with its obligatory destruction of colts not deemed worthy of perpetuating their type.

Very few breeders of riding horses keep stallions in France, since any entire kept privately must be checked by government stud inspectors, and may have to prove his worth as a show-jumper or racehorse before being permitted to cover any mares. Instead of keeping a stallion of his own, the small breeder in France, who generally owns no more than five or six brood mares, and often fewer, looks to the excellent entires owned by the state, and headquartered at one of the twenty-two large French government studs scattered throughout the country. These stallions are bought as three-year-olds at the stallion shows and sales which are held every autumn.

Considerable numbers of these carefully-selected entires are maintained, and at one big stud farm, for instance, there are more than seventy Normandy half-bred stallions alone. The main breeds represented at the government studs are the Normandy half-bred, the Thoroughbred, the Anglo-Arab and the Trotter. Shortly before the covering season is due to begin, the many stallions of the various breeds leave the main studs with their handlers to stand at smaller depots, and in August they return to the main studs.

In Spain, horse-breeding is overseen by the army, which maintains the stud books of all breeds in conjunction with the *Sindicato de Ganaderos*, or Stockbreeders' Union. The army authorities are also in charge of the government stud farms, which are situated in various regions of the country, with the largest and most impressive stallion station, the *Deposito de Sementales*, being located in Jerez de la Frontera, Spain's sherry capital and centre of a great horse-raising area. At the Jerez *Deposito* are to be found Arabians, Andalusians, a few Thoroughbreds and a number of excellent cross-breds.

However, the Spanish horse-breeding authorities, while doing everything in their power to keep standards as high as possible and to make available to private breeders the best stallions they can find of the various breeds, do not dictate policy to individual breeders. If a private breeder wants to use one of his home-raised colts as a stallion, or wishes to acquire an entire from another breeder to use with his brood mares, he is quite free to do so. The eligibility for the appropriate stud book of the resulting progeny will depend purely on whether its sire and dam were themselves registered or not. However, an army inspector will examine the youngstock before they are actually entered in the stud book to make an official grading which will depend on their conformation, temperament, action and freedom from hereditary unsoundnesses. The most sought-after classification is *sobresaliente*–outstanding.

The Spanish army horse-breeding officials, all of them cavalry officers, keep a close watch on private stud farms in the hope of spotting an outstanding young stallion prospect. When they find one, the government will make an offer for the young entire, and it will always be a most competitive one, since there is no question of the state acquiring a stallion it wants for a price below its real market value. Most of the state-owned stallions leave the stallion stations for the covering season, and go with their handlers to the farms or ranches of private breeders, who hire them for a nominal fee. The results of this policy on Spanish horse-breeding have been very rewarding, and the dedication of the Spanish army officers to their task exemplary.

The Spanish government also maintains its own stud farms, where it breeds from carefully-selected mares. The colts from these establishments are rigorously tested to see if they are worthy to stand as stallions at the state *Depositos*. Andalusian colts, for example, after they have passed the tests for conformation, temperament, action and soundness, must cover twenty-four miles of rough and varied terrain under saddle in no more than two-and-a-half hours. When the ride is finished, the temperature of each colt is taken, along with its pulse and respiration rate, to determine its grade of fatigue. Then, after twenty-four hours' rest, each horse has to complete a course of ten jumps to see whether, after the efforts of the previous day, he still retains the desired energy and agility. Depending on the joint classification of these two tests, the colts are used as stallions by the government or are rejected.

PART SIX

The history of riding

The history of riding

One of the earliest records of man riding a horse is an engraving on bone, found at Susa from the third millennium BC. Horses had been hunted as food from early times; then, as primitive agriculture began to be practised in the more fertile valleys of Persia and Turkestan, they were domesticated, probably to draw sleds and the first wheeled vehicles. The onager was ridden before the true horse; at Ur it is shown with a nose ring, and also with a strap round its jaws, kept in place by another behind the ears, like a drop noseband. The onager is very fast–it has been paced by a jeep at 35 m.p.h., but horses are more docile and stronger, and bovine means of transport were changed to horses as sources of food and water in the valleys and steppes were reduced, causing man to become nomadic. A mounted man could hunt fast-moving quarry and also raid the fertile valley dwellers. The first time man employed some means other than human power was an immense step in world history.

Most of Europe was swamp or heavily forested, so it was on terrain suitable for man and horse that riding developed. A figurine of a horse ridden by a groom in about 1580 BC shows that riding had spread to Egypt by this time, and a clay figurine from Mycenae about 250 years later proved that it had also reached Greece.

Legend and myth surround horses, such as the centaur, half man and half horse, which was how a mounted man must have appeared to the incredulous, who might never have seen either a horse or a rider. The psychological as well as the physical advantage over the man on foot is still exploited, and history and literature constantly remind us that 'to turn and wind a fiery Pegasus' is heady stuff. Pegasus means, literally, 'from the water'. This association of the horse with the sea was based on fact, for invaders and colonists transported horses by ship, and they had to swim ashore. Ancient sculpture shows them rising from the waves; Poseidon and the Diascuro are famous instances, and Pegasus was depicted with wings because the speed of the horse indeed suggested that it was winged. The Emperor Wu of China sent an expedition to Turkestan to obtain 3,000 horses to improve the local breed of coarse, Mongolian type, and in order that he might be suitably conveyed to the other world he bargained for two 'heavenly horses' every year. Sacred poetry claimed that they issued from a pool and were dragon horses, so they were depicted rising from lumpy waves like an eiderdown. These were the horses that well authenticated records say sweated blood; the only explanation is that this was caused by wounds from parasites.

There is a remarkable document, the Kikkuli Text of about 1360 BC, which gives detailed instructions on the training of horses, and the Hittites used clay tablets to record information on horsemanship. But until Xenophon's time most of our knowledge comes from bas-reliefs, cave drawings, pottery and sculpture. In the British Museum there is a statue of a Libyan woman riding an Arabian horse with typical high set tail and dished face. We know that Jacob found horses in Egypt, and the Bible has many references to horses used in war. The Egyptians hunted lions, gazelle, ostriches and the wild ox, and in Asia they also hunted tiger, so the horse must have been well trained and obedient to face its natural enemies. In Babylon and Nineveh the people boasted that their children were taught to ride, shoot and tell the truth. Darius, who wrote his own obituary, stressed in it the fact that he was an accomplished horseman.

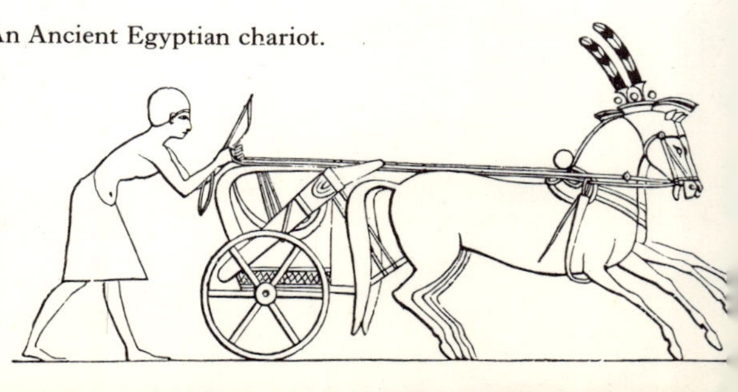

An Ancient Egyptian chariot.

A hurdle race at Cheltenham.

Runners in the paddock
before the Irish Grand
National at Fairyhouse.

The Japanese have become keen horsemen in recent years. Here their 'Derby' is in progress.

Out of the starting gate and away. Races in the United States are generally run on dirt track courses.

Riding was at first considered suitable only for soldiers and archers, but before long to be carried on a fine horse instead of sitting in a chariot was thought fitting for kings and rulers, their rank emphasised by skins and embroidered cloths spread on the horse's back. Bits were of bone or ivory; crescent shaped, with holes drilled for attaching the reins. Later, the working of metal enabled bits and horse furniture to be made of bronze, iron or silver.

Bits, bridles, buckles and bells have been found in royal tombs, and horses as well as slaves were sacrificed, not only in the superstitious belief that they would continue to serve their master, but because the horse was regarded as a comrade. This association of horse and man has become a symbiosis resulting in taboos against eating horseflesh. Sometimes this taboo was enforced to maintain the supply of horses for use in war; the Council of Calcuith forbade it on the grounds that 'no eastern Christian would do so'.

The Scythians and the Parthians were skilful horsemen and warriors who could fire a cloud of arrows at speed. The Parthian arrow was fired backwards in retreat. The Scythians introduced horses to the Etruscans, those early inhabitants of an area around Rome, where they used a simple, ringed snaffle; they also took horses to Ireland. A Scythian horse reconstructed from a mound burial in the Altai mountains is an animal of quality carrying a saddlecloth held in place by a breastgirth and fillet strap and wearing a plumed mask.

It is difficult to establish the size of riding horses because the thick winter coats of cooler regions caused them to be represented as bigger and heavier. The Arabian horse was favoured because of its speed, courage and mobility, and was sought after for interbreeding with coarser northern types. Assyrian drawings portray this type, and show the troops of Asurbanipal mounted on far larger animals than the mere ponies of their adversaries the Elamites. One method of selecting the fastest animals for breeding was to keep a herd without water and then release it to a known source, noting those that arrived first. Domesticated horses had long manes, encouraged to assist riders to mount; the 'wild' horse had a stiff, upright mane, but there is apt to be confusion about this because manes were shorn as a sign of mourning to show the owner would need to mount his horse no more. Horses were taught to stretch out their legs and lower their backs, and also to kneel. Kneeling horses are found in statuettes and on ceramics in tombs, awaiting their master's return to the saddle. The Stoic philosopher Posidonius reported that Iberian cavalry in 90 BC taught their horses to crouch, and also that they carried two riders, one of whom dismounted to fight on foot. The advantages of riding were so great that men rode asses, onagars, elephants and camels; two archers rode one camel – the one facing backwards could cover a retreat. On a coin, Alexander the Great is shown attacking King Porus of the Punjab, who is mounted on an elephant.

The first book on equitation still available was written by Xenophon, born in 430 BC, the son of a man of the equestrian class in Athens. Members of this élite corps submitted sons to years of training and to strict examinations, and they had to be wealthy in order to provide two horses, which could cost nearly as much as a house.

After serving in the Peloponnesian wars, Xenophon joined a mainly Greek army under Cyrus, son of Darius II, and after it was betrayed into defeat he was elected to lead the 10,000 in their two-year retreat from the Tigris to the Black Sea. Among the books he wrote after his retirement to a country estate near Corinth was *Hippike* for his two sons, also in the cavalry. He learned many things from the Armenians and Persians, who had longer tradition of horsemanship than the Athenians; the first 'horseshoe', a piece of cloth covering the foot and tied above the fetlock for protection among ice and rocks in the mountains; mounting 'Persian fashion', which was what we call giving a leg up – sometimes a servant or prisoner bent to offer his back to his master. This, Xenophon thought, would be useful for the wounded or elderly. He also recommended working horses downhill: 'There was no fear of horses becoming unsound as Persians always run their races downhill and their horses are as sound as any Greek.'

His advice on buying a young horse remains valid today, for he said that after seeing the horse ridden it should be put back in its stable and later brought out again to find out if it was still obedient and sound. Grooms should lead horses from either side to prevent their becoming one-sided, and the proper way for a man to treat his horse was to visit the stables frequently and to know more than his servants.

Xenophon was a true horseman and the great masters still pay him tribute. The horse must learn, he said, to associate man with everything he likes: food, rest and relief from flies. As a horse cannot be taught by word of mouth but by association with reward and admonition, what is done by compulsion is done without understanding. He is best rewarded by being dismounted and allowed to roll in the sand and then return to his stable, and he will learn to be obedient and to jump. He quotes a book by Simon of Athens which must have been destroyed. One of his maxims was that if a man buys, feeds, trains and exercises wisely he will own famous horses and be a famous horseman. 'I do not approve of a seat as if a rider sat in a chair but rather as if he were standing upright.' Although padded cloths were used to sit on and for protection in battle, horses were mainly ridden bareback, and the 'double back'–the well-muscled ridge that Xenophon stressed was so important–was as essential as 'a neck that rises like a cock, not sticks out like an oar, for it is not by arching the neck but by stretching it that horses try their powers'. The pasterns should be sloping, 'not straight like a goat', which jars horse and rider. Broad, short loins enable the horse to raise its forehand; the *levade* and the turn, and the agility of a polo pony, were imperative in battle. Feet, too, and their care, came in for much comment; hard hooves that rang like cymbals and a well developed frog were particularly important in long campaigns before the days of shoes. The frieze on the Parthenon well illustrates riding at this period.

Xenophon believed in bits with a simple mouthpiece on which various rollers could be fitted, some smooth and some, called 'echini', the name for seahorse, roughened like that creature. A bit of that date found in Athens is exactly as he described it. The horse is 'always after the part that is getting away from his mouth. The bit should be like a chain so that he drops it from the bars, and varied to prevent his becoming accustomed to the feel of a particular one.' Not the bit but the hand was what mattered; coaxing rather than harsh measures should be employed, and variety in training should include circles, transitions, halts, and also cross-country work and jumping walls and ditches. But showing off was not to be tolerated. 'The cavalry leader, by making his horse prance and rear, slows up those behind' and destroys a harmonious performance. The horse was bestowed by man for his happiness, he maintained, and the gift must not be abused.

Mahommed puts the same words into the mouth of Allah. 'The horse shall be for man a source of happiness and wealth; its back shall be a seat of honour and its belly riches, and every grain of barley given to it shall purchase the indulgence of a sinner and be entered in the register of good works.'

There are no more books on horsemanship in existence for two thousand years, but a lot is known about riding in various parts of the world. The Greeks colonised southern Italy,

Xenophon, the founder of modern horsemanship.

taking thousands of horses to Sybaris and Taranto, and it is related that their horses danced to music. The Greek word for riding ground–*harena*–is still the 'arena' of today. Horses and riding were widespread in France and in the Iberian peninsula; Greek and Phoenician traders took them throughout the Mediterranean basin and beyond, and tributaries to the river of knowledge turned horsemanship from simple transport and a means of war into relaxation and an art in itself.

Hannibal and his son Hamilcar were great horsemen, the latter defeating the Romans with Spanish cavalry at Cannae and continuing to North Africa to join the King of Syria. The Romans were prejudiced against saddles until AD 380, when Theodosius' cavalry had real saddles with trees and were demonstrably successful. A hundred and fifty years later the Venerable Bede said that saddles were in use in England. Attila the Hun was believed to be the first to use stirrups for his troops, and again their success led to the stirrup coming into regular use. The growing discoveries and craft of working metal had developed the primitive 'hipposandal', or fitted overshoe, into a plate punched for nails; bits, armour and weapons also became widespread.

The first knights were a bodyguard of noblemen at the Court of King Chosroes, and royal princes were taught riding, polo and chess. Genghis Khan established an empire from China to Poland, based on hordes of mobile guerrillas on tough, shaggy ponies, the sweat from their bodies freezing solid into an insulated shell. Messengers could travel over two hundred miles in twenty-four hours and be replaced at strategically placed staging points. The Mongolian herdsman of today has changed little, riding pony or yak with equal facility, vaulting into the short-stirruped saddle and using a twisted halter of hide. Sometimes a small bit dangles under the pony's chin, and the lassoo is a loop at the end of a long pole.

Later, the raiding hordes of Turkestan trained their horses to exist on balls of barley and sheep's fat, which could be carried on the saddle during long raids. Other tribes prepared their horses by galloping them and reducing their food for about two weeks, then staking them out in the open after throwing icy water over them, and finally galloping them again, often with weights of 300 lb. Those that survived were considered hardy enough to be used on raids.

Tremendous distances were covered, not only by the Crusaders and those engaged in wars. Charlemagne's kingdom stretched from the Pyrenees to part of Italy, and his six daughters, riding astride, often accompanied him. One was the mother of Alfred the Great, who rode to Rome at the age of four, and in various visits many new ideas and methods must have been exchanged. Charlemagne copied the oriental custom of parks or paradises–enclosures for hunting game. Thus did royal and feudal parks spread in Europe. As swamps and forests gave way to agriculture, larger horses could be bred and fed, and these became the foundation stock of the Great Horse or *destrier*, able to carry a man in full armour. William the Conqueror dubbed his son 'Ridere', or knight, from *rad*, an Anglo-Saxon word meaning 'to be conveyed'; the French *cavalière* later produced cavalier and cavalry.

Feats of horsemanship are recorded, such as one performed by Jelel el Din whose army of 60,000 was defeated by Genghis Khan. When only 700 remained he mounted a fresh horse, plunged over a 20-ft. bank into swiftly flowing water, swam ashore and was pursued as far as Delhi, where he escaped. The great Hippodrome at Constantinople, begun by Septimus Severus, was completed by Constantine the Great, and equestrian acts were a change from gladiators and Christians. The bronze horses brought from Chios stood there until the city fell to the Crusaders, when they were taken to their present site at San Marco, Venice. Fame could lead to power, for Basil, a Macedonian slave, was so successful at taming fierce horses that he came to rule men, and a stable boy, Philadorus, entertained the Byzantine public by galloping round the arena standing on his horse's back and playing with a sword.

Breeders used to bring their horses to the great rulers to choose the best, and it was the wild colt, dismissed as unrideable by the courtiers of Philip of Macedonia, that the boy Alexander pleaded to be allowed to own. His father agreed, on condition that Alexander could ride it then and there. To everyone's amazement Alexander turned it away from the shadow which he alone had noticed was frightening it, mounted it and rode back in triumph. The great horse carried his master to victory in many battles and was finally killed. He had a town named after him.

The interchange of horses, through purchase, gifts and the tides of war, continued, and the Arabian was highly prized for riding and for racing. Good mares were syndicated, for Arabs believed that it was the female line which was dominant. The King of the Yemen took 80,000 horses from Syria; King Bahram had a stud of 40,000; while the Caliph Motassem had as many as 130,000. The Saracens spread both horses and their style of riding throughout North Africa and into Moorish Spain, and as a result to the Americas, where their influence can still be traced. There is a theory that the horse with knee action, the ancestor of the hackney and the trotter, originated in hilly areas where survival depended on ability to lift the feet at speed over rough and stony ground, whereas the horses of desert and steppe could safely stretch out from birth in a flat gallop – the forerunners of the Thoroughbred with its low, daisy-cutting action. Most horses were taught to amble, and a man was paid to train a nobleman's horse to amble. The word comes from the French *armer* – to arm against the bit. Various broken paces still exist, such as 'tripling' in South Africa and pacing in America in preference to the two-time diagonal of the true trot. The 'amblynge nag' and the palfrey were comfortable conveyances; the hack, from an Andalusian word *jac* or *haca*, evolved into a showy animal used for social purposes such as display and courting.

By the fifteenth century riding horses had become definite types suited to special purposes, and those from famous studs, especially those with certain brand marks, fetched very high prices indeed. The Great Horse, led by a squire until mounted for battle or mock combat, was trained in curvets and careers so that men encased in armour could hurl themselves at each other and try to unseat their adversaries with a long lance. To be unhorsed was a disaster; in jousts and tournaments it was considered sporting to let a competitor remount. These contests were quite literally 'knockout', and sometimes fatal. Richard Coeur de Lion handicapped his knights, although this word, from a card game called 'hand i' cap' was not used until Samuel Pepys wrote of it – it was used in connection with racing.

Time hung heavy at courts between battles, and sport was the main recreation, women often taking part. Hawking was considered ladylike, and deer and boar were rounded up so that nobles could administer the *coup de grâce*. But there were tough and accomplished horsewomen, such as Anne, daughter of Louis XI, who was the equal of any man, both in statecraft as Regent and in hunting wolves and wild boar. Mary of Burgundy, who was killed when her horse fell on her after jumping a ditch, was said to have been the first to place one leg over the pommel; others say that Catherine de Medici saw an opportunity of displaying her shapely legs on the first side-saddle. Diane de Poitiers, the mistress of Henry II, was renowned as an expert on horses and hunting; she performed equestrian feats thought to be too difficult for women, and rose at 3 a.m. to ride for hours to keep fit.

One of the greatest horsewomen in the world was Queen Isabella of Spain, a descendant of Alfred the Great and John of Gaunt, and whose daughter married Henry VIII. She rode from the age of three; at ten she scorned the mule that etiquette thought suitable and rode a blood horse, and she went to her coronation on a white horse and dressed in white brocade. She helped her husband in all his campaigns, riding hundreds of miles in extremes of heat and cold, wearing armour over a light dress. Between campaigns she hunted, and also reorganised the great stud farms of Andalusia, encouraging the Arabian blood which Pliny said the Moors

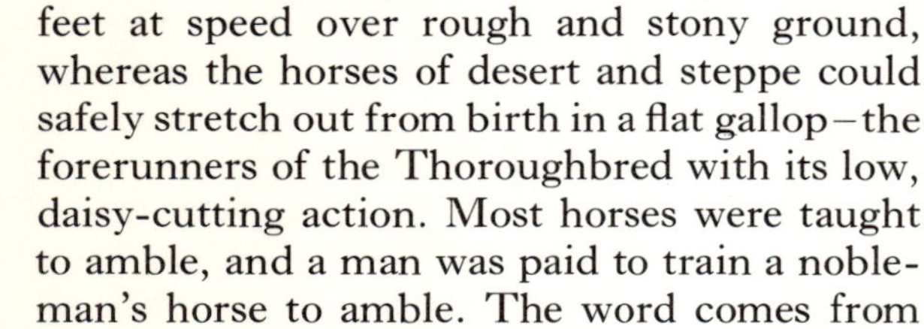

The King of France,
unhorsed at Bouvines in 1213.

A suit of armour with 'Lamboys', presented by the Emperor Maximilian to Henry VIII.

cherished. She rode to Mass on a horse with a bridle and saddle of gold, and wore a shirt embroidered with pearls, yet she offered to sell all her jewels and her Crown to finance Columbus' expedition and thus was responsible for the spread of horses to the New World as well as to many parts of Europe. Her son married Margaret of Austria, her daughter Juana married Margaret's brother, whose son, Charles V, ruled an empire unknown since the time of the Romans. Her influence on the equestrian world was thus incalculable. When her daughter, Crazy Jane as they called her, arrived in the Netherlands, she astonished everyone by riding '*a la gineta*' on a high-peaked Moorish saddle. The Spanish *caballero* or horseman was expected to ride both with very short stirrups and also with the long, *manège* seat.

Poor women, if they rode at all, sat sideways or rode pillion, while women of rank were carried on a whirlicote to protect their formal clothes, resting their feet on a board and standing sideways. Sitting sideways on the rump of an ass or pony is still seen in Ireland, and is used by peasants in some Mediterranean areas. Greek maidens rode naked to enable them to bear children easily; their legendary ancestors, the Amazons, killed their sons at birth. The word Amazon means 'without breasts', a meaning overlooked by later artists and writers who represented them as buxom wenches. History consists largely of wars and therefore of men, but women of rank at least had the chance to ride and ride well.

The days of the juggernaut were numbered, for the invention of the crossbow and of firearms brought about a revolution, triggered off by the relatively unimportant battle of Pavia. A handy horse was now needed to turn and manoeuvre while the clumsy weapon was loaded, fired and reloaded, and an assailant was outwitted by feints, sudden attacks from the rear, and quick retreats to avoid pistol shots. To do this the horse had to be controlled without giving away the rider's intentions. From this necessity modern dressage was born, and it was in Naples that the renaissance took place.

Political and economic circumstances combined for a time to make Naples the nucleus of an emergence of the arts after the Dark Ages. Count Fiaschi was the first to open a school and write a book on what was called 'the secret weapon'–the art of the light, supple and manoeuvrable horse and rider–in 1539. He used combined aids and a long switch, and also the voice and music. His book shows sketches of movements at the trot and canter, as well as devoting a section to farriery, which by then had become an important craft.

After him came Grisone, the most famous of all the School of Naples, which later spread to Rome. Grisone's book was translated into English by Blundeville; Queen Elizabeth I sent the Earl of Leicester to study his methods, and pupils came from all over Europe to learn this new system. No less than four of his pupils came to England as Master of the Horse, to advise on royal studs and give demonstrations. Some of his methods may now seem crude, and his bits look horrific, but they were revolutionary at the time, as was his understanding of the mentality of the horse. A contemporary writer recommended that a disobedient horse should be locked in his stable for forty days, then mounted with long spurs and a 3-ft. long iron bar ending in sharp hooks, which were dug into the horse's quarters while an assistant applied a hot iron under the tail. The English writer, Gervase Markham, also described such a rod with rowels, called a peerch.

Grisone used volts, serpentines, the *caraglio*, or 'snail' exercise of diminishing circles, and high school airs such as the *levade*, with emphasis on lightening the forehand by putting more weight on the haunches. The 'ready' or managed horse had to be balanced in order to make rapid starts, stops and turns, and part of the *manège* was sloping. The bits he illustrated had long cheeks and high ports; the mouthpieces had balls, rollers, bells and revolving cylinders, with the object of teaching the horse to be behind the bit or to anticipate its action, rather than the use of brute strength, and he shows the first known curb chain. No doubt these bits were severe, but like the 'Western' bits based on the Moorish, the horse learned to respond to the first touch and to neck reining, and was undoubtedly handy.

Less well known to posterity was Grisone's pupil Pignatelli, because he only wrote a few notes and devoted himself to training horse and rider. Some circus riders had come to Naples from Constantinople to try their luck, and Pignatelli was so impressed by their humane ways of training their horses that he adopted their methods and studied how they obtained such subtle movements. A writer describes Pignatelli's pupils managing their horses, the rider sitting calm, sword in one hand, reins in the other, by a slight movement of heel or bridle hand conveying something unseen by his opponent but causing the horse to dance aside, rear up, stop, pirouette, turn almost on a plate, and take the enemy by surprise. Thus horsemanship, using almost imperceptible combined aids, rather than just a strong seat, was all-important. At first work on two tracks was intended to confuse the enemy, but it was later used for suppling movements and to obtain obedience to the leg.

Pignatelli charged high fees, and pupils were expected to study for several years. It was he who introduced the pillar, hitherto used only to exercise horses during a siege, and it is said that he first made use of a convenient tree on his training ground facing ruined Pompeii and Vesuvius. He discarded the cumbersome bits, replacing them by the 'simple canon' used by his circus friends, for 'the intemperate hand of the rider breeds hardness, not the bit'. Thus he began the free forward movement suited to the lighter, high-couraged blood horse, restoring Xenophon's precepts with the help of saddles, stirrups, shoes and some of the experience sifted from 2,000 years of valuable, if unstable progress.

His most famous pupil was de Pluvinel, who became riding master to the young prince, later Louis XIII, and whose sister married Charles I. Pluvinel's book, in the Socratic form of question and answer, is as tedious in style as a recorded interview, but immensely illuminating on the advances that continued in France. Pluvinel made more use of the leg than merely heel and spur; his major aims were grace and suppleness, still a feature of French riding. He used two pillars instead of one, and his pupils, horse and man, worked between them using a cavesson (*cauzan*) of soft leather, for he never put a bit into a horse's mouth until it was schooled with the cavesson. Young horses were broken in this way, then bridle and saddle added; next a light rider without spurs, and finally an experienced rider to produce a horse '*bien dressé*'; perfectly obedient to hand and leg and accustomed to guns, trumpets and shouting. Coarse aids were condemned because they could give away the rider's intentions, and 'he should know by instinct whether his horse defeated him from ignorance, high spirits or lack of memory'.

Pluvinel was a stickler for detail, and even grooms were criticised over the cut of their boots. His royal master was not exempted from having his equipment checked, 'because your life depends on it, Sire'. It was difficult to persuade the gentlemen of the court to discard their heavy armour and to hold their arms lightly instead of with a stiff wrist, but by introducing tilting at the ring, the quintain and the carousel, and inaugurating a college of arms to teach young riders, he gradually weaned them from their old ways. He obtained 'the best horse in the world, Bonite' for the young Prince, who preferred riding to his other studies and won a watch in one of the new competitions, and cleverly ensured a feminine audience for, he said, 'there is nothing like ladies to obtain a graceful performance'.

Pages held armfuls of switches, and a swishing noise was made with flat strips of leather to speed the horse round the pillar. The main paces used in training were the trot and the gallop, which was probably the canter, so called because it was the lolloping pace of the Canterbury pilgrims. Ambling was considered a fault, but usually meant cantering false or disunited, and the *passage* was not what is now known, but the Spanish Walk.

In Vienna, equestrian history was being made when the riding hall was erected in the Josefsplatz about 1570. The Emperor Charles V, one of the finest horsemen of his age, could assemble and use the best from Spain, Portugal, Italy and the Netherlands. The Lippizaner stud was first based near Trieste and consisted of Spanish and Moorish stock with some Italian and later one pure-bred Arab–Sigslavy. When Trieste was ceded to Italy the stud was moved to Piper in Austria. Without interruption the classical airs of *haute école–capriole, courbette, ballotade, croupade, levade* and *mezair*–have been taught and practised, as well as the airs on the ground, the *piaffe, passage* and pirouette. No rider is considered fully trained for at least ten years. The newly-revived art soon took root in Germany, where a remarkable book on riding was published in 1588 by Master of the Horse Lohneysen, another pioneer of the use of a horse's natural aptitude by kindness and understanding.

England produced perhaps the greatest horseman in the world, William Cavendish, Duke of Newcastle. Because of him the word cavalier came into the English language. He taught Charles II and Prince Rupert, both brilliant horsemen; the latter once owed his life to the fact that his horse was 'ready' and jumped an enormous hedge that all his pursuers refused.

'I have ridden since I was ten,' he says in his book *New Methods of Dressing the Horse*, 'have read all the Italian, French, Latin and English books ever writ, good methods and bad; ridden with masters of all nations and tried their systems. I have spent thousands, spoiled many horses, and been a very long time learning the art.' When Charles II fled from England he was accompanied by Newcastle, who opened a School in Antwerp which, with Vienna and Versailles, was one of the pillars that supported equitation through the centuries. His power over horses was said to be miraculous. He himself explained it by saying that he worked according to nature, that is, by studying the mental as well as the physical qualities of the horse, 'for he hath imagination, memory and judgement; I work on this which is why my horses go so well'. The rider is as good as he rides, was his maxim; presumptuous fellows get falls but 'I have never known a good horseman thrown'.

His critics and opponents have called him harsh, mainly because he required obedience before anything else, but after all, is that not still a requisite? Until it is perfectly dressed, he maintained, he never knew a horse that would not rebel before he would go freely. Great resistance was rare, but no man could find subtler ways to oppose the rider than the horse, who must learn to accept the rider before he can love him. Readiness–the *bien dressé* of de Pluvinel, grew into dressage, and although no longer a matter of life and death it is still essential before a more advanced performance can be required. Newcastle uncompromisingly stated that the unbalanced horse at best required the services of a bone surgeon, while the ready horse could turn, circle, perform the *sorpiger* (serpentine), pirouette, *terra a terra*, and go through fire and water. Shakespeare's 'round' he called the *corvet*, but the *capriole* was 'yerking out behind'. Many Englishmen contended that horses so trained might start 'dancing and fancy tricks' in the middle of a battle, but Newcastle replied that it was hard enough to obtain such airs at will, and after a campaign of several days it was unlikely that any horse would perform them of its own accord.

'Sitting is one thing, there are thousands of things in the art' he insisted. 'Anyone can ride from London to Barnet, for sitting in a saddle is easily learned.' But the prophet was not honoured in his own country, though outside it he was and still is recognised as a master of his art. He believed in suppling the horse's shoulders, using a sliding rein like the present Market Harborough, which continued in use in Europe and has now been revived as a training aid. He rode with one rein attached to a soft padded cavesson, sometimes tied to the saddle and sometimes held in the hand, using the outer rein as a rein of opposition. The second rein was attached to the curb bit and used independently; thus he employed the separate actions of the double bridle, obtaining flexions at the poll and jaw, the basis of later orthodox schooling, and his suppling the shoulders was a preliminary to shoulder-in. He used a simple bit, like Pluvinel and 'the master Baptista' (undoubtedly Baptista Pignatelli), with a tight noseband to keep it in place, and adapted the French saddle still used on ceremonial occasions by the classical schools. He used the voice as a training aid, and a long switch which lasted him six months; long spurs used discreetly, he believed, were preferable to bruising the horse's sides by constant use of blunt ones, and he stressed that fitness and proper shoeing were all part of the horseman's art. Above all it was his vast understanding of a horse's nature ('I would not hurt his mouth nor anything about him if I could help it; forgive him his mistakes that next morning he may know you have mercy as well as justice') which enabled him to impose the discipline that produced 'miraculous' results. His cultural background and mastery of languages helped preserve his accumulated experience for posterity.

Blundeville and Markham he considered 'more scholars than horsemen', but to Markham is owed the purchase and importation of the Arabian sires mated with mares which were part of the Queen's dowry and became the foundation stock of the English Thoroughbred, whose descendants spread all over the world. Charles II deserves credit for initiating many far-reaching changes; he was a devotee of racing and took part in contests at Newmarket clad in his shirtsleeves, an activity not entirely approved by Newcastle, himself more a *manège* man. Political and military considerations combined to make many historians decry Newcastle's reactionary ideas, still incorporated in classical teaching. Markham, for instance, wrote at the same time: 'To stop, thrust him on and make him coutch his hinder parts'. The old advice to the armoured knight, to keep a strong hold on the reins in order to remain in the saddle, died hard, and is still used as a lifeline by the inadequate rider.

The progress of educated riding was like a seesaw. The best survived; oases in a desert of ignorance. Spain and Italy declined in importance although great horsemen remained, though in France Pluvinel's standards were maintained and advanced. Comparison of school movements of those days – the D-shaped turns at the end of the career and the clover leaf of *voltes* connected by a long stem for the gallop – with those shown in de Saunier's *Art de la Cavaletie* of 1761 shows that the oblong *manège* of today had come into use, subdivided by quartermarkers, and instructions were given for changes and counter-changes of hand, and *demi-voltes* and lateral movements for which combined aids as we know them must have been required. The correct leading leg was understood, and the paces of the loose horse studied and analysed, as well as the correct posture and deep, flexible seat of the rider, taught once again without stirrups. His illustrations of bits show a revolution in thought as well as design, for the mild snaffle with the mouthpiece increasing in thickness toward the outside and two joints, still known as the French snaffle, or one with cheeks and a single joint much like the Fulmer, were used, although other bits were also shown, including some with curb chains a few of which were twisted.

It is impossible to cover the history of riding in detail in a few thousand words; one can only note the milestones and landmarks along the way. France maintained the great School of Versailles, producing a line of *écuyers* of unrivalled brilliance; Vienna and Antwerp combined technical skill with Xenophon's 'never lose your temper but rely on kindness, patience and admonishment when he disobeys'.

In Europe, great studs existed to breed suitable horses, mainly for the army. In England, Henry VIII ensured a supply of horses by a law obliging dukes and archbishops to keep seven trotting stallions for the saddle; marquises and bishops five, down to one for clergymen whose wives owned a velvet bonnet. Sport and games played a large part in the story of riding, not necessarily toward improvement. James I wrote to his son, who was temporarily at Newmarket for the improvement of his health, that 'the honourablest and most commendable games a king can use are on horseback, such as tilt, ring and low riding'.

Polo was the national game of Persia and India, but did not reach Europe until the end of the nineteenth century, being brought from

India, like the gymkhana, by the British Army. In central Asia the popular game was *buzhaski* (*buz*, a goat and *kashidan*, to pull). It was played by teams of twenty or more, and to add interest several teams could play at the same time. Whips with wooden handles were carried, but only for hitting opponents, not their ponies; the area was unlimited, but single goalposts stood about half a mile apart with a circle about fifty yards in diameter round them, and rocks and precipices were desirable hazards. The ball of goat-skin was thrown in, as in rugger, picked off the ground by a rider without dismounting and tucked under the knee. Regional variations were a game of the thirteenth century, *Tchigan*, a kind of mounted tennis; a Kirghis game, *kyz-kun* (kiss the girl), and *papakh-ayum*, or grab the hat. Tent pegging, cushion polo and polocrosse are modern versions, to which have been added a variety of gymkhana events.

The wild goose chase originated in Ireland, and became the steeplechase and the point-to-point. At first one rider tried to overtake another across natural country and obstacles; gradually riders matched favourite horses on the established lines of flat racing, and later a number of competitors were despatched at once. Such races resembled a cross-country event, and in order to keep the course secret men with flags bobbed up out of ditches as the rider approached. But one result was that more thought was given to fitness, stable management and weight. Owner-riders, or their grooms and valets, were the jockeys over fences. Jockeys, from the Romany word meaning a whip, were often Jews or coloured boys, expected to win by fair means or foul; poorly rewarded, and on one occasion refused Communion because of their occupation. Flat racing became the sport of kings, which also improved the riding horse, for famous sires often served local mares. Hunting, steeplechasing and flat racing caught on in North America and were soon of a high standard; Australia and North America started rodeos; easily organised with the raw materials ready to hand. Horses and riding reached South Africa mainly from Java and the Dutch East Indies, but until recently horse sickness was a great handicap in this part of the world.

The two distinct styles of seat continued, depending to some extent on terrain and economic circumstances. The rancher and cowboy found the long stirrup and *manège* type seat suited to long hours in the saddle, and to lassooing a horse or steer; the style of saddle was convenient for a man needing to carry his worldly goods tied on to it. The American Indians, terrified when they first set eyes on a horse, were quick to adapt to riding, and evolved a style suited to their way of life. They vaulted on to the horse's back at the run and trained their horses to gallop while they hung by one foot, the horse's body between them and their enemies, firing under the horse's neck or even under the belly. The same feats are practised by some North African tribes in a 'fantasia', and were demonstrated by the Egyptian police in this century. The Bedouin carried a lance and rode with a slightly longer stirrup than the Moors, who used swords and rode only mares.

Many of these peoples used simple bits or none at all, except where the fearsome mediaeval bit of the Saracens that passed to Spain was still preserved, more as traditional ornament than from conviction. The cavalry of Chad might have been lifted straight from the fifteenth century. The American Indian, lacking the resources of metal for the buckles and decoration, adapted the bit to the bent-back type seen in Western films, with a port and single curb rein resembling the Kimblewick. But the ability to stop, turn and gallop without severe bits was noted by a British cavalry officer. 'He puts his horse to the gallop leaning much forward and clinging with his naked legs and heels round the horse's flanks, his stick brandished in his hand. Then he checks the pace, turns right and left, pulls up in his bitless

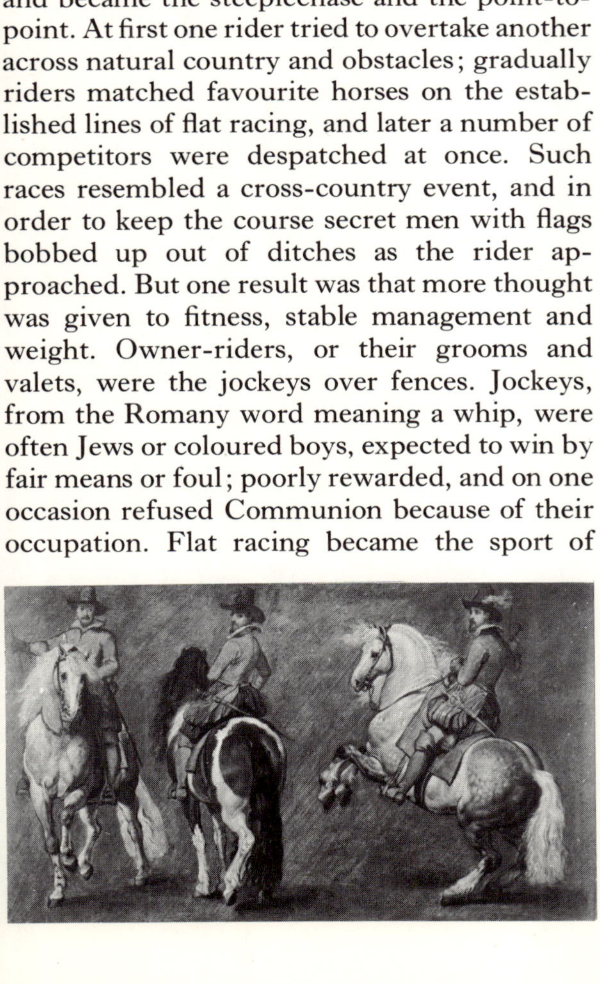

A delightful study by Rubens.
reproduced by gracious permission of H.M. the Queen

halter and exhibits if not the power of flinging his horse upon his haunches like the Turk and other bit-using orientals, more control than the English dragoon with his heavy bit. The Arab horse in our hands is hot and inclined to pull. With the Bedouin some screw is tight where ours is loose. If a stallion breaks loose in camp and the scene is chaos, he gets a halter and shows no trace of anger, where the English groom would bestow a job in the mouth.'

As northern Europe became industrialised the teaching of riding grew more sophisticated. Hanover and Brunswick (later to merge into Germany), Poland and Russia all had great riding schools. In Vienna alone there were twenty. The original Spanish Riding School had been rebuilt, modelled on Versailles, the first known instructor being von Regenthal, followed by no less than four of the great von Weyrothers. A tablet had an inscription ending in 'for the instruction and training of young noblemen and to train horses for school riding and combat'. Fees were high and entrance exclusive, except for the Emperor's pages and foreign noblemen who could get a permit. Others were allowed to ride from noon until three, called the hour of the cavaliers, and one who took advantage of this, Count Wilczek, 'when he at last broke into a canter made onlookers feel positively giddy'. The public thronged to see the daily rides, especially when the archdukes were present, and mounted displays and carousels were held, one of which was performed entirely by ladies.

From its declared standards the School has never deviated. The four-year-old stallions from its own stud are introduced to simple free forward movement under an experienced trainer, and only in the second year are asked even to become flexible, supple and obedient. When mature they pass from the *Campagne* or lower school to the high school, to develop absolute balance and dexterity. Their special abilities are studied for concentrated training in particular airs. But every movement is a natural one, seen in the young horse at play, and no artificial movement is ever, or has ever been, taught. Pupils have the exceptional advantage of learning on a trained horse, taught by an experienced instructor, so that they can know and feel what they are trying to achieve. Work on the lunge without stirrups gives the deep, supple seat that will not disturb the horse's equilibrium and can give him the right help, for the two must form a harmonious whole. Only by self-discipline can the rider expect to obtain control of the horse through the speech of the aids.

The School pays tribute to Guerinière, chief instructor under Louis XIV at the riding school in the Tuileries after fifteen years at Versailles. To him is owed the best in modern riding: the free forward movement, fluidity, controlled impulsion and accuracy of dressage, and the versatility and courage of the *militaire* which became the three-day event, yet without sacrificing the classical tradition. Rejecting force and cruelty, still prevalent in spite of great horsemen, great books and great schools, Guerinière aimed at a partnership of horse and man and demonstrated it convincingly; he achieved it by introducing lighter and better horses, mainly from England, and inventing better school movements, particularly shoulder-in, *renvers* and *travers* (head or tail to the wall) and suppling work on two and four tracks. Inadequate translations of his *Ecole de Cavalerie* (none is yet available in English) led to confusion and misunderstanding of the interpretation of two or four tracks, but no one disputes that he furthered the control of the quarters by the independent use of the leg, and stressed its primary importance in transitions. He also modified the system of pillars and cavesson, redesigned the saddle to one still in use, and increased jumping and cross-country work, using a mild bit.

The brilliant Comte d'Aury modernised the cavalry school at Saumur, opened a hundred years earlier, at which the élite Cadre Noir of officers specialised in *haute école* and produced a line of great *écuyers*.

International exchanges took place so that new and better methods were disseminated. One of Guerinière's pupils was invited to appear before George II, who pronounced him one of the most elegant horsemen he had ever seen. Montrose was taught in France, and his Highlanders on ponies conquered Scotland. The Duke of Wellington studied at Angers.

The British were individualists, and with the mixed blessing of a climate where riding was possible out of doors all the year round, and their growing industrial and maritime wealth, tended to despise any 'foreign tricks', as they deemed *manège* work. They careered across country with varying degrees of skill and complacency, on superb horses, and saw no need to be taught, having ridden from the cradle. One squire, watching a display of dressage, remarked that if his own horse started behaving like that he would shoot it. It was a matter of pride that for every continental riding master there was a British Master of Foxhounds. The English hunting saddle was claimed to be one reason for British superiority (a cheaper edition was made for export to the Colonies!); the *manège* type was thought impossible to fall off and thus somewhat unsporting. That 'Gentlemen must stand up in their stirrups at the gallop to avoid friction on the bottom' was customary, even inevitable, but the stirrup bar was often behind the centre of balance so that the rider 'sat as if in a chair', in Xenophon's words.

The increasing size and speed of horses led to indifferent riders resorting to the spade bit and similar instruments of torture; the high-spirited horse was starved or bled – the latter at one time a sovereign remedy for all equine and human ills. The reluctant horse was encouraged by the application of herbs, such as arsemart, under the saddle, and too much was left to ignorant and superstitious grooms. The 'desperate animal ignoring all aids' was believed to be the result of original sin rather than faulty training. This is not to suggest that equally bad riders did not exist elsewhere, nor to deny that a great tradition of fine horsemanship was built up in the eighteenth and nineteenth centuries which is still hard to surpass. Sport and steeplechasing helped to produce a supremacy, both of horse and rider, still in evidence in cross-country phases, while those fewer civilians who rode in continental Europe usually learned in the *manège*, where horse and rider are more disciplined. To this, and perhaps to the heavier, stronger type of horse bred for the army, may be due the advantage in dressage of the German, Russian and Scandinavian competitors, for such horses demanded a stronger use of back, loins and legs. Possibly, too, they owe more to the Viennese School, whereas other nations inclined to the French. But there is no doubt that countries previously isolated from international competition, such as Canada, the United States, Australia, the Argentine and Brazil quickly proved that horsemanship is an art that can be acquired by those who have the dedication, the will and the understanding.

In England one voice crying in the wilderness was Richard Berenger, Master of the Horse to George II although of French descent. He translated Bourgelat's *Le Nouveau Newcastle*, wrote a valuable history of horsemanship and a work on riding. Perhaps he did not contribute anything highly original, but he wrote in English, boldly and succinctly. Union was his watchword: co-ordination of horse and man which 'many talked about but few could define'. There were two kinds of riding, he said, the useful and the ornamental. Few attempt the second, but some knowledge is needed even for the first, for 'who would venture

De la Guerinière, *Ecuyer du Roy* to Louis XIV.

alone in a boat who knew nothing of wind or tide, yet mankind thinks it can climb on a horse's back and ride'. Had he lived a few centuries later he would have seen the trekker refuting this, and the holidaymaker drifting out to sea on an airbed. Prejudice gave rise, he wrote, to theories that the methods of French, Spanish and Italian riders of repute, each adopting postures peculiar yet fundamentally the same, were sometimes better and at other times desirable. It was assembling and distributing the horse's strength equally, thus helping to lighten its forehand – in other words, balance – that were the purpose of training. He is one of the few who mentions cavalletti, and that in early times boxes of clay were used for the same purpose. He also describes as a training exercise a trench dug so that the horse had to follow its course.

Berenger believed that too much use could be made of the senses of hearing and seeing and not enough of feeling. Too great use of the first two resulted in dependence on memory; sound should be used sparingly and as an element of surprise. He also recommended variety in lessons to help keep the horse's attention, and

The Duke of Marlborough.

in acknowledging Pignatelli he emphasises that many of his own exercises stemmed from him and his successors. Using the whole *manège* he calls 'changing large', while using part of it is 'changing narrow'. He also describes 'putting the horse under the button' or the *oberstreicher*; this is stretching the rein hand up the horse's neck without change of weight or position to ensure there is no increase of pace. He considered the switch an aid through sight, for later it had to be replaced by the sword, and the trot the most useful pace for work to obtain equilibrium. Both horse and man needed to improve their carriage by planned exercise, and he described the correct posture of the rider. Horses were imperfect from weakness, bad conformation, want of courage and sloth, but disobedience was due to lack of skill by the rider, ignorance, bad temper and incapacity. He calls the fork the 'twist' and the horse's chin 'the beard', and he also writes of lunettes for impetuous horses, which must mean blinkers although the word was also used for half-shoes or grass tips.

Many books were published in the eighteenth and nineteenth centuries which throw light on riding, both good and bad; most concerned the cavalry rather than the civilian rider. One criticised the way that riding instructors were appointed, which was to appoint anyone who volunteered and had an eye on the 'perks', such as an NCO without qualifications but an appreciation of the leather breeches provided. Dress for riding is the subject of a book in itself; it was colourful, picturesque and, in portraits, unpractical, although informal clothes were more drab but more comfortable.

By this time not only Europe but North America was developing fast, and all kinds of equestrian activities acquired regional and national idiosyncrasies. But every country used the hack, and the richer ones had particular animals on which to appear in society. The park hack was a light type with presence and extravagant paces, especially the ability to canter at almost walking pace to enable the carefully dressed men of fashion to ride alongside carriages or be seen by those on foot. For this the *manège* seat was adopted, with the tip of the toe just reaching the stirrup, and one hand

A nineteenth-century view of the horse, taken from contemporary engravings.

on the hip when not sweeping off the hat. An exaggerated cheek and single rein were obligatory. The covert hack was used to ride to the covert side before changing on to a hunter.

In France the Limousin was popular for display; the equivalent in Spain was the 'courting horse', with mane and tail beribboned and like a fleece. In the evening promenade the young *caballero* was to be seen greeting his friends, the horse stirring up clouds of dust as the fashionable dishing action brought its huge feet down in almost the same place. Ladies rode specially trained horses, accustomed to the side-saddle, for women of rank and those whose husbands or fathers could afford for them to ride expected them to be decorously covered from head to toe 'for ladies can fall in all positions and there is absolutely no saying what might happen'! A special horse, special clothes, a special escort and a special teacher were required, and the first known bra ('a handy little contrivance of twill called the bust bodice') was recommended for 'amazones'. Needless to say, long before the first world war emancipated women from the long skirt and the male escort, there were many who were bold and dedicated horsewomen in spite of the view expressed by one man that young ladies should recognise that in the hunting field they belong to the third class.

In the nineteenth century there emerged 'the great trio', one of whom was Baucher, a virtuoso and a genius whose intuitive skill enabled him to produce his horses, usually English Thoroughbreds, in a month. They performed quadrilles without apparent aids and only infinite adjustments of weight; it was said 'like clockwork but with great sadness'. He carried lateral flexions to such a degree, obtaining these at first from the ground, that his horses could move freely forward with their heads at right-angles, and the *passage* was performed with such a short stride that tremendous elevation was obtained. As a civilian his only shop window was the circus; he never appeared out of doors. He was born at Versailles and later studied at Milan before returning to France. The authorities at Saumur were so impressed that they sent officers to study with him, but some of his methods were unorthodox and he was turned down as an instructor. In his *Dictionnaire Raisonné d'Equitation* he admitted it was difficult to put his methods into words, and the sharp spurs he used were not always successful when used by his pupils. He used the hand before the leg, and the intense criticism he evoked was a tribute to his brilliance. The Cirque d'Éte was thronged by fashionable society and the stars feted and lionised. Baucher was killed when one of the huge chandeliers of the Cirque Napoleon fell on him as he was mounting. After his death one of his women pupils, Pauline, bought two of his horses and became a star of the State Circus in Russia; another pupil, Caroline Loyo, performed in London on a side-saddle.

The second of the trio was James Fillis, who although English spent most of his life in France and then became chief *écuyer* at Leningrad. His *Breaking and Riding* is still in print, and was one of the first books to devote a part entirely to jumping. He too performed in a circus, and could make his horses canter backwards and change legs at every stride, carrying collection to an extreme to suit the enclosed areas he had to work in. He disagreed with Baucher on many points, although himself admitting the difficulty of analysing some movements, which before the days of photography was understandable. He was a disciple of Guerinière; the Germans he considered stiff, partly due, he said, to their being taught without stirrups but with their toes higher than their heels 'which would make even a Frenchman stiff' and they were obsessed with 'bracing the back' as advocated by Müseller and probably necessary on the type of horse they rode.

Fillis used a double bridle for jumping, employing the curb only between fences, leaving the horse's head entirely free during the leap and picking up the snaffle rein on landing. He said the legs should be used to obtain a 'bascule'. Obstacles were usually some distance apart or this would have needed sleight of hand. Nevertheless, both Baucher and Fillis made a contribution to new facets of riding.

One early twentieth century innovation often taken for granted was clipping, introduced to England by the army, who saw it in use in the Peninsular war. Crude methods such as scissors. razors and singeing had been used on the heavy winter coats, but it is hard to imagine now the problems of fitness, stable management and ills and chills before shears and then clippers became standard equipment. By the time someone had ridden miles for a vet, who also had to return on horseback, many animals were dead or on the way to recovery. At that time the farriers' knowledge had advanced little, if at all, from that of the Spanish Albitares, who had learned their medical lore from the Moors and Arabs, whose early books show that they were advanced in scientific matters, even being aware of hormone tests for pregnancy.

Like the great renaissance in Naples, the second was also conceived in Italy, circumstances one more contriving to produce the climate and the man. Caprilli rebelled against the orthodox 'straitjacket' of *haute école* applied to the cavalry, of which he was an officer. His 'new' method crystallised the accumulated experience of centuries into the principle that the rider must abandon his instinctive urge to preserve his own equilibrium with gravity and adjust it to that of the horse in motion. This had unconsciously been done by horseman of the past 'in balance', but Caprilli analysed the problem and concluded that at speed, especially during the acceleration of take-off, the weight of the rider's body must be equally distributed over the horse's centre of balance, and the angles of the body–waist, knee and ankle–closed to act as shock-absorbers, the base being the foot and the stirrup. To achieve this, the rider's body had to be in advance of the perpendicular, and a shorter stirrup used so that the rider's seat was out of the saddle throughout the parabola. It was this 'forward' position that was hotly disputed for a quarter of a century, as everyone had believed that a rider had to lean backwards during descent in order to remain in the saddle. Caprilli, and his pupil Santini, disproved this for all time by convincing successes in open competition, and the Italian cavalry adopted the 'forward seat' even down the famous slide at Torre del Quinto. He objected to stylised riding and maintained that non-intervention, widely adopted and adapted in the first part of the new century, was the natural method.

It was coincidence, perhaps, that at about this time Tod Sloane brought a hurricane of change into racing with the ultra-short stirrup and crouched position. Although he brought the 'monkey on a stick' seat to Europe he did not invent it. Centuries earlier observers had commented on its use by untaught boys, but it was an American trainer, Huggins, who proved that horses ran faster if thus ridden. It was his custom to attend minor race meetings and to buy likely winners, and at these meetings the jockeys were black 'boys'. But the horses he bought and brought south failed to maintain their form, until finally he bought the boys as well as horses. Not until he let the boys ride in races did they achieve spectacular success, and he realised that this position on the horse at speed distributed the rider's weight where it handicapped the horse least, and he instructed his stable jockey Sloane to adopt it.

Caprilli's system has been modified over the years because of modern competitions involving speed 'against the clock', and of the course builders' increasing skill in setting problems of difficult combinations and courses. Lateral movements and rapid transitions are now essential, comparable to the athletic qualities demanded in the School of Naples to outwit the bullet. As it is the rider who walks the course beforehand, it must be the rider who decides the tactics and dominates the partnership, though this must be done tactfully and without disturbing the combined equilibrium.

A horse equipped and ridden Western style.

The success of Continental riders in jumping events at Turin and the International Horse Show in London in 1907 gave British riders cause for thought. It was Lord Baden-Powell, Inspector General of Cavalry, who, after visiting Vienna and Saumur, recognised the need for a school of equitation in Britain; Netheravon, later moved to Weedon, was opened. In the United States Fort Riley was established at about the same time.

Horse shows had been held for over a hundred years. One, held in 1864 at Islington, included ridden classes and 'leaping', and in the heyday of the horse at the turn of the century agricultural shows had classes for hunters, hacks, polo ponies, children's ponies, and jumping. At first there were no proper rules; and considerable chaos sometimes ensued.

In the present century various bodies, such as the Fédération Equestre Internationale, the British Horse Society and the Pony Club, help to maintain and to improve standards of riding and of horse management. But in the present era of an explosive increase in the popularity of equine pursuits, there is still cause for concern. In spite of qualified instructors, the example of the successful, infinitely better horses, and the camera and television for all to see and learn, there are far too many who try to train their own horses without acquiring the basic principles of horsemanship.

In this day of do-it-yourself and of instant everything, it is still impossible to achieve instant horsemanship, for it is the one activity requiring the participation of an animal larger and stronger than man in harmonious partnership with him. Perhaps another revolution is due, and computerised simulators will enable the novice rider to acquire the supple and independent seat once taught on the lunge and between the pillars.

It is to be hoped the *beau plié* and the permissive rosette will not prove more beguiling than mastering an art. 'Every science and every art has its own principles and rules that lead to new discoveries and to perfection,' wrote Gueriniѐre. 'Why should horsemanship be the only art for which practice alone is needed?'

PART SEVEN

Methods and techniques of training

Chapter one

The mentality of the horse

In order to get the best out of any horse, it is necessary first to understand it; genuine communication is possible only through a knowledge of its mental processes.

In spite of a long history of association between the horse and man, many of the primitive instincts the horse developed in the wild have stayed, and still affect its behaviour today. It is important to look back at these early behaviour patterns and the environment that fostered them. The basic physical structure and the mental faculties of the horse have remained virtually the same for 5000 years. The most important characteristics of the horse can be summarised as follows: herd instinct, need for security, the following instinct, love of routine, laziness, excitability, nervousness, sensitivity and courage.

The herd instinct is the strongest of all the horse's basic urges, as its survival in pre-domestication days largely depended on the herd grouping. A horse alone stood little chance of survival. Because of this, horses will naturally want to be with others of their kind. The effects of this urge can be greatly reduced by training, but many horses are reluctant when asked to leave a group, and it is instinct that lies behind most 'nappy' horses' behaviour. The instinct can, however, usefully be exploited during training: a horse will, for example, jump with more enthusiasm if asked to do so at first towards its companions rather than away from them.

Linked with this instinct to be with its fellows is the horse's need for security. In the domestic horse of today security is centred on the stable–or field–in which the horse lives. Horses are generally fairly greedy animals, and there is no doubt that their usually being fed at 'home' also contributes to this. Horses kept on their own are seldom happy, for company also represents security. Horses suddenly isolated will pine not just for companionship but for the security it brings; horses moved to new surroundings will take some time to settle down and regain the temporary loss of confidence occasioned by the move. This is true, though less markedly, even if the horse is in company and is handled by humans it knows and trusts. A horse frequently subjected to strange accommodation–as, for example, a show-jumper travelling to different parts of the country during the summer–will become more adaptable and be less affected by these changes.

When the position of a training area is being considered, the factor of security needs to be borne in mind. If it is placed too close to the stable a horse will find concentration difficult; it should be situated far enough away from the stables–and preferably the proximity of other horses–for distraction to be avoided. It is advisable, however, for a horse, particularly if it is young, to be allowed to become familiar with the area where lessons will take place before they actually begin. Calm is of prime importance in successful training, and only with security will there be calm. A horse in a completely strange field will be looking round, not concentrating on its trainer.

The following instinct of the horse is easy to identify. Like all herd animals, horses were accustomed to looking towards a leader: they are, almost all of them, naturally 'following' animals. This creates, nowadays, a need for a leader that can seldom be satisfied by another horse, though even trained horses will, when out to grass, show tendencies to lead or to follow. In any group of them it is easy to see which horse is 'boss'. It is to man's advantage that most horses have a natural inclination to obey, for it makes them surprisingly co-operative. This becomes less surprising when one realises that in his mastership of the horse–a domination sometimes taken undue advantage of–man is filling a need in the horse which makes his training of it considerably easier.

A pattern of life which has rhythm, a regular routine, is important to the horse. It encourages security, calm and well-being. It is for this reason that so much emphasis is placed on the importance of regular feeding times and regular exercise. However, an occasional interruption of routine is no bad thing, as the stabled horse without enough activity or any variations in it is likely to become bored. Boredom is a dangerous state of mind which is liable to lead to bad habits and a bad temper.

The love of routine is linked to the horse's natural laziness. In its wild state it spends almost all its time browsing–as, indeed, horses out at grass will do–unless it is disturbed and excited or frightened. Only its willingness to co-operate makes it possible for man to persuade the horse to perform with the amount of energy he requires.

In spite of these characteristics, horses are frequently both excitable and nervous. The horse's ability to run away when danger threatened enabled it in the past to survive predators more crafty or more aggressive than itself. In order to become instantly alert to possibilities of danger, its nervous system had to be highly tuned, and when startled and alarmed now its natural reaction is still to flee. A horse with a rider it trusts can be prevailed upon not to; but it is no good shouting at or abusing a frightened horse, for this will only make matters worse.

Fear is also likely to make a horse excitable, as are conditions that remind it of its wild state–such as galloping–or which it finds unnatural or strange–like being asked to jump. Part of the excitement of hunting and racing comes from the simulation of running with the herd. Obedience produced by training will get over much of the excitability, as the horse will learn to behave calmly in situations that are naturally exciting to it.

It is perhaps strange that in spite of the–already somewhat conflicting–traits of laziness, excitability and nervousness, most horses are also possessed of great courage. Without this courage it is unlikely that they would ever be able to trust humans to the extent that they do.

One important characteristic which is fundamental to our ability to train the horse is its sensitivity. Because it is a non-aggressive animal originally equipped to run away rather than to stay and fight, it has a relatively low pain threshold. In other words, it is very sensitive to touch. Without this sensitivity the subtle leg aids used so unobtrusively on a highly-trained horse would be impossible to achieve. In the early stages of its training the horse responds to aids because it wants to get away from the touch of the leg. This is even more true of the discomfort caused by a bit. The mouth is much more sensitive than most other parts of the body, and a mild bit should always be used until horse and rider have reached a stage of training where the correct use of, say, a double bridle will improve performance. It is less ridiculous than might be supposed to suggest that a horse which pulls badly in a severe bit should be ridden in a bit as mild as a rubber snaffle. With the pain–from which the horse is trying to escape by running away–removed, many horses stop pulling and fighting and will calm down.

As far as training is concerned, perhaps the most useful attribute the horse has is its memory, which is extraordinarily retentive, though some horses are undoubtedly more intelligent than others. As with other animals capable of being highly-trained, there are several points to remember. The first is the importance of making sure that what is being asked of the animal is completely understood and within its capabilities at that time. An imperfectly understood lesson will be wrongly remembered; an animal either mentally or physically unprepared may not be able to carry out a particular exercise. Cause and result must be closely related in time if the result–reward or punishment–is to have any meaning. Thus a horse that has just performed well should be rewarded immediately; one that is being wilful or obstinate must be punished at the time of its bad behaviour. Horses should, incidentally, be sparingly punished. Unsatisfactory behaviour most often results either from imperfect communication or from some outside factor affecting the horse's concentration, calm or confidence. Occasionally, though, it must be admitted, they are just badly behaved, and then they must be treated firmly. A horse always allowed to have its own way will eventually become totally undisciplined and difficult to handle.

First steps in training

The early training of a young horse is made much easier if the horse has been accustomed from the start to being handled. Anybody breeding a foal and hoping to train it would be well advised to spend a good deal of time with the foal during the first three years of its life. A prospective trainer buying a three-year-old will also need to devote time to its general handling before any formal training is undertaken.

On these early days the whole future of a horse can depend. It is important to establish from the outset a basis of trust on the part of the horse for the person who is handling it. If the horse learns that it is not going to be hurt or frightened by anything the trainer does, the later lessons will proceed much more smoothly. It should be taught to accept without fear or fuss the presence of the trainer, learn to have its feet picked up, to be groomed, to be led in hand. At this stage it should ideally also become accustomed to the presence and the absence of other horses, and be encouraged to begin to learn concentration and obedience.

All these lessons should be taught gradually and calmly. Unless the horse is deliberately wilful–and horses occasionally can be, at any age–discipline and obedience should be established without any use of force or demonstration of anger on the part of the trainer. A gradual progression of obedience and respect will help the horse in all its later stages of training. If the horse is handled in this way, it should, when the time comes, learn to accept the bit and eventually the saddle and a rider, without undue difficulty.

LUNGEING

The practice of lungeing is in general use in most of the horse countries of the world as a part of the basic training of the horse. The degree of emphasis placed on this exercise and the manner of its execution naturally vary according to the trainer. Some will use the lunge line only for gymnastic exercises designed to supple and strengthen the young horse prior to his being backed; others will rely on lunge work to a far greater extent, regarding it as the foundation for ridden schooling both on the flat and over fences.

Lungeing is not, of course, entirely confined to the training of the young, unbroken horse. It has considerable value in the retraining of the older, spoilt animal and is also used to supple and relax the schooled horse before he is worked under saddle. The essentials of lungeing can be simply described. It is an exercise where the horse circles the trainer, contact between the two being maintained by the line connecting the trainer's hand to the ring fitted on the nosepiece of the horse's cavesson, while the horse's pace is controlled by the trainer's voice and the movements of a long whip held in his other hand.

The principal objects of working on the lunge are:

1. To promote the gradual formation of muscles in the young horse without these being developed in opposition to the weight of the rider. By working the horse in both left- and right-hand circles, equal muscle development on either side of the horse can be ensured.
2. To supple the horse laterally by the equal stretching and contracting of the dorsal, neck and abdominal muscles on each side, made necessary by the horse's body being bent on the circle.
3. To encourage a tensioning of the spine, induced by the horse stretching his head and neck forwards and downwards. This causes the back to round and the hind legs to be further engaged under the body. The proper engagement of each hind leg is made easier on the circle, as the inside hind must be brought further under the body if the horse is to maintain his balance.
4. To increase the flexion of the joints as a

1

5

2

6

This sequence of photographs demonstrates progressive stages in work on the lunge. (1) The young horse is first led round the trainer by an assistant, who then (2) moves to a position inside and slightly behind the circling horse. (3) A good, free trot with the horse moving well forward. (4) Crossing a low grid of logs or cavalletti is a valuable balancing exercise as well as a useful introduction to jumping: the horse is led before being sent down the grid at the full length of the lunge rein. (5) A jump in excellent style over an improvised obstance. Note the sympathetic contact maintained by the lunge rein. (6) The horse stands square at halt on the circle.

3

4

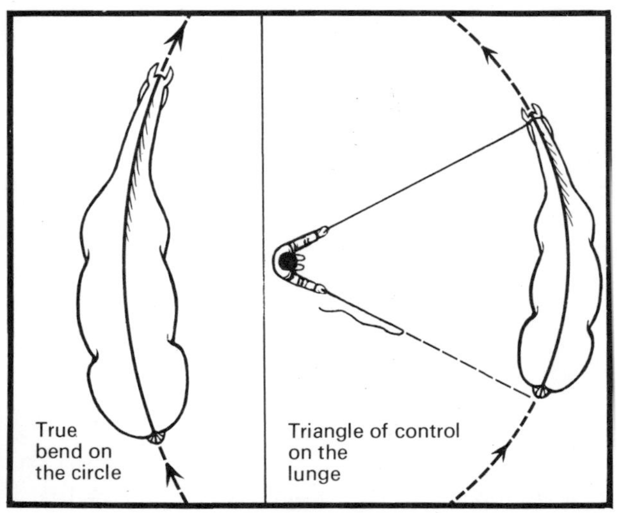

result of greater and more supple muscular development.
5. As far as is possible to flex the spine of the horse and to correct his natural asymmetry.
6. To improve the horse's balance as a result of the increased engagement of the hocks.

These are the physical benefits of lungeing; the mental discipline is equally important, particularly the development of the horse's powers of concentration. Properly carried out, lunge work helps to produce a calm horse accustomed to the *habit* of obedience. It is in this exercise that the horse learns to obey the first of the aids, the voice.

The other vitally important lesson that lungeing teaches is the essential requirement of the riding horse: free forward movement. The urge to go forward is as much the result of a trained mental attitude as it is a disciplined physical movement.

At a later stage of his training the horse may be given his early jumping lessons on the lunge, unhampered by the weight of a rider. Lungeing is also a useful way of exercising a horse who for one reason or another cannot be ridden, or for settling an over-fresh horse before the rider mounts him. Fifteen minutes on the lunge is frequently enough to take the edge off an unruly horse and so avoid the ignominy of being bucked off.

The equipment required for lungeing consists of a cavesson, a lunge line of tubular web material fitted with a swivel hook to fasten on the cavesson rings, and a long whip with a thong. In due course a pair of side-reins will be needed. These are attached first from the cavesson and then from a snaffle fitted both to it and to rings placed at varying heights on a body roller. In some cases trainers dispense with the roller and prefer to attach the side-reins to a saddle. The disadvantage of a saddle is that the height adjustment of the side-reins cannot be made as precisely as it can when a roller is used.

An essential part of lungeing is the availability of an enclosed space in which the horse can work without distraction.

The methods used to teach the horse to work on the lunge will vary according to the particular preferences of the trainer, but whatever methods are employed it is first essential to teach the horse to move freely forward when being led in hand. To make sure that the horse understands what is required of him an assistant frequently leads him round on a circle until he has learnt the words of command. The horse can then be asked to perform work on the circle.

A LIGHTWEIGHT LUNGEING CAVESSON

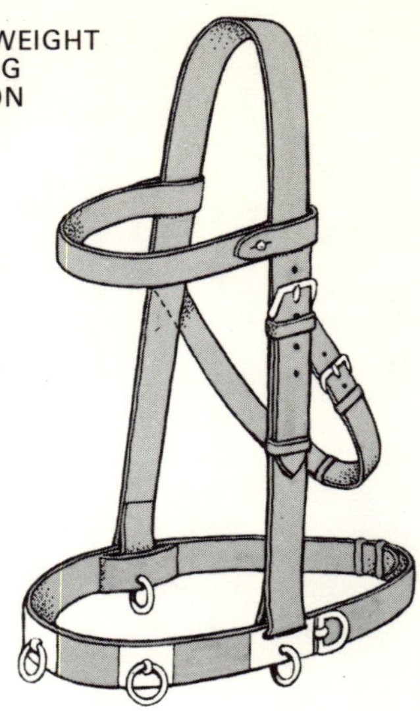

The side-reins are brought into use as a method of encouraging a steady and correct head carriage. Whether they are fitted to the cavesson or to the bit, however, they should always be adjusted at full length until the horse takes up the slack and begins to seek out the bit and to 'accept' it. Only when he does this is it safe to work–for periods of only a few minutes–on shortened reins.

When a horse is being lunged the side-reins take the place of the rider's hands, and the whip acts as a substitute for the legs in pushing the horse into contact with the bit. Only a few trainers lunge a young horse directly from the bit, though the French method is used to place the horse 'on the bit'. It is, however, a method which really belongs to the more advanced stages of the horse's training, and it should be practised only by experienced horsemen.

Specialised aspects of lungeing apart, in general the exercise should be regarded as an essential and valuable part of every horse's training programme.

LESSONS UNDER SADDLE

Once the horse has been backed it should continue its lungeing lessons, both with and without the addition of the rider. By this time the lungeing process will have become familiar and the horse will know what is expected of it. It will be helped to get used to carrying a rider if it first has to do so during lessons that it already understands. Only when the horse has fully accepted the presence of the rider should he begin to take any active part, gradually increasing his participation until the horse is responding to the rider as well as to the trainer on the ground.

The objects of working the horse on the lunge apply equally to the elementary exercises carried out under saddle. In addition, once the horse has learned to accept the rider's weight it must learn to carry it properly as well. It must also, of course, be taught the elementary language of the aids so that it can understand what it is being asked to do. All horses, not necessarily just young ones, greatly benefit from proper schooling. A well-schooled hunter, for example, will not only be a much more pleasant ride, but a much safer one than an unschooled horse; it will be more obedient, more comfortable. will both get less tired and tire its rider less. A well-schooled horse will perform whatever is required of it to maximum effect with minimum effort; the reverse is true of an unschooled animal.

Impulsion is the most important factor of all in schooling—without it, none of the schooling exercises can be properly performed. The horse should first be introduced to the school at the walk, the rider concentrating on riding into each corner rather than allowing the horse to cut across the corners. From the outset the horse should be encouraged to respond to the lightest possible aids; if it does not respond to a light use of the legs the aids should be reinforced by light taps with a schooling whip (which is long enough to be used without the rider altering the position of his hand on the rein) rather than by letting the aid degenerate into a kick.

The horse should be taught to move forward at walk, trot and canter and within each gait to vary the pace. The variations are the ordinary walk (or trot or canter), and the collected and extended walk. At ordinary walk the horse moves freely forward with long strides, a light contact with the bit being maintained. For collection, the stride is shortened by a strengthening of the pushing force of seat and legs on the part of the rider encouraging a greater engagement of the hind legs under the horse's body, and by a slight closing of the hands on the reins to contain the increase in impulsion. In the extended walk the driving aids are maintained but the horse moves forward on a released rein.

The trot is the most valuable pace for schooling. The ordinary trot should be energetic, with plenty of drive from the quarters, the horse covering with its hind feet the tracks of the forefeet. In collected trot the forehand is lightened with the increase in impulsion, the stride is shortened and becomes more vigorous. When the extension at trot is asked for, the horse will lengthen its stride and increase speed in response to the increased driving force of the rider's legs while a light contact with the bit is

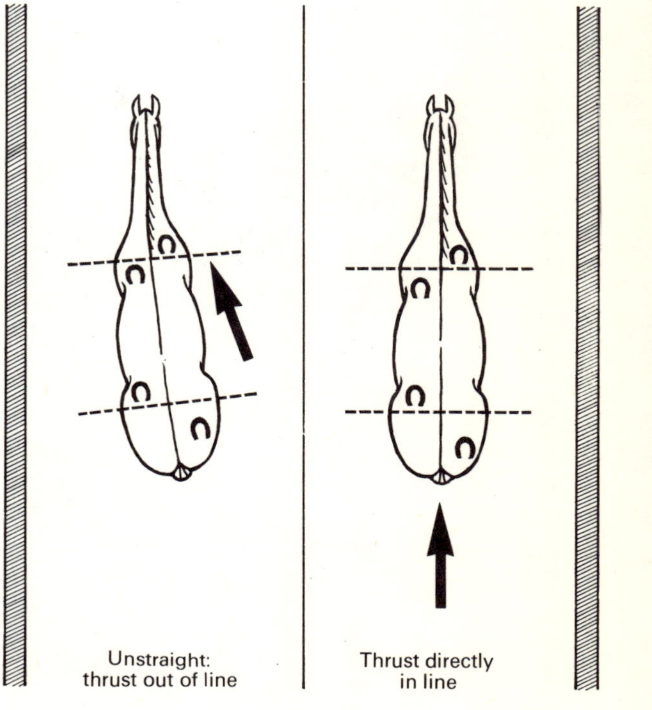

maintained. Both collected and extended trot should be requested only for short periods.

The rider can either rise with the trot or keep to the sitting trot. The latter should only be introduced when the horse's back muscles are sufficiently developed and it has become reasonably well balanced.

Cantering in the school should only take place when the horse's action is controlled, even and active at walk and trot, and when it has learned the necessary adjustment to balance, necessitated by the presence of the rider's weight, by being asked to canter while out hacking. It is easiest to teach the horse to lead on the correct leg on a circle–that is, with the inside leg–by asking for the transition to canter with the horse bent very slightly in the direction of the movement. It is important to ask the horse to canter from a controlled trot with good impulsion, not to allow it to trot faster and faster until it eventually falls into canter. If the horse strikes off on the wrong lead, it should be brought back to trot and then asked to canter again when balance and impulsion have been re-established.

A young horse being schooled under saddle.

Half-circle to the right at walk.

Circle right at sitting trot.

Transition from trot to canter.

At all paces, the fullest possible use should be made of the school, the direction being changed by riding diagonally across it and circles of varying sizes being made.

The horse will benefit at a fairly early stage by the introduction of poles laid on the ground, over which it should be asked to walk and trot, and later on by the use of cavalletti. There should be no interruption in the rhythm of the horse's stride, and it is important that the poles, which should be placed down the long side of the school, should be correctly spaced and adjusted to the stride of each horse. At this stage the poles should not be considered as an introduction to jumping, though one of their objects is to lay the foundations for it. They are introduced to help the horse develop suppleness, maintain an even length of stride, improve its balance and pick up its feet. When the horse has become thoroughly accustomed to negotiating the poles when they are laid in a straight line, they can be placed on the circumference of a circle, thus combining their own functions with those of work on the circle.

During all its early training, the horse will be learning to move freely forward, to move straight, to develop evenly and to become supple and balanced. Work on the circle is the best way by far of achieving all these ends.

There are several general points to be borne in mind during the early stages of training. All horses–and young horses in particular–will feel exuberant and have natural outbursts of high spirits at times. Obviously, it is preferable if these do not take place during a lesson, and it is therefore advisable to allow the horse a period at liberty every day. If the horse is being kept out at grass it will have this opportunity in the normal course of events; stabled horses should also be allowed an hour or two in the paddock when they can work off their high spirits and relax. Secondly, concentration is an acquired art; only a little concentration should be asked of the horse at first, and as its training advances work in the school should be broken up by a relaxed ride. Lessons should always be kept short, and be brought to an end before the horse shows any signs of becoming bored–boredom prevents concentration and encourages disobedience. An intelligent trainer will in fact use the hacking periods to advantage as well: riding the horse up and down hills and trotting it over uneven ground are just two examples of ways in which the lessons of the school can be extended to help the horse develop its muscles, balance and wind.

ELEMENTS OF RIDING

In considering any stage of training of the horse it is necessary also to consider the possible effects of the rider on the horse's movement. A horse at liberty will at an early age learn how to balance itself effectively; it will be able to vary pace and speed, to change direction and to turn, without losing balance. The addition of the weight of a rider on its back, however, will necessitate considerable adaptation on the part of the horse, which will need understanding and assistance from its rider.

In order for horse and rider to become a successful combination, the rider must work as closely as possible *with* the horse. One important aspect of this is for the rider to develop a well-balanced, supple seat. The correct position of the rider resembles that of a man standing, with bent knees, rather than one sitting in a chair. The head should be held up, looking out over the horse's ears rather than down at its withers or at the ground. The back and shoulders should be held straight, but not stiffly. The seat should be positioned deep in the saddle, with the legs hanging down naturally so that the heel falls in a straight line with hip and shoulder, and the toe of the foot faces to the front at a slight upward angle. The rider should push down and forwards into the centre of the saddle, gripping with a relaxed, not a tense, grip with the thighs. Gripping with the knees and lower leg will tend to push the rider up out of the saddle rather than down into it.

It is important for the rider to develop a seat that is secure, supple and completely independent of the reins. Flexibility is obtained by relaxing the muscles, balance by the rider adapting his own distribution of weight to the shifting centre of balance of the horse.

There are various exercises which help the rider to achieve a correct seat. Riding without reins (while the horse is exercised on the lunge, for example, or in the school) will ensure that no dependence is placed on the reins in order to maintain balance; riding without stirrups encourages the rider to develop a strong seat and to move with the horse as it moves. As the rider's balance improves and his own muscles develop, he should ride without stirrups not only at the slow paces but for school movements and jumping as well.

Because of the importance of a harmony in balance between horse and rider, the rider should understand how the horse's weight is naturally distributed and how its centre of balance alters as it changes pace. Only rarely is the horse's weight evenly distributed over all four limbs. When the horse moves at speed or is extended, its weight shifts to the forehand as it covers the ground with longer, lower strides, and the centre of balance therefore moves forward. With increased collection the stride is shortened, the forehand lightened, the quarters and hind legs engaged further under the body, and the weight and centre of balance thus move to the rear. The rider should at all times be positioned as closely as possible over the horse's own centre of balance, and it is for this reason that the different seats have been developed for different riding activities. A jockey will ride so that his weight falls as far forward as possible, while a dressage rider will use a seat with much longer stirrups–which also helps the subtle, accurate application of the aids–and his weight falling further to the rear.

A full understanding of the aids and how they should be applied is also of prime importance. The aids are the language which the rider uses to communicate to the horse what he wishes it to do; imperfectly understood and incorrectly applied, they limit rather than improve this system of communication. An inexperienced rider should be given instruction in the use of the aids on a trained horse which will respond when the aids are correctly applied, for he will then be able to feel the result of his application of them. An untrained horse will quickly become confused by unsure use of the aids and will find it increasingly difficult to learn the language.

The natural aids are the legs, hands, body and voice. The legs encourage impulsion and the development of free forward movement, which come from active engagement of the quarters and hind legs, and also control the movement and position of the rear part of the horse. The hands guide the forehand and the positioning of the horse's head and neck. Used in combination, they encourage an increase in collection or extension. The combined use of leg and hand should, however, be restricted to a horse reasonably advanced in its training and understanding of the aids; a young horse should at first be given clear and simple instructions of either leg or hand, used separately.

Equal action of both the rider's legs will increase impulsion and direct the horse straight forward. A pressure applied unequally will encourage the horse to bend round the inside leg, used on the girth, while the outside leg is applied behind the girth and controls the quarters.

The use of the hands is divided into the 'simple' reins, the direct or open rein and the indirect or contrary rein, and the more advanced 'reins of opposition'. The use of the direct rein should first be taught to the horse, and the indirect rein (neck rein) introduced at a later stage.

The weight of the rider's body, as has been explained, has a considerable effect on the balance of the horse, and can therefore be used as an additional aid. The voice, used either soothingly on an excitable horse or more sharply on a lazy animal, is generally also effective, particularly as it is one aid to which the horse will have become accustomed even before its formal training begins.

An unfit rider will be less able to apply the aids correctly and to use the advantages of a correct, secure and supple seat than will a rider who is fit. Similarly, the horse must be allowed gradually to develop its muscles and to learn to balance itself properly. Most horses, left to themselves, develop unevenly. As a rule, because of the way they lie in the womb, they are likely to develop stronger muscles down the right side of the back, a tendency which is encouraged by their usually being led in hand from the near side when young. They should learn to lead equally well from both the near and off sides in order to avoid this muscular block being further built up. At a later stage in training, work on the circle will help to reduce this natural stiffness. It is perhaps advisable initially to teach the horse by working in the direction it finds easiest (i.e. on the left circle), but as soon as it has understood what is required, work should proceed on the right circle in order to encourage evenness. In the same way, most horses will at first be more comfortable to ride on one diagonal than the other at the trot, but the diagonal should be changed both during school work and while hacking. All these factors will help contribute towards making a horse that is evenly developed, well balanced and responsive.

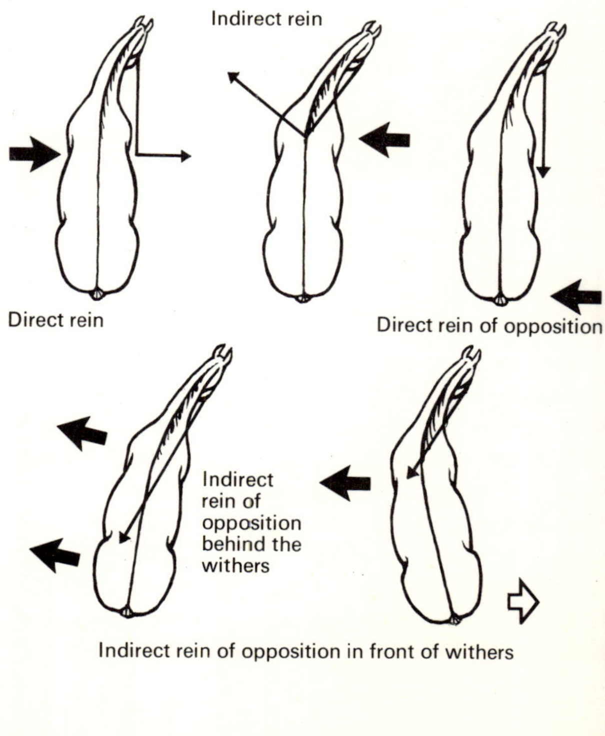

Chapter three

Advanced training

DRESSAGE

Advanced dressage, often called *haute école*, defines not so much a particular set of movements as the refined execution of them. It is their expression rather than their nature which classifies them, and the most essential ingredient of this expression is lightness.

What is lightness in this context? A painter, Max Liebermann, has said that 'art is doing without'; and the lightness that turns equitation into an art is the result of eliminating from movements executed by the horse any force or effort that does not directly contribute to them. Whatever additional effort the horse makes is a resistance which detracts from lightness. Thus lightness is contingent upon the rider's knowledge and tact as well as on the horse's skill and training, which will successfully have straightened it and developed in it a permanent urge to go freely forward. It is only possible to achieve lightness once these important preliminaries have been firmly established.

The urge to advance is called impulsion, and has been achieved when, rider's hands permitting, the horse will advance on its own, regardless of the nature of a prior movement, and when the slightest pressure of a rider's leg will shift its haunches. Straightness comes with the disappearance of all the horse's undesirable natural asymmetry; once this has been achieved the hind legs will follow in the exact track of the forelegs and the haunches function uniformly, causing the horse's weight to be correctly distributed. Thus a picture of harmony is produced: harmony between haunches and shoulders, haunch and haunch, shoulder and shoulder.

There is no precise distinction between 'ordinary' training and dressage, or between 'ordinary' dressage and *haute école*, though there are some stages of advanced training in

A dressage horse during a test moving forward on a circle into the centre of the arena.

Deauvillois in action: relaxed, supple and with excellent impulsion.

dressage which only those specialising in this branch of equitation will need to study. It may, however, be interesting to others—both as non-specialist horsemen and as spectators—to know something of what is involved.

Before any attempt is made at collection, the horse's paces must have attained regularity and maximum development. There is no real relation between trot and canter, and work to to improve both paces may therefore go forward along parallel, largely independent lines. Work out of doors is preferable to work in the school for this, as the natural impulsion will be greater outside, the horse can be kept on a straight line for longer, and the going offers a variety not available in the school. For instance, it should be possible to trot the horse up a slope, with the consequent flow of weight to the forehand and relief to the quarters; the neck will stretch and the hind legs be engaged much more than on the flat. It is also better to work out of doors towards achieving even strides of the trot by sitting more often to the less frequently used diagonal.

The remedies for crookedness in a horse are work on the circle, shoulder-in and counter-canter. A horse that is naturally bent to the left, for example, requires work particularly on a right-hand circle, when he will bend round the acting right leg, on the right shoulder-in, the true canter on a right-hand circle and the counter-canter track to the left. The spine is straightened by exercises such as these both on the circle and on elements of curves, and is suppled longitudinally by others.

Before going any further let us define the much misused term 'collection'. Collection is a combination of *ramener* and *mise en main* accompanied by lowered haunches and engaged hind legs. *Ramener* signifies a narrowing of the angle between head and neck, the poll remaining at the apex; the *mise en main* is a relaxation at the jaw accompanying the *ramener*. In this way the horse is collected 'above' as well as 'below', along a topline compressed by tightening the natural undulations of the spine.

Ramener and elevation of the neck are ob-

Adjutant le Rolland of the Cadre Noir school at Saumur, a superb horseman.

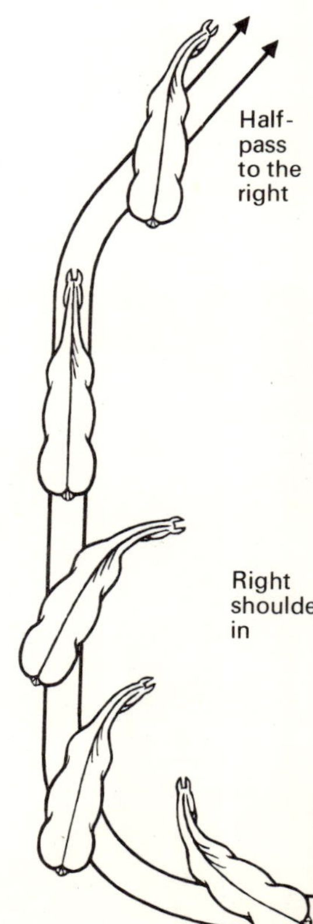

tained progressively and simultaneously by the advance of the entire body towards the head. With a low neck the *ramener* is easily–all too easily–attained, for there is a tendency to overbend; while raising the neck excessively results in a hollowing of the back and misplacement of the hocks. It is essential to work towards *ramener* without haste; do not try to raise the head more than is necessary in order to avoid overbending. Once the quarters are supple and the hind legs increasingly engaged, the haunches will place the forehand in a higher position. This 'relative' elevation is important.

Collection itself is characterised by the flexibility of both the front and the back of the horse, making it work on 'short bases' to give a flexible equilibrium with great multilateral mobility. This improves both the horse's paces and his changes of direction and speed.

As soon as exercises have made the horse generally reasonably flexible the *ramener* may be introduced. It may now be necessary to deal with a 'contracted mouth'. Any resistance–whether or not it originates in the mouth–will cause contraction there, and thus a relaxed mouth usually indicates the absence of resistance. In the same way that the mouth conveys the general state of the horse, a rider able to finger the mouth into relaxing well will himself help to do away with any incipient resistance more successfully than attempting to do so by using other aids. Of course this is only possible where the resistance neither borders on revolt nor comes from any permanent rigidity of the horse.

The half-turn on the forehand may be helpful at this stage in a horse's training, as are local encouragements by hand, known as flexions. While the mouth prepared by them will doubtless yield more easily, abuse of this method nullifies the effect by limiting it to the mouth itself, without the rest of the body being affected by the benefits of the flexing actions.

The relaxation is shown by a tongue movement similar to that produced by swallowing, and makes the bit ride up and fall back into place with a soft but unmistakeable noise. A slightly opening mouth is enough for this gesture which, quite slow, ought to be devoid of nervousness. The well-trained horse's mouth is mobilised in this way at the rider's very first prompting and ceases as soon as the prompting hand ceases to act.

Before trying the direct flexions leading to the *ramener*, one practises the lateral flexions, which with most horses are indispensable. Their purpose is a gradual suppling of the sides of the head at the point where it joins the neck. By increased tension of the direct rein and yielding of the opposite one, the horse's head is drawn sideways, the trainer being careful to ask for only a little at a time and to keep the head straight on its vertical plane so as to avoid a twist at the poll.

The work on two tracks supples and develops muscles which play a minor part in travel on only one track, increasing the engagement of the hind legs, loosening the shoulders and improving the horse's balance through swift changes between straight and oblique travel and movement from right to left at the half-pass through counter-changes of hand on two tracks. If this sentence covers many lines, the exercises cover a lot of ground in training!

This is particularly true of shoulder-in, which may be performed in various ways. We shall deal here only with that used in training, not for test or show, because it is the most easily stopped and restarted. The horse, uniformly bent in its entire length, moves forward in the direction of its convex side. The size of the circle determines the degree of the bend, and must be dictated by the stage of training the horse has reached. Detaching the horse from this circle by use of the inside rein, which takes on an intermediate rein effect, a few steps of shoulder-in are performed. To stop the movement, all that is required is to ride the horse forward on the circle, where it travels again on one track but still holds the bend for an instant return to the shoulder-in.

In the half-pass the horse is kept straight, except for a slight *placer* (lateral flexion of the head) causing it to look where it is going. The main difficulty here is to maintain the impulsion and to obtain the engagement of the inside hind–that is, the near hind in half-passing from right to left. If the horse has been trained to engage its hind leg at the prompting of the rider's leg on the same side, this engagement of the inside hind can be controlled. Almost the entire value of the half-pass lies in this, for without proper engagement of the inside hind the outside one will cross so little as to rob the exercise of its purpose, which is to improve engagement by bending hip, stifle and hock joints.

The half-pass may be performed either on straight lines, with the head to the wall, the tail to the wall, or on the diagonal. It is particularly interesting when practised on the circle, where the horse may be placed in one of two positions: the forelegs travelling on a circle either inside that travelled by the hind legs, or outside it. In the first case the exercise is called haunches out, and in the second, haunches in. Using both these methods, the trainer increases at will the crossing effort of the hind legs over that of the forelegs–haunches out–or of the forelegs over the hind legs–haunches in.

Work at the canter is at first performed at the horse's natural speed, where it can hold a satisfactory balance. The pace is slowed only gradually after the gait itself has been straightened in work on the circle, chiefly at the counter-canter.

When tackling flying changes of leg one must beware of two common faults: an exaggerated slowing down of the pace, and too imperative an intervention of the aids. The flying change can be simply taught: counter-canter in a circle, keeping the horse at a steady pace but tightening the circle to the limit of the horse's ability, then neatly but smoothly reverse the aids, thus demanding a strike-off on the other lead. Repeat the same exercise with increasingly clear demands for a definite change of direction towards the centre of the circle. After a few days of tuition the changes will come easily. Do not, however, ask the horse for changes of lead at every fourth, third or second stride until he has completely mastered single changes.

Single changes are first asked of the horse from the outside to the inside lead on the circle, then on straight lines, and eventually on the circle again but now from the inside to the outside lead. At this stage of training one also goes back to working on transitions, for if the horse is able to strike off correctly and calmly from a walk, canter three strides and return to a walk, he is ready for flying changes at every fourth stride. When the inserted strides at the canter come down smoothly to two, then to one, the horse is ready for flying changes at every third, then every second stride. For changes at every stride, return to the circle and canter on the inside lead: circle once or twice and then ask for a change of leg followed immediately by a second change. Achieve just one change at every stride (track to the left, track to the right), and the horse may be considered as good as trained to them.

The pirouette at the canter is a combination of two movements–that of the pace proper and that of the pirouette itself–and should be worked out in two parts. The prerequisites are a correct and ready performance of a pirouette at the walk and the ability to obtain a single calm stride of canter from walk. In pirouette, the horse's hind legs mark time in the required pace while the forelegs circle round them.

After one full pirouette at the walk, request a second in the same place but this time inserting a single stride of canter during it. Very gradually three or four strides at canter are inserted. This achieved, a half-pirouette at the canter is requested in a corner, at first using the walls of the school as a frame and to encourage the horse to position itself to execute on the spot the pirouette, a movement it already knows, remaining throughout at the slowed canter. The rider should intervene only to maintain the pace and indicate the direction of the half-pirouette. Improvement comes with half-pirouettes being achieved ever further away from the wall, to begin with entering the movement from a half-pass. Only when the half-pirouette is performed easily, away from the walls, from a straight line on a single track, unimpaired in pace or lightness and without any rush, is the horse ready for the complete pirouette.

Passage and *piaffe* are both derived from the trot. Like it, they consist of alternate contact of the diagonals with the ground. Only the times of suspension (both of the total and of each diagonal) are longer. *Passage* is a slow, shortened, very collected, elevated and cadenced trot, while *piaffe* is a collected trot on the spot.

Passage may be induced by slowing the trot and, by burdening the resting diagonal, lengthening the time of the other's suspension, the hands doing the burdening by a neck-reining action, the legs reinforcing the trotting action of the horse. Similarly, the *piaffe* may be reached by slowing down the *passage*.

It is in fact preferable to teach these two movements together. Working into *passage* by slowing the trot, and then into *piaffe* by taking off from the trot, facilitates the transitions between the two airs, which are more difficult than either of the airs themselves. There are several ways of doing this, but the usual method of training is to work the horse along the wall, with the trainer on foot using a longish whip. Without keeping the horse from advancing, the length of stride is gradually reduced by the hand holding the reins, while the other maintains the limbs in activity by light touches of the whip to the top of the croup, the chest or the legs. When, still advancing a little, the horse begins to give a few regular strides of *piaffe*, the lessons are continued under saddle. Initially a a second rider may be used, the aids coming first from the trainer on foot only, then from rider and trainer together, and eventually with the rider alone determining the movement. Only two or three strides should be taken strictly on the spot, and the horse should always be ridden forwards on completion of the exercise.

Some of the movements discussed here are difficult ones which only a specialist dressage partnership need know. Others, however, are quite simple to achieve if the necessary groundwork in the horse's training has been thoroughly understood and carefully carried out. It is worth remembering that successful advanced dressage is as much a matter of executing simple movements superlatively well–with maximum lightness, straightness, balance and suppleness–as it is of performing the airs of *haute école*.

JUMPING AND EVENTING

It is no paradox to say that the training of a jumper or eventer takes place primarily on the flat, not over fences. The best of training cannot give a horse jumping aptitude where nature has denied it, but by exercising its body, teaching it skills and making it manageable, training can develop whatever capacities the horse does possess. Thus elementary training on the flat is the same for hunters, jumpers, event horses or those specialising in dressage, though the emphasis placed on the various exercises will differ according to the career the horse will eventually be asked to follow.

To begin with, you cannot do much with any horse that does not obey readily; and since submissiveness is a matter of physical and mental interaction you must first gain, then keep, the horse's trust while developing it to make it able, not just willing, to obey. The kindest horse becomes nappy if overtaxed, and the most able jumper stops and runs out if allowed to get the upper hand. Body and mind are thus interrelated, and you must know your horse in order to know how best to use it.

Although there cannot be an exact system of training uniformly applicable to all future jumpers or eventers, there are some general rules for any successful training programme. Never ask for an effort or movement unless the horse has been prepared for it; never extend any particular lesson beyond an hour–preferably divide the lesson into two or three short sessions instead; never end a lesson with a tired horse; never pursue two aims at once.

Any horse worthy of being considered trained must reach the level of the FEI horse trials dressage test, which includes only those movements fundamental to the general training process. The jumper and eventer require less collection than the dressage horse, though they must be able to engage their hind legs properly, and have a slightly convex crest and a nearly vertical head carriage to allow the bit maximum effect and the hand action to be conveyed along a properly rounded topline all the way to the quarters. Perfect straightness is not required of them, provided they do not deviate on a straight line; but instant, unhesitating forward movement at the slightest prompting of the legs is essential.

Transitions in pace must be effected effortlessly, without the horse fighting or evading the hand. Since this obedience is just as important for changes of direction as for transitions, mounted exercises over fences should not be attempted until smooth control of speed and direction has become second nature.

Although one trains concurrently on the flat and over fences, we will deal here with each part separately. A word first about conditioning, an adjunct to training which in preparing a jumper, not to mention an eventer, plays a very important part. It is a complex process, consisting of outdoor work to develop muscle and wind, and a diet in keeping with the work expected and with the horse's particular temperament as well as supervised physical care, from daily grooming to the sophisticated science of 'interval training'.

Whenever we speak of lessons, we mean a session either acquainting the horse with new movements or, with those already familiar to it, improving their performance. Since any lesson exacts a sustained mental and physical strain on the horse, they should be kept short, even if it means repeating them fairly often. What we call work is performed out of doors, without the necessity of clock-watching. Its chief object is to develop muscle, wind and balance, though during the periods of work we can also usefully check up on the progress made during lessons. Initially work consists of long stretches at the walk, and rather short but regular ones at the trot increased progressively to eight miles or so. Eventually the horse should be able to maintain a good, controlled gallop for between three and four miles. Good muscles and sound limbs are second in importance only to perfectly functioning lungs, heart and digestion. The walk develops all the muscles without tiring them; the trot adds sturdiness to both muscles and joints; and the gallop increases lung power.

Let us deal with lessons on the flat first. Once the horse's confidence has been safely gained, forward movement becomes our chief occupation. In taking over a new horse, green or not, the wise horseman returns to basics and works back up through the successive phases. The first rule should always be not to act with legs and hands at once. When the legs act they must do so neatly, and both hands advance to where contact with the mouth is lost. Conversely, the legs must become inert when the reins tighten for a slowing down of pace or for halt. As progress is made, legs and hands will remain in permanent contact, but both will always still yield slightly before the other pair acts. This includes hand yielding to hand and leg to leg, as when a single hand changes direction or a single leg shifts the haunches, the other having previously given way. It makes the lesson so much easier for both rider and horse.

Even a spurred heel may be incapable of obtaining correct forward movement from a horse that leans on the leg, back or rears. These are great evils, requiring strong remedies that may be beyond the competence of the average rider. For him we recommend the use of 'rigid reins', which is not only an uncomplicated method but also cannot blemish the horse. Two wooden sticks, ideally polo sticks, are strapped on to the upper eye of a curb bit fitted with its chain on a tighter link than usual, the sticks being cut to a length permitting the rider comfortably to hold their opposite ends when seated normally in the saddle. From the halt his legs alone request the forward movement, and if there is a defence the hands move ahead vigorously, driving the horse's head and neck forward. This advance of the weight will carry the horse along with it. Rearing under these conditions becomes impossible, and after a few days of treatment the legs alone are almost always enough to persuade the horse to move forward. The only skill required is to combine the action of the rigid reins with that of the legs. Feeling powerless to fight this combination, the horse submits morally, which is what matters; it is never enough to parry a defence against his disobedience–it must be destroyed.

Once forward movement is safely assured, we look for lateral mobility of the haunches. Half-turns on the forehand and the single leg having acquired meaning, we proceed to half-turns on the haunches, and when the legs are sufficiently in control both to prompt forward movement and to shift the haunches, the hands in their turn may begin to exercise control over the effects of the legs.

A puller will act in one of two ways. He either lets himself be carried by the hand, a matter of balance, or he will fight, trying to evade the hand by stretching out his neck and increasing his speed. In both cases the horse must be refused the contact he requires. Intermittent hand action–half-halts in the first case, closing and opening the fingers firmly but gently with

The Irish bank at Hickstead

H.R.H. the Princess Anne on the steeplechase course at Badminton.

Richard Meade and Laurieston at Munich.

The individual gold medallist in the dressage at Munich, Liselotte Linsenhoff.

An inviting indoor jumping course.

steady hands for the second–will do the trick.

We also teach the horse to yield to the opening rein, to neck rein and respond to the direct rein of opposition, and both during lessons and, most particularly, out of doors, exercise it in neck stretching. First at the walk, at the halt if need be, but eventually at a rising trot the neck is stretched when prompted by the hands sliding gently along the reins from the horse's withers to the rider's chest. This exercise is particularly useful because it helps the horse to find and improve his new balance under the rider: the hands not only permit but encourage the forward movement prompted by the legs; the confidence the horse gains in a smoothly-acting hand causes it to seek contact with the hand. It supples and muscles the back as well as the neck–particularly at the base–and in the right direction; rounding the entire top-line in a convex line it promotes the engagement of the hind legs; and the overall streamlining produced leads to paces covering more ground, without which one cannot even begin to think of collection.

There are three further useful exercises the horse should learn. Shoulder-in leads to the desired degree of collection, supples the horse in all its length, sets it increasingly on the haunches and also gives it assurance in lateral travel. Bent on a circle, the horse describes it once or twice under good impulsion, then leaves it to execute a few steps to the side. The inside rein (left on a left circle for the left shoulder-in) acts as a counter-rein of opposition behind the withers, the legs maintain the forward movement while the horse always remains bent. Just a few steps should be executed, then return the horse to the circle, do another few side steps and back to the circle, and so on.

Rein back is an excellent way to improve engagement of the hind legs, though it must be put off until the rider's legs are sure to obtain instant free forward movement. If not, since its entire value depends on the transitions from walk to rein back to walk, the exercise might do more harm than good.

Counter-canter is best first taught out of doors, where no walls impede us from accustoming the horse to cantering without a change of lead on widely-looped serpentines and without any constraint by the aids. As ease in doing this increases the loops can be tightened, and the only real difficulty is for the rider not to disturb the balance of his horse.

These exercises, carefully and correctly taught, will enormously improve the balance, suppleness and obedience of the horse.

CAVALLETTI

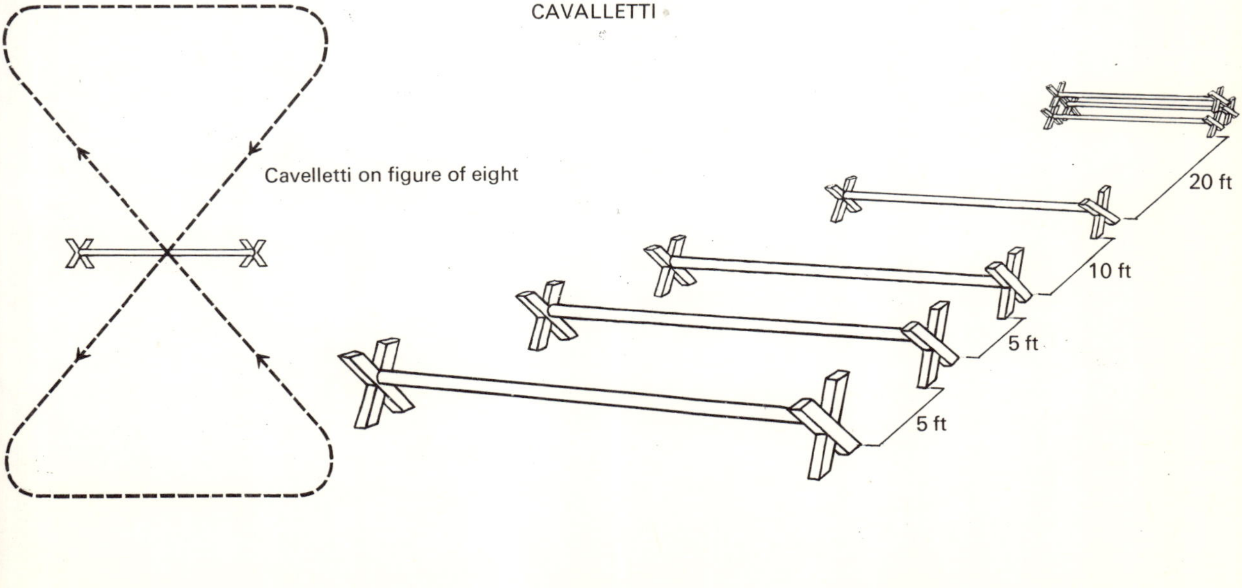

Now one may start work on the lunge line over cavalletti. The cavalletti (round, not square-cut poles) are set to about eight inches for work at the trot and separated, at least at first, by one of the horse's normal strides. This distance is gradually increased to lengthen the stride. Before approaching the cavalletti the horse should be sufficiently warmed up for the trainer to be able to maintain an active trot. From a single pole and then just two, he will graduate to five or six. While only two poles are in use more than one stride should separate them, so the horse does not mistake them for a single 'fence'. By doubling the distance between them we oblige the horse to take another stride.

Work over cavalletti under saddle should be done at a rising trot to relieve the horse's back. It is an exercise that improves balance, works the back muscles in the proper way, stretches the neck and develops the horse's shoulder movement.

The worst faults in a horse–whether these arise out of ignorance or vice–are cured on the flat; but it is over fences that the worst fault of the rider is generally apparent: his tendency to make his horse jump too often and too high. Wherever possible, the horse should not be expected to perform under saddle anything it has not already done at liberty before, and most exercises should be performed at the trot. Such are the general rules. I personally do not recommend the use of a ground-line or guard-rail before a fence except perhaps with an exceedingly awkward horse or one at the very outset of its training, for it will teach the horse to judge its fences by the foot rather than by the top, a bad habit later in competitive jumping. The fences used for training should be wide along their length and have thick, brightly-coloured poles.

We want the horse to use its neck correctly, to raise its shoulders and to stretch properly over a spread. Depending on its natural form over uprights and spreads respectively, emphasise work over one sort of fence or the other. All poles on uprights should be in the same plane and in sufficient numbers to keep the fence from looking hollow. Low, wide oxers make the best novice spread fences; once these are raised, watch out again for hollowness, and make them more inviting by building them in an ascending line. If the horse fails to wrap itself properly round uprights, work more at the trot and practise neck stretching over the fences. A horse not stretching properly over spreads should be encouraged to do so by having a pole positioned on the ground at a suitable distance on the landing side.

To help the horse get going, one might let three or four cavalletti precede the fence. They should be set up at maximum height (about one and a half feet), and the distance between them adjusted accordingly. They will also encourage the horse to take off at the right distance from the fence itself.

The next stage is to combine the two kinds of fences, first by setting up an upright followed by a spread. First these should be separated by three normal strides, this distance gradually being reduced down to one stride and then to just one very short one, to teach the horse to make maximum use of its abilities without relying on speed. Once this exercise is being successfully performed reverse the positions of the fences and repeat it.

Once the horse is going happily, quietly and freely at liberty, lessons under saddle can begin. The first fences, at a height far below the horse's scope, should be approached slowly. Return to walk between each jump, trotting or cantering only for the last few strides before each obstacle, for by doing so you will oblige the horse fully to use itself, and will have increased control over him at the same time.

When none of this work poses problems work on training courses can begin. The speeds required in competition can be practised, for the horse must be able to approach a fence with long or short strides to order, depending on the nature of the fence itself and its relationship with the rest of the course. So variations of speed at

Captain Raimondo d'Inzeo and the Irish-bred Bellevue. The wall stands at 7 ft. 2 in.

the canter are now requested, both on the flat and over fences. Begin by counting the normal strides taken between two points, and then either rein in and push with the legs after each stride, to shorten it, or with legs pushing and hands rhythmically accompanying the movements of the neck, lengthen it.

The horse should be taught to jump both upright and spread fences at an angle. Construct jumps that can be approached from either side, and ride the horse through a figure of eight with the fence at its centre. The loops should at first be narrow enough for the horse to reach the fence almost straight; as the loops widen, the angle of approach widens too. This exercise works wonders with horses that have a tendency to run out always to the same side. If you have a horse that runs out to the left, approach the fence from the left so that it is almost impossible for the horse to turn away to that side. To make doubly sure, stay on the near lead for your right leg to keep the haunches left and oppose a change in that direction.

If your horse is skilled in turning in the air it will land over a fence facing a new direction, which will save precious seconds in a speed competition. This turn is made during suspension. If the rider wishes to turn left on landing, he should turn the horse's head squarely to the left during the jump, tightening the left rein while the right hand yields and the left leg acts well behind the girth.

Introduce your horse gradually to different kinds of obstacles. The first banks, open water, lake and drop jumps he is asked to negotiate should cause no trouble if they are small and easy and an experienced horse gives the novice a lead. If hunting can be looked upon as a means rather than an end in itself, a little hunting can be useful at this stage.

Wherever jumping lessons take place–in an enclosed space or across country–they should at first be kept to a very moderate speed. For every obstacle there is a preferred speed–preferred because it allows the horse to jump with the minimum of effort. The question of speed is very important in both show-jumping and eventing, and merits careful attention. In competition the correct speed for any particular fence cannot always be maintained as there are generally other speed achievements to be considered, but in training the horse must be brought up to confident, easy jumping by being taught at ideal speeds. Only when these have been so successful that the horse is really happy and relaxed should the speeds be varied.

To summarise, therefore, the basic training of a horse on the flat develops his physical capacities and the self-confidence he needs both in himself and in his rider, for it establishes a mutually comprehensible language which enables the rider to convey his meaning to the horse clearly and precisely for instant obedience. Training over fences improves the horse's form, teaches it to make better use of its natural abilities and gives it practice in coping with increasingly complex obstacles. Stable care and feeding must allow the horse to sustain his efforts without coming to any harm and, very importantly, the horse's breeding, aptitudes and conformation must give it enough natural qualities to enable it, with training, to perform satisfactorily.

PART EIGHT

Equestrian sports

Chapter one

Racing

FLAT RACING

'It is difference of opinion that makes horse races.' Just when the first difference of opinion produced the result predicted by this remark of Mark Twain's will never be known, though there was horse racing in ancient Egypt, the Greeks certainly had a word for it and no self-respecting Roman circus would have been without ridden horse races, in addition to the more spectacular chariot variety. The first historical record that exists of mounted racing is as long ago as the year 624 BC, at the thirty-third Olympiad.

The English Thoroughbred horse surpasses all others as a quality racehorse, and to a certain extent the suitability or otherwise of a climate for breeding Thoroughbred horses has determined the popularity of the sport in any country. The breeding of racehorses was for a long time dominated by English studs and English Thoroughbred strains, and an enormous export trade all over the world developed. Nowadays, however, the strength of the competition from horses bred elsewhere, particularly on the Continent, and in the United States, has resulted in the Thoroughbred's more truly international ancestry. French and Italian stallions, in particular, have been so influential that it is now unusual to find a winner of any of the English Classic races without some European bloodlines in its breeding background. Today the spotlight is on American breeders who, through a policy of purchasing the best Thoroughbreds wherever they are available, are currently riding the crest of a wave. Four of the five English Derbys run between 1968 and 1972 were won by American-bred horses. In spite of this reversal of trends, it is still true to say that England and Ireland are the greatest exporters of high-quality bloodstock.

The first record of a race in England dates from 1174, in the reign of Henry II, and took place at Smithfield. Until the time of Charles II, horses used for racing in Britain were home-bred horses known as the racing 'galloway', sometimes the 'hobby horse' or 'running horse'. It was not until the seventeenth century gave way to the eighteenth that the importation of the three famous Arabian horses from the east took place; three horses which were to have a profound effect on racing all over the world, for they led to the establishment of the Thoroughbred.

While the Thoroughbred was evolving, so too was the organisation and control of the sport of racing. The British royal family has always taken a keen interest in racing; Charles II was himself principal arbiter when any disputes arose, and the aristocratic status of this 'sport of kings' has always been maintained. Queen Anne was responsible for Ascot racecourse, which was laid out in 1711 and has remained the most fashionable of all the British courses. The days are now past, however, of hard-riding amateur jockeys, many of them peers, who often rode against each other on horses of their own breeding.

In the early days of racing the most usual distance for a race was four miles, and races were frequently run in heats. It was unremarkable for contestants to carry as much as twelve stone in weight. Under these conditions close, exciting finishes must have been the exception rather than the rule, and it may well have been lack of spectacle as much as any humanitarian feelings about exhausted horses that led to a gradual reduction in distance and the abandonment of heats, though this method of racing did persist well into the nineteenth century.

The English Derby, still regarded as the world's greatest test for a three-year-old over a mile and a half, was in fact first run at Epsom in 1780 over only a mile. It is the three-year-old races that have become most popular on an international scale. In Britain, apart from the Derby, there are four other Classic races: the

St Leger (first run in 1778), the Oaks (1779), the 2,000 Guineas (1809) and the 1,000 Guineas (1814). In France, which now holds a position of considerable eminence in the racing world, there are the Poules d'Essai and the Prix Royal Oak, equivalents of the Guineas and the St Leger respectively; the Prix Diane (equivalent of the English Oaks) and the Prix du Jockey Club (like the English Derby), both run at Chantilly; and an additional Classic, the Grand Prix du Paris, which is run at Longchamp in June.

The British model for three-year-old racing has been closely followed in Europe, unlike the two-year-old tests, of which there are fewer on the Continent than in Britain. In the United States, on the other hand, the emphasis is placed on precocity, with some two-year-old races being held in January. Racing was first introduced to the United States in 1664 by the commander of the English forces invading what is now New York. Commander Nicolls laid out a two-mile course very near to the present-day Belmont Park course, still the best known internationally of the North American courses.

The pattern of meetings is similar on the Continent and in Britain, meetings being held for perhaps four consecutive days. In France, most major meetings are held at weekends, with big races being run almost exclusively on Sundays. In the United States, on the other hand, they may continue for anything up to a hundred days, the average being between fifty and sixty. American racing is usually conducted on dirt-track courses, unlike the turf to be found all over Europe, though with the increase in international competition turf courses are now sometimes to be found running inside the dirt tracks. The most famous Classic races in the United States are the Coaching Club Oaks, the Belmont Stakes and Withers Stakes, all held at Belmont Park, the Pimlico (Baltimore) Preakness Stakes and the Kentucky Derby.

Racing is also a popular sport elsewhere in the world. Australia and New Zealand have a particularly high reputation, and breed horses of as great quality as those produced by France, the United States and Britain. The Melbourne Hunt Cup, run at Flemington Course over two miles, is among the world's most famous handicap races. The Randwick course at Sydney is a well-known one, as are the Auckland and Christchurch courses in New Zealand.

The British introduced racing to other parts of the Commonwealth, too; it is popular in Canada, where the oldest fixed event, the Toronto Queen's Plate, was first endowed by William IV in 1836 and continues to be sponsored by the reigning monarch, and in spite of breeding difficulties caused by the climate, racing thrives in India. Most of the countries in Latin America both breed and race Thoroughbreds. The courses at Buenos Aires, Santiago and São Paulo are particularly well known, and the Grand Premios Brasil and São Paulo is one of the top international races.

Racing in Britain is controlled by the Jockey Club, founded in 1751. It sanctions racecourses, approves programmes, licenses officials, jockeys

Mill Reef, 1971 winner of the English Derby, lying in fourth place at Tattenham Corner.

and trainers and enforces its own rules. Initially it was set up to organise racing at Newmarket, still the headquarters of British racing, though its influence quickly spread to other centres. Similar institutions in other countries have used it as a model.

In the United States each state has its own organisations to regulate racing within that state, though the Thoroughbred Racing Association of the United States is perhaps the most important of the bodies responsible. It was in the United States that mechanised starting stalls, now a familiar sight in the racing world generally, were first used.

All over Europe racing is controlled to varying degrees by state authorities, and in some countries breeding too comes under government supervision. In communist states both horses and racecourses are owned by the state, and in spite of the curious anomaly of competition that this situation creates, racing is a popular sport in many of the Iron Curtain countries.

Air travel has solved problems of transport to such an extent that the trend is now towards international breeding of racehorses, though because of the difficulties of differences in seasons, climates, tracks, training methods and so on truly international racing seems unlikely to develop. The Washington International, held in the United States in November, for example, can apparently never be considered more than an interesting exercise. The newly-sponsored Eclipse Stakes, the King George VI and Queen Elizabeth Stakes (now, thanks to sponsorship, more valuable than the Derby) and the Champion Stakes, revitalised with Levy Board money, provide in England the principal opportunities for a meeting between Classic horses. For European owners and breeders the world's richest race, the two-mile Prix de l'Arc de Triomphe, which takes place at Longchamp in October, overshadows

An unfortunate start to a race at Atlantic City, Wyoming.

A keen, calm and beautifully balanced canter down to the start.

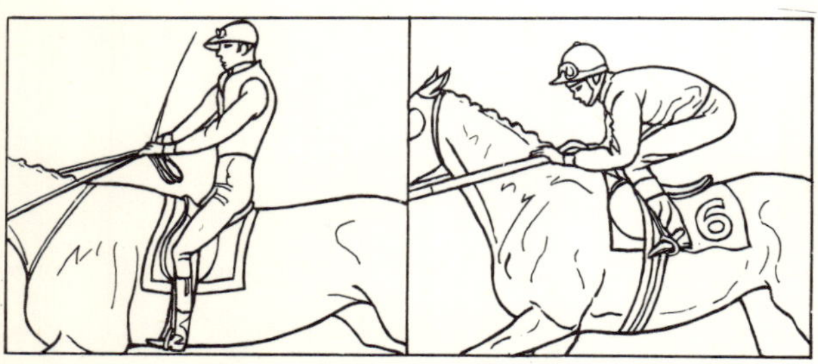

all other tests for three-year-olds and older horses.

It is impossible to discuss flat racing without making some mention of the financial aspect. What was once a sport conducted for and by participants, has now grown into a multi-million pound industry, in which the interests of bloodstock and racing for its own sake have become merged with those of spectator-entertainment and gambling. Modern racing derives its principal support from gambling, so it is hardly surprising that it should now be geared very largely to the interests of the betting public, and more particularly, perhaps, to those who rarely visit a racecourse. Enormous sums of money are offered by the big Classic races, and the total turnover of the betting industry is now huge.

Brigadier Gerard has almost become a legend in his own time.

STEEPLECHASING

If it is true that the British have a tendency to over-sentimentalise about animals–particularly horses–the trait was never more evident, nor more played on by Press, radio and television, than in the days following the Kempton Park Christmas meeting in 1966. All the crowds had gone, the stands stood silent, no bookie shouted nor tic-tac waved, but in the racecourse stables there stayed one horse who had been the focal point for every eye at the Sunbury track on the day after Boxing Day. Then he had faltered and failed for the first time in two years of top-level sport, for only the third time in his last thirty races. Now, with 5 lb. of plaster encasing the injured leg that had brought his downfall, Arkle remained the centre of attraction not only of the racing world but, it seemed, for the entire population of Great Britain and Ireland.

He was the latest–and greatest–in a long line of famous 'chasers in a sport whose history in Great Britain is charted through a course of revered names . . . Prince Regent, Golden Miller, Manifesto, back as far as the aptly-named Lottery, winner of the first of the Grand National, a race which today still provides 'chasers with their most gruelling test–and their richest reward.

Many countries in Europe–and some in other parts of the world–stage races over a variety of obstacles, but these are often fairly innocuous constructions, bearing little relation to those seen today on English and Irish courses. There are exceptions, notable among them the famous Gran Pardubice Steeplechase run in Czechoslovakia each October. The course for this four-mile event is really more like that for a severe cross-country test, incorporating numerous natural jumps, brooks, hedges and ditches. One comprises a hedge six-foot deep and six-foot wide, with a ditch eighteen-foot wide and eight-foot deep on the landing side. Not surprisingly, there have been occasions when every starter in this epic contest has come to grief.

But despite a few races of this sort, nowhere in the world has steeplechasing risen above the level of a minor sport except in Great Britain and Ireland (its spiritual home) and, to a lesser extent, in France and the United States.

Paradoxically, though steeplechasing in France is of lesser stature than its English cousin, the prizes to be won by French owners–at least on the principal jumping courses, such as Auteuil near Paris–are of considerably greater value, a reflection of the higher prize money available for all forms of French racing, thanks to the vast rake-off the sport receives from the official Totalisator, which provides the only legal form of horse race betting in France.

Often called the 'winter game' in Britain, where it reaches a climax in March, the steeplechasing calendar bears a different look across the Channel. The Grand Steeplechase de Paris, French equivalent of the Liverpool Grand National, is run in flaming June, as is the big hurdle race, the Grand Course de Haies d'Auteuil. And Auteuil itself takes a December to February break from racing each year.

But in Britain, though the crowded season now extends from late July right through to the following June, the important segment has always been linked with the foxhunting season. Indeed, in their earliest form steeplechases were private hunting matches, a sort of take-your-own-line test across country, with no cup or prize, only a large wager staked between participants.

One of the most famous was the Co. Cork clash between Messrs O'Callaghan and Blake, who settled a friendly difference of opinion as to the relative merits of their respective hunters by racing the four and a half miles from Buttevant Church to where the spire of St Leger church could be seen peeping above the trees. Hence the adoption of the name 'steeplechase', and after the first of them was over the winner was richer by a 'hogshead of claret, a pipe of port and a quarter cask of old Jamaica rum'.

This celebrated affair took place in 1752, and references to similar tests across country appear with increasing frequency in contemporary newspapers and journals as the nineteenth century approaches. In England, the first recorded cross-country event with more than two starters took place in 1792, when Charles Meynell, son

The three leaders coming over Becher's Brook, by far the most formidable obstacle on the gruelling Grand National course at Liverpool, for the second time.

of the famous sportsman Hugo Meynell, beat Lord Forester and Sir Gilbert Heathcote in an eight-mile Leicestershire gallop from Barkby Holt to Billesdon Coplow and back.

Around this period, too, it seems that hunting men of the day amused themselves with what are sometimes referred to as 'wild goose chases', in which one rider set a course across country pursued by others, whose object was to catch the leader but, by the rules of the game, perforce following the same line.

Something similar was practised by out of season or out of scent sportsmen in the United States during the late eighteenth and early nineteenth centuries where, particularly in the state of Maryland, the events became known as 'pounding races', apparently because the aim of the leader was to choose the stiffest obstacles so as to put down–or pound–his pursuers.

In the United States, jump racing has moved forward on lines different from those in Britain, splitting somewhere along the way to become virtually two sports: the hurdle races and steeplechases, held mainly for variety as extra events at some flat meetings; and races over timber at so-called hunt meetings. In the former the obstacles for both hurdle and steeplechases are soft brush fences, varying only in size, but the timber races are decided over much more formidable fixed fences of the post-and-rails type.

These hunt meetings started as amateur affairs, not much more than point-to-points (which are still staged, though not on the same scale as in Britain), though they have since become professional and are held under the sanction of the National Steeplechase and Hunt Association.

A well-bunched field takes a hurdle at Hurst Park.

The best-known race at a US hunt meeting is the Maryland Hunt Cup and, though trans-atlantic traffic in steeplechasers is rare, the American horse Jay Trump crossed the ocean to win the Liverpool Grand National in 1965 after having previously been successful in two Maryland Hunt Cups.

The hunting field of the eighteenth century was an obvious ancestor of steeplechasing in Britain. But before any infant can struggle into the world two parents are normally necessary, and it was flat racing, then fast becoming more popular and better organised, which provided the second half of the equation in this case.

At many flat race meetings of that day it was customary to include a race for genuine hunters . . . but run on the flat. However, it seems that, as with present-day point-to-pointing, race organisers experienced difficulty in ensuring that such events were contested by 'genuine' hunters–and not dominated by flat racers in search of easy pickings. Then, in 1811, the Clerk of the Course at Bedford devised a plan to achieve this desirable aim–a plan so beautiful in its simplicity that it was only astonishing nobody had thought of it before.

Reasoning that hunters were accustomed to galloping over hedge, ditch, rail and gate, whereas flat racers kept their feet as near to the ground as possible, he encouraged the one and deterred the other by decreeing that jumps should be erected for this particular race. So a quartet of fences–'4 ft. 6 in. high with a strong bar at the top'–was put up before the two starters (from eleven subscribers . . . were some frightened off?) went to the post. Each obstacle was leapt twice during the three-mile run and none caused concern to either contestant, both certified as having been 'in at the death of foxes in Leicestershire'.

So steeplechasing moved on a stage. Now 'made' courses began to replace the hell-for-leather gallops across any stretch of land that happened to be available. But it remained a spasmodic sport, and the first real sign of organisation did not appear until the arrival on the scene of one Tom Colman. A racehorse trainer since before the days of Waterloo, in 1830 Colman was proprietor of the Turf Hotel at St Albans, and the plans for the first St Albans Steeplechase are said to have been hammered out round one of that establishment's dining tables by a group of young officers from the local garrison.

Whoever was responsible for initiating the trial race, held on 8 March 1830, Colman himself was the driving spirit when the venture was repeated every year from 1831 to 1839. So he was the first to put the sport on a regular reliable footing.

The new race was by far the most important of its type throughout Britain, drawing the best horses and the best riders. Among the latter one Jem Mason, destined to achieve the highest rank in his profession, caught the eye when he won the 1834 renewal on The Poet. Mason, in his heyday one of the dandies of London–'the finest natural horseman in England'–was on his way up the ladder of success when, in the same race three years later, he began his association with a 16 hand bay named Lottery, the first of the great steeplechasers and winner of the 'Grand National' of 1839.

There had in fact been previous 'Grand Steeplechases' held at Liverpool, but the 1839 one is now universally regarded as the start of the long series. The conditions included a proviso that no rider was to 'open a gate or ride through a gateway or more than 100 yards down a road', which suggests that an element of take-your-own-line still existed in the early made-up circuits.

Each horse carried 12 st. (the race did not become a handicap until 1843) and there was the usual stipulation 'gentlemen riders', quite meaningless in modern eyes since there can be little doubt that the majority of jockeys were paid for their services. Fees were not laid down by rule as they are today and varied with the quality of the rider, but the five guineas received by one or two top men, such as Mason, compares favourably with the £13·50 a time paid to all 1972 riders.

Today nothing in racing comes outside the rule book. Steeplechase fences, excepting water jumps, must be at least 4 ft. 6 in. high, with twelve obstacles or more in a two-mile 'chase (the minimum distance) and another six for each succeeding mile; hurdles, stuck into the ground so as to slope away from runners, must measure 3 ft. 6 in. from bottom bar to top, with eight to the minimum two miles and at a proportionate rate for longer distances. In 1839, however, individual racecourses set their own standards, and the Grand National obstacles (twenty-nine of them in the four miles) were mostly innocuous 2 ft. banks with gorse on top. There were three that occasioned considerable respect–two brooks with sizeable fences in front and both taken out of heavy plough, and a loose stone wall reckoned to be near 5 ft. in height. The first of the brooks was not long in losing anonymity. It is now renowned throughout the racing world as Becher's Brook, named after Captain Becher, one of the few true amateurs, became the first to taste its waters after leading the way until then on his mount Conrad.

But Lottery and Jem Mason met no trouble at any of the jumps, winning with the greatest of ease after jumping the last hurdle (and it was literally a hurdle) 'as flippantly as though he were just starting', in a recorded time of 14 min. 53 sec. Nowadays the winner usually completes the 4 miles 856 yards in about 9½ minutes.

The new race was an unqualified success, as it continued to be in each succeeding year, but with the sport as a whole the picture in the 1840s and 1850s was not so rosy. The Jockey Club, by then the respected and all-powerful controllers of flat racing, refused to take cognisance of steeplechasing, and the lack of a governing body of its own led to the young sport being infiltrated by many whose activities had caused them to be barred from the 'legitimate' Turf.

Those amateur sportsmen who considered themselves its founders were rudely shouldered aside by an extremely dubious type of professional. Surtees describes the meetings of this period as 'generally crude, ill-organised things'. Every sort of chicanery and villainy was rife. Away from the immediate vicinity of the judge's box all power drained from such stewards as there were. Clearly, something had to be done before the sport disappeared beneath the mire.

Happily, there were still honourable men who considered it worth saving. One of them, Fothergill Rowlands, took a decisive initial step in 1859, when he set about organising a race at Market Harborough for which he sought the backing of various hunts so that it, and the meeting at which it was run, would benefit from the authority and reputation of the Masters and hunts concerned. Support was not immediately forthcoming, but in 1860 thirteen hunts subscribed, and the race became the first National Hunt Steeplechase. After fifty itinerant years travelling the courses of England, Wales and Scotland, the race settled at Cheltenham in 1911 and has been run there ever since.

The movement against the disreputable elements gathered strength, and articles and correspondence in the sporting Press of the day led to the formulation of regulations–sometimes referred to as the 'Market Harborough Rules'–

under which, it was claimed, 'fraud will be defeated'.

Shortly after this Messrs Weatherby–for long the secretaries of the Jockey Club–were acting as stakeholders for at least some National Hunt races, thus lending further respectability and authority, while from 1863 a Jockey Club sub-committee dealt with certain steeplechase matters. This small body did not have full powers but from it, in 1866, evolved the Grand National Hunt Steeplechase Committee, which was to retain its somewhat unwieldy title until 1889, when it became, more simply, the National Hunt Committee.

Whatever its title, the new organisation soon began to bring order out of chaos, one of its early and effective moves being to extend its recognition only to those meetings acknowledging its power, while debarring any who competed at meetings which it did not recognise–as the Jockey Club had done for flat racing.

So steeplechasing moved into safe, strong hands, hands that were to control its course for the next hundred years until, in 1968, the two controlling bodies–the Jockey Club and the National Hunt Committee–became one, so that all racing in Britain now comes under a single authority, as had been earlier recommended by the Benson Committee of Inquiry.

When the National Hunt Committee was born it controlled a sport still not far removed from the hunting field. The separation was gradual but definite. For example, by 1885 the whole Grand National course had been railed and, more important, it was all grass for the first time. Even before this, the Thoroughbred was gaining ascendancy over his less aristocratic brethren and, as the sport became more 'respectable', so the half-bred hunter faded from the scene. Increasing popularity and opportunity led to the development of breeding for jumping, though with the longer time-lag between conception and reward this does not–and never can–approach the scale of commercial flat race production.

This passage of years between foaling and fruition results in steeplechase breeding being an even less exact science than that for flat racing, since, though certain names and bloodlines appear again and again in pedigrees of leading performers over fences, their immediate sires and dams have frequently reached advanced years–or may even have died–before their progeny can advertise their potential. Because nearly all male steeplechasers are geldings, a sire can only be judged on what he has produced or on his flat racing achievements, and the latter is often not a reliable pointer.

Thanks to Levy Board funds and private sponsors, prize money opportunities for leading steeplechasers have reached levels undreamed of in the first seventy-five years of the NHC's existence. Until several years after the second world war the Grand National still offered the only worthwhile target. The Cheltenham Gold Cup, instituted in 1924, provided a true level weight championship test over $3\frac{1}{4}$ miles, but when the great Golden Miller scored his record five wins in a row from 1932 to 1936 he won only £670 each time for Miss Dorothy Paget. His Grand National victory of 1934 netted £7,265.

Now, however, steeplechasing has a Pattern of Racing system of its own and the whole calendar is peppered with valuable events. Sponsorship of the Gold Cup led to a prize of £15,255 for the 1972 winner, Glencaraig Lady. Three weeks later Well To Do won £25,765 in the part-sponsored Grand National. Nor have hurdlers been neglected. The Schweppes Gold Trophy, one of the season's hottest handicaps, was worth £9,698 in 1972, while the top weight-for-age test, the Champion Hurdle (started at Cheltenham three years after the Gold Cup) carried a prize of £15,648.

More money in prizes has meant a rising demand for horses and rocketing prices. Though still far from reaching the figures paid for their flat racing brothers (the lack of breeding potential has its natural effect), few eyebrows are raised at bids of £5,000 to £10,000 for horses with any sort of form, the current record of 16,000 guineas having been paid for the hurdler, Major Rose.

But one of the fascinations of the 'chasing game lies in the ever-present prospect of finding a future star, perhaps foaled in some rural Irish backwater, bred for the jumping job but without too much of the look of success about him to push his price up to impossible levels.

POINT-TO-POINTS IN BRITAIN

Why point-to-point? From February to May each year not far short of 200 hunts throughout Great Britain put on their annual race meetings, called point-to-points, at which all the events are confined to horses hunted during the preceding season. Usually one race is for members of the promoting hunt, a couple are for horses from neighbouring packs, one is for ladies only (normally open to any hunt) and there may be two open races for men, one for moderate horses and one for the cracks from far and wide. And all these races are also called point-to-points, though in fact this is just what they are not.

Hunt racing courses may vary from round to oblong, from oval to triangular, even (though rare nowadays) to a figure-of-eight, but all have one attribute in common: horses competing over them must end up where they started.

It was not always so. Originally a point-to-point race was just that: it was run from point A to point B, and first there the winner. It was then a more accurate description than 'steeplechasing', with which sport point-to-pointing shares a common ancestry. Both grew out of the eighteenth-century hunting field, from the private matches with which sportsmen of that day amused themselves out of season – or out of scent. They were headlong cross-country gallops to prove their hunter better than the other fellow's, sometimes just for the hell of it but more often with a hefty wager at stake.

This single sport became two, one branch developing into the professional sport of steeplechasing and hurdle racing we know today and the other retaining its associations with hunting and amateur riding. For much of the nineteenth century the dividing line was blurred. The Lonsdale Library volume on steeplechasing gives 'about 1885' for the first point-to-point. In his book *The Continuing Story of Point-to-Point Racing* Michael Williams quotes a considerably earlier date, the Worcestershire Hunt's first fixture on 2 March 1836, three years *before* the first Liverpool Grand National and in the heyday of the St Albans Steeplechase.

Probably that Worcestershire point-to-point was much like many a minor country meeting of the day, and perhaps 1866 might be better considered the year when the parallel paths began to diverge. This was the year of the formation of the Grand National Hunt Steeplechase Committee, a body born out of a desire among honest supporters of steeplechasing to bring a Jockey Club type of control to a sport with a poor reputation. In time it was redeemed, but the new authority's early efforts were wholly directed towards extending its discipline to those areas where there was none, to meetings in business to attract the public to 'made-up' racecourses and to bring profit to promoters often not too fastidious about where it came from.

Taking the last fence in excellent style, this horse went on to win his race at the Zetland hunt point-to-point.

The 'professional' half kept the new body fully occupied in its initial years and it may have been quite pleased to leave the hunts alone, happy in the knowledge that their little affairs were untainted by any of the scandals which had blighted steeplechasing, and knowing, too, that most Masters of hounds had sufficient honour in their own country to be able to maintain order and reputation.

In addition, point-to-points in the last quarter of the nineteenth century were very local, almost private meetings. Races were run for the diversion of competitors. Often the 'points' remained a mystery until the day of the race when, after being first mustered near the intended finish, runners would then be shepherded by road to the start, five or six miles away. After the starting signal, the 'course' followed would be up to the individual, the only stipulation being that no rider, on pain of disqualification, was permitted to open a gate or ride more than one hundred yards down a road. (A similar condition had been included in the rules for the first Grand National.)

The race – and there was probably only one – was restricted to members of the hunt, who rode the horses on which they regularly followed hounds and wore the clothes they wore to the meets. Racing colours were coming in, however, first being worn in open races and then gradually superseding sashes and armbands in other events, though at many fixtures right up to the 1930s hunting dress was still demanded for the members' race.

As the twentieth century advanced so point-to-point meetings began to look at least something like those enjoyed today, though courses were still almost exclusively across natural country, perhaps with one or two hedges strengthened, but none being specially built. Flags to mark jumping points, yes; wings, never!

Hunts still paid scant attention to the needs of spectators, though the social aspect was gaining sway as early as 1900, when many Masters and committees were taking advantage of the day to give a 'Thank you' lunch to the farmers over whose land they had hunted all winter. The farmers' pre-racing lunch, the big marquee with its rounds of beef and pickles, the barrels of beer – and the teas later for the wives and children – was an important feature for the next fifty years, and so, gradually, the point-to-point became more than just an end-of-season get-together for hunting folk, the shift from a semi-private frolic to a public entertainment gathering momentum with the advent of the motor-car age.

For as the city-dwellers, enjoying the new-found pleasures of a countryside now within easy reach, discovered the point-to-point, so the hunts discovered that their annual race-meeting provided an outstanding opportunity to raise what was becoming increasingly hard to find: money. Thus the paradox – the motor car, whose invention helped render horse transport obsolete, played its part in ensuring the survival of one of the oldest equestrian sports.

The North Cotswold point-to-point.

Since taking over control of the sport from the Masters of Foxhounds Association in 1934, the National Hunt Committee (now the Jockey Club) has never countenanced admission charges for people. For cars it is a different matter and, though the appearance of the first Baby Austin at meetings in the late twenties may not have been entirely welcomed by die-hard traditionalists, crammed car parks were soon to bring smiles to the faces of hunt treasurers everywhere.

The car itself has led to some redesigning of courses, 'viewability' now being the prime consideration. Natural country, on the way out from the first war, gave way to standardised fences of birch after the second – apart from the far West Country, where banks held on until the sixties. 'All fences in view from the car-parks,' claim the advertisements, and in many cases it is true.

For his money the visitor can go anywhere

A polo match at Zimmerman, Minnesota.

Pato, a popular game in Argentina.

American pacers in action on the Hollywood Park track.

Close on the heels of the leader, a trotter pulls out to overtake.

(apart from the parade ring and sometimes a small segment round it), and see everything. Rails are few and often the only ropes near the finish, where the crowd could overflow on to the course. But otherwise the racegoer can wander where he pleases: to a fence to experience the crashing, shouting thrill at close quarters, to a hillside vantage point, to bookmakers or Tote to back his fancy, to the car park for a pre-race picnic, or a between-race drink with friends. It is this intimate, near-to-the-heart-of-things atmosphere which has always been the charm of a point-to-point, and always will be, even in a changing age.

The principal requirement for competing horses, most of whom, as in steeplechasing, are nowadays Thoroughbreds, is that they must have been 'regularly and fairly' hunted – and possess certificates to prove it. To qualify for such a certificate a horse must usually go out with hounds at least ten times. Some do go out to hunt; others merely to qualify. The subject has long been a controversial one. 'He calls hunted fairly a horse that has barely been stripp'd for a trot within sight of the hounds', wrote Adam Lindsay Gordon many years ago, and so it probably always will be. Responsibility for certificates used to rest with the Master, but under a recent change of rule he merely agrees that a horse has been hunted with his pack; the owner confirms the 'regularly and fairly' part.

Ladies' races are becoming increasingly popular.

Horses must also be five years old or over. Many start at the minimum age and, if they make the grade, move on, perhaps via hunter 'chases (the races staged for qualified hunters at National Hunt meetings from February to June) to bigger things. There have been many examples. Limber Hill, Gold Cup winner of 1956; Halloween, twice successful in the King George VI 'Chase; that fine Scottish 'chaser Freddie; and, more recently, The Dikler – all these horses went to the top after starting in hunt racing; many others succeeded at lesser levels.

Then there are those who will always be point-to-pointers, familiar figures season after season, sometimes kept to the one sphere by their own limitations, often because their owners like it that way. And finally there are the old 'chasers and hurdlers, no longer fast enough to win round Plumpton or Nottingham, but able to continue to give useful service (and probably enjoy themselves into the bargain) in point-to-points. With their experience they provide ultra-safe conveyances for young men, and women, first learning the art of race-riding.

In fact, until 1972, when the Jockey Club at last relented, the point-to-point was the only outlet in Britain for lady jockeys. It is still the only place where one can see them riding over fences.

Ladies first started competing after the first world war – against men. Many did so well it has been suggested that it was in self-defence that the men persuaded the Masters of Foxhounds Association, then the sport's controllers, to segregate the sexes after 1929, though the ladies may have brought it on themselves by pressing for more 'ladies only' races. One of these soon became customary on every card. Under the existing rules they were mostly adjacent hunts' events, but from 1963 meetings were permitted to stage two open races – one for men, one for ladies – and from 1967 ladies won additional opportunities when a rule change allowed them to compete against men in members' races, provided they carry the full 12 st. 7 lb. in weight, though a basic 11 st.

is set for ladies' races.

At all meetings the five or six races held take place over a minimum of three miles, with the rules specifying not fewer than eighteen jumps for that distance (from at least eight separate fences). Standards are laid down (the lowest height for fences was recently raised to 4 ft. 3 in., only 3 in. lower than that on National Hunt courses) and inspectors appointed by the Jockey Club see they are maintained, though a frequently-heard complaint concerns the variation in stiffness of fences.

Courses also vary. Some are sharp, others galloping; some are flat, others undulate. Post-war financial problems have, however, brought a reduction in variety, since more and more hunts have adopted a shared-course policy so as to cut expenditure. Thus some venues – like Larkhill in Wiltshire or Tweseldown in Hampshire – are used as many as four, five or six times in a season, and two or three hunts to a course is now very common.

Garthorpe, the course near Melton Mowbray in Leicestershire, is usually used four times, the last being the 'Grand Finale' meeting of the Melton Hunt Club, always a great draw since, with every race open to club members' horses from anywhere in the country, it provides a fine end-of-season championship between the best from the various areas. Other point-to-point 'classic' open races which often produce such clashes are the Worcestershire's Lady Dudley's Cup, first run in 1897 and now at Chaddesley Corbett; the Heythrop's four-mile Lord Ashton of Hyde's Cup at Stow-on-the-Wold in Gloucestershire; and, longest of all at four and a half miles, the Ralph Grimthorpe Gold Cup at the Middleton's meeting in Yorkshire.

Classics these may be, but prestige only is gained, since the rules forbid any prize in this all-amateur sport worth more than £40 to the winner. For non-open races the maximum is £30, ludicrously low compared with the £20 top limit allowed when the National Hunt Committee first took control in 1934.

Many point-to-point owners regretted that the committee of inquiry into point-to-pointing, set up under Lord Leverhulme in 1970, did not recommend any increase in prize money – in fact came down firmly against it. However, the committee's recommendations for a new 'command structure' for the sport, since acted on, do to a marked extent meet the widespread view that the authorities were out of touch with the rank-and-file supporters.

The Jockey Club and MFHA now have equal representation on a new Point-to-Point Liaison Committee which administers the sport, acting on recommendations received from the next down the line – an MFHA sub-committee on which representatives from various interested bodies sit, including a newly-formed Point-to-Point Secretaries' Association. With many changes still to come, not least the establishment of a register of half-bred horses (due to operate from 1974) as well as the re-examination of the whole financial structure of the sport forecast in the Leverhulme Report, is is to be hoped that this Secretaries' Association will provide owners, riders and meeting organisers with a voice powerful enough to be heard in the Jockey Club corridors of power when any changes are discussed in the future.

The first fence in the Adjacent Hunts' race at the Heythrop point-to-point.

Chapter two

Show-jumping, eventing and dressage

SHOW-JUMPING

Show-jumping is a newcomer to the world of equestrian activity. It was first mentioned–in a French cavalry manual–towards the end of the eighteenth century, and not until the 1860s are there records of organised show-jumping competitions. In Ireland, which roughly a hundred years before had seen the first steeplechase, the Royal Dublin Society's show of 1865 included a high-and-wide leaping competition; the following year there was a jumping competition in Paris, but this was nearer to cross-country than to show-jumping, for after a preliminary parade indoors the competitors were sent out into the country to jump over mostly natural obstacles. It is probable that in the USSR, where there was strong French influence, there were also jumping competitions at this time.

On the Continent, jumping quickly became widely spread, particularly in Germany and France, where jumping competitions were included in the 1900 Olympic Games, held in Paris. In Turin in 1901 officers from the German army, and possibly also from the Swiss, were invited to take on their counterparts in the Italian army at jumping. The era of international show-jumping had begun, though it is unlikely that any of the pioneers of the sport envisaged its spectacular growth.

A show-jumping competition, at whatever level, is no better than the course over which it is jumped. One can have a field of the best horses in the world, but if they are given a course of too-big fences and impossible distances something like chaos can ensue; if they are given too easy a task, not asked enough to test their ability, the competition will be dull, resulting in a mass of clear rounds and then a 'steeplechase' against the clock to decide the winner. On the other hand, a course of good, solid fences arranged in an inviting way can produce an interesting competition even from inexperienced contestants.

Sophistication of course-building has come about to a considerable extent over the years. At one time the courses at almost all British agricultural shows were very much of a pattern, and that an extremely dull one: six or eight fences along one side of the ring, a similar number along the other, and a water jump or triple bar in the centre.

Slats which rested on the top of the fences were easily dislodged, with subsequent penalisation, so that horses had to jump well over each obstacle. Time was unimportant, and riders were perfectly entitled to circle a few times before a fence in other to get on the right stride. It was not unusual, after jumping the upright fences, for a rider then to undo his martingale before asking his horse to stretch out over the water. Each round could, and frequently did, take minutes to complete, and the wonder is that it ever caught on as a spectator sport at all.

Nor were slats confined to minor contests. They were in use at the 1912 Olympic Games, with penalties given for touches, knockdowns, or landing on or within the demarcation lines of spread fences; and foreleg mistakes cost twice as many faults as those of the hind legs. Even though it was the custom to have a judge at each fence there was still so much room for error that there must have been frequent dissatisfaction.

Gradually the snags were ironed out, especially after the formation of the FEI (Fédération Equestre Internationale) in 1921 which brought some uniformity of regulation, although slates were still in use in Britain after the second world war and 'touches' are still penalised in some national classes in the United States. Generally speaking, however, penalisation is now clear-cut, with four faults given for each fence knocked down, be it with forefeet, hind feet, nose or tail. Only the water still presents problems, for the adjudication still depends upon the human eye, though the use of a plastic tape which is encouraged by the FEI in international competition, indeed is mandatory at CSIO shows (official international jumping shows), at championships and the Olympic Games, at least ensures fairly reliable evidence in the event of a dispute.

The ease with which a show-jumping competition can be judged is, to a considerable extent, a reason for its popularity. Four faults for a fence down, three for the first refusal, six for the second and elimination for the third. Given a digital clock – happily on the increase – to see whether a horse has incurred time faults, any spectator is in a position to see for himself how one horse has done against another.

All that remains is for a course to be built that will produce a competition the spectator wants to sit through. Accusations of lack of imagination are frequently levelled – and with reason – against course-builders everywhere, yet to a large extent the wonder is that they manage to bring in as much variety as many of them do, for basically there are only four different types of fence: the upright, the parallel, the staircase and the pyramid. In spite of this limitation, an imaginative and knowledgeable course-builder can use his basic material, with the many variations of each type, to produce a track capable of getting the best out of a horse at any stage of its development. This of course is, or should be, the aim – to encourage a horse to jump to the best of his ability, not to trap him into making mistakes, which may not only make for an ugly round and unsatisfactory competition but could ruin a horse's confidence for the future.

Having established the class of horse he is building for – novice, Grade B, or top-grade international – the course-builder can then start his plan. Even for a novice the course should not be too easy, for one should assume that the horses have all had reasonable schooling at home. There used to be a tendency to build such easy courses that the most minor of competitions ended in a timed jump-off, but it does a green horse little good to have to scamper round against the clock in competition with others who have gone clear.

For top-class horses the builder uses not only the size of the fences, but also the distances between them, between the various elements of a combination fence, and the position in which they are sited both in relation to each other and to the perimeter of the ring. Although the length of a horse's stride will clearly vary, at a strong canter it will on average be about 10 ft. Fences more than 80 ft. apart are considered as unrelated, which is to say there is sufficient distance between them to make fundamental changes of pace and stride.

If fences are within 39 ft. 4 in. of each other they are considered to be part of a combination fence, which normally consists of two or three parts though it can be more. The distance between fences is what makes them easy or difficult to jump, although this problem has to be considered in conjunction with the shape of each fence. Easy distances between two uprights would be 26 ft. for one stride and 35 ft. for two; between two sets of parallels they would be 23 ft. and 34 ft., with variation for

A perfect take-off at the water jump at the Dublin Horse Show.

mixed combinations. So given a distance of, say, 29 ft. the rider would have to choose between two short strides or one long one, and his answer would depend upon his horse's ability, length of stride and courage.

How a fence is built will have considerable bearing on how well it is jumped. Thus a flimsy upright with three or four poles will be less inviting, and will be knocked down more often, than one which looks solid from top to bottom. A triple-bar will be cleared more often than an upright of the same height; a fence four feet wide is more difficult than one of twice the width. To some extent the difficulty of a fence is apparent rather than actual – how often has one seen horse after horse making heavy weather of a course and then, as soon as one horse jumps well clear others follow immediately in his wake. The maxim among many event course-builders is to 'frighten the rider but not the horse', and this applies in the show ring as well.

A horse measures the fence to be jumped from where it is in contact with the ground, so if there is no ground-line a fence is more difficult to assess. If there is a false ground-line it is that much more difficult: if, for example, a set of fairly sparse parallel poles has a bush in the centre a horse may use this as his guide when in fact it is set several inches behind the first set of poles to be jumped. Incidentally, although parallel poles are so-named, usually they are not truly parallel but have the front poles slightly lower than those behind.

An obstacle that has frequently caused a great deal of trouble is the water-jump. This used to be far more true than it is now, mainly because the majority of water jumps were so badly built, with the water either scarcely filling the bottom of the pit or alternatively slopping out beyond the tape. In recent years the standard has improved considerably, and mistakes at this obstacle now tend to happen because a rider, having jumped the upright fences with forethought and precision, reverts to a gallop and a prayer when faced with a stretch of water. When a pole is put over a water jump it usually proves the easiest fence on the course, simply because attainment of the necessary height almost guarantees that the width will be cleared also.

Nelson Pessoa in action at Harwood Hall, Upminster.

Clearly the size of the fences will depend upon the type of competition: if it is against the clock in the first round they will not be too difficult, although at the same time they should not be so insignificant as to give too great an advantage to a horse whose only asset is speed. Jumping is, or should be, the essential element of the sport at all times.

It is significant that when British riders first went abroad after the war they found it almost impossible to keep pace with the continentals, and that the first post-war international victory was by Harry Llewellyn and Kilgeddin in a puissance competition in Rome. Now Britain tends to produce few real puissance horses, which is perhaps why Swank's record, set in 1937, still stands. The puissance is the only type of competition in which jumping alone,

Harvey Smith and Summertime,
a highly successful combination.

Both horse and rider escaped unharmed from their fall in Copenhagen.

The Mexican, Jesus Portugal, on the Derby Bank at Hickstead.

irrespective of time, is the deciding factor. Starting with perhaps six fences, a puissance is habitually reduced to two, usually a wall and a triple bar, which are raised for each new round until all save the winner have dropped out. This is sometimes considered an unfair competition, it being thought a poor way to reward a horse jumping clear by making him then jump again over bigger fences until he reaches his limit. True puissances are rarely jumped in Britain, usually being limited to a set number of barrages, probably three, with all those still clear dividing the spoils.

The FEI rules offer guidance to course-builders for individual competitions; they are more restricting when it comes to Nations' Cup courses. Thus, except for indoor shows, a Nations' Cup course must be about 867 yards long, with thirteen or fourteen fences between 4 ft. 3 in. and 5 ft. 3 in. in height, including at least one double or one treble, but not more than three doubles or one double and one treble. There must be a water jump of not less than 13 ft. This still leaves a course-builder with sufficient scope to erect a course which will favour his home team's particular strength, but it does maintain a certain uniformity in these most important of normal international competitions.

The most difficult courses of all are naturally reserved for the competitions which carry the greatest prestige, the Olympic Games. They have increased in severity to such an extent that the course built for the team competition in Mexico was widely considered to be both unjumpable and unfair. A comparison of heights shows that in 1912 the maximum was 4 ft. 6 in., with most fences at 3 ft. 7 in., which the noted equestrian historian Schmidt-Bodner rated a difficult course at that time, while in Mexico the maximum was 5 ft. 3 in. and the lowest jump, the first, stood at 4 ft. 7 in. with a 4 ft. 11 in. spread. It was not just height that made the course so difficult, but also the placing of the fences. Combinations are, by their very nature, the most testing of obstacles, and in Mexico the treble was placed just five strides after a big triple bar, 4 ft. 11 in. high with a 6 ft. 6 in. spread. The treble consisted of a 4 ft. 11 in. wall and two sets of parallels, 4 ft. 11 in. by 5 ft. 7 in. and 4 ft. 11 in. by 5 ft. 11 in. One of the three elements – usually the final part – was knocked down no fewer than 108 times, and there were also many refusals. It was here that Marion Mould and Stroller came to grief, and many experienced riders decided that the only way to jump it was deliberately to knock a brick off the wall in order to be correctly placed for the two parallels.

The winning Canadians, with 102¾ faults, had by far the highest score ever, the previous 'top' being Germany's 68½ faults in Tokyo. Ever since the war there has been a gradual increase in the score of the winning team. For many years there has been debate about the merits of having a set Olympic course, and the arguments in favour of producing a surprise at the last minute seem to have little merit. No one, surely, would suggest that the Jumping Derbys, such as those at Hamburg or Hickstead, are any the easier to win for being held over virtually the same course year after year. A rider may know what he and his horse have to do to win, but they still have to do it.

COMBINED TRAINING

Three-day event competitions are perhaps the most demanding and rewarding of any equestrian activity. In order to do well, a horse must combine speed, stamina, obedience and considerable jumping ability; his rider must be knowledgeable and expert in three distinct branches of horsemanship–dressage, cross-country riding and show-jumping. The object of combined training is to test the all-round quality and versatility of the horse.

Combined training is the modern English term for what the French have always known as *concours complet*–the complete test.

It all began–as do so many things of value concerning the horse–with Xenophon in 300 BC. He instructed his cavalry commanders thus: 'The duty will devolve on you of seeing, in the first place, that your horses are well fed and in condition to stand their work, since a horse which cannot endure fatigue will clearly be unable to overhaul the foeman or effect escape; and in the second place, you will have to see to it that the animals are tractable since, clearly again, a horse that will not obey is only fighting for the enemy and not for his friends.'

Over the centuries the great horsemasters evolved systems of training and riding horses which were adopted in whole or in part by cavalrymen, who placed the emphasis upon endurance. Long-distance rides for chargers have been staged for more than three hundred years, the severity of the test being governed by the ratio of distance to speed. Perhaps the classic example was the ride from Vienna to Berlin, held on 2 October 1892. It attracted an entry of ninety-eight officers from the German, Austrian and Hungarian armies. There were rest areas and compulsory checkpoints all along the distance of 370 miles, and the winner–Count Wilhelm Starhemberg, a lieutenant in the Austro-Hungarian Royal Hussars–completed the journey in less than three days. Riding the 15·2 hands horse Athos, a Hungarian half-bred, his time was 71 hours and 26 minutes. The last of the forty-two finishers arrived only fifteen hours later. The distance was mainly covered at the normal military trot interspersed with walking and leading periods.

Ten years later the Brussels to Ostend ride, though only a little over a hundred miles, ended in disaster. Only half of the sixty starters reached their destination, and sixteen horses died. The inference must be that the horses were not fit, and in the modern three-day event there are still–though not very often now–horses whose preparation has not included the vital process of getting gradually and thoroughly fit. The introduction of compulsory veterinary inspections during a three-day event has done much to reduce the casualties of earlier competitions.

The French developed the endurance idea with *Raids Militaires*, including one from Biarritz to Paris–a distance of some 450 miles. In Germany the Emperor's prize was awarded to the winner of an overnight journey of some hundred miles, for which ten or eleven hours were allowed. The United States Army followed suit, though it was the French again who developed the first true precursor of the modern three-day event, the *Championnat du Cheval d'Armes*, which was staged by the cavalry around Paris in April 1902. It was concluded by a show-jumping test–as it is today–in order to keep the general public in touch with the proceedings and thus ensure their support.

H.R.H. the Princess Anne on Columbus performing an excellent dressage test.

Like other competitions at this time, the *Championnat* was an entirely military affair. It was open to one officer from each regiment of the army. Dressage was held in a military *manège*, a steeplechase over the course at Vincennes, a long-distance roads and tracks phase of fifty kilometres through the surrounding countryside and finally the show-jumping, held at the Grand Palais in Paris. It was won by Lieutenant de Saint-Phalle riding a Thoroughbred mare, Marseille II.

This event was so successful that it became an annual affair which has continued up to the present day, only being held over during the war years. It was designed to produce an all-round horse and rider, reducing the tendency to over-specialise, and aimed at improving the type of horse being bred and the standard of riding and training. *Haute école* and *reprise libre*, included in the first test, were abolished. Horses were also examined after each phase and awarded points according to their condition, and finally–in 1922–the roads and tracks phase was shortened and a cross-country course introduced. In the same year, too, the event for the first time occupied three full days.

In Sweden a military three-day event was held at the cavalry school at Stronsholm in 1907; Belgium followed suit in 1910; and Switzerland held their first event at St Gall in 1921. Then the United States joined in, and the Indian Army School at Saugor also ran tests for officers' horses on similar lines. The growing international interest was partly responsible for the introduction, at the 1912 Olympic Games, of equestrian events. They were organised, and–deservedly–won by Sweden, all the competitors at this time still being drawn from military sources.

It was not until after the second world war that civilian participation in three-day events began. The 1948 Olympics were held in England, with the three-day event being based at Aldershot in Hampshire. Up to this time the British had on the whole held aloof from the sport–the officers were busy hunting and playing polo, and were suspicious of everything connected with dressage, which was quite unfamiliar to them. But the post-war era produced a worldwide outburst of enthusiasm for horses and everything to do with riding for pleasure, at a time when the horse's working contribution was beginning to dwindle in importance. A diminution of the military monopoly quickly followed.

The new interest in horses was particularly strong in England, and it was fostered by the Duke of Beaufort. He had been present at the Olympic Games in 1948 and subsequently gave Badminton Park, his home in Gloucestershire, to be used for what is now perhaps the most famous three-day event in the world. The first event was held in 1949 and was open to civilian riders and, what is more, to women too. Nobody took much interest in this at the time, and certainly none seemed to expect the enormously high quality of performance that has put lady riders in the front rank as three-day event competitors.

Let us look briefly at what riding in three-day events entails. In the words of the FEI, it constitutes 'the most complete combined competition, demanding of the rider considerable knowledge in all branches of equitation and a precise knowledge of his horse's ability, and of the horse a degree of general competence resulting from intelligent and rational training.'

Combined training competitions are usually held over three days, though the various phases of the competition have been modified to produce one-day events and Pony Club trials as well. The one-day events were started in the early 1950s as a nursery for the three-day event, and they have grown enormously in response to an increasing demand during the twenty years since they were established. The cross-country phase has had to be reduced in length, and the steeplechase and roads and tracks phases also curtailed, but the principles remain the same and the aim is still that of a complete test for horse and rider.

In any combined training competition there are three phases: dressage, speed and endurance, and show-jumping. The dressage test is designed to show that the horse is educated, balanced, supple and obedient to ride, and though many of the movements included in the test form part of the schooling programme of any horse, others set out to demand from horse and rider a greater degree of schooling and a

Miss J. Wofford and Kilkerry of the USA on the steeplechase course at Munich.

more polished performance than the average hunter, for example, would acquire.

The second part of the competition, the speed and endurance test, consists–in three-day events–of several phases. There are two on roads and tracks, these being separated by a steeplechase course. Finally comes the greatest test of all: the cross-country phase, a gruelling course of some thirty obstacles spread over between four and five miles of open country. The whole speed and endurance section of the competition will cover some twenty miles, the roads and tracks being taken at a brisk trot or easy canter, the steeplechase course at a good gallop and the cross-country also needing to be ridden at speed, allowing too for the individual horse's ability over solid, and sometimes tricky, obstacles. Only bold, clever horses with confidence in their riders and obedience to them, and being at the peak of fitness, can hope to complete the second day with bonus points in hand and no undue exhaustion.

The third day involves a show-jumping course designed as a final test of fitness of both horse and rider. In spite of the previous day's exertions, horses are expected to be able to complete a relatively modest course; an exhausted or unduly stiff horse will find even show-jumps of this kind difficult to negotiate.

It was the Duke of Beaufort who first realised that the tests involved in three-day events should ideally suit British-bred and trained horses and their riders. Any good hunter who is temperate enough to perform a dressage test and jump a course of artificial fences in addition to galloping across country tackling big fences at speed can succeed in combined training, provided he has sufficient quality. Most natural horsemen who have been following hounds since childhood can master sufficient dressage, in the hands of a good instructor, to satisfy the judges. Strangely enough, the show-jumping phase seems to be the bugbear of many riders, perhaps because the necessity to 'see a stride' and to know whether a horse is right or wrong in his approach is more important in the ring, where the fences come thick and fast and there is little time or space to correct balance and adjust stride, than it is across country, when there is often the width of a field between each jump, and related fences are the exception rather than the rule.

In all countries where horses and riding are a natural feature–for driving and riding in the past and for pleasure riding now–the demands of versatility in achievement of combined training are a particularly popular challenge; every year the sport makes phenomenal advances in popularity all over the world.

A French competitor, M. Bentjac, enters the fearsome leaf-pit at Burghley.

COMPETITION DRESSAGE

As with almost every other form of physical activity, the art-sport of dressage has long since taken its place in the competitive sphere. Essentially, dressage riding continues to exist, as it began, as a pleasurable activity with two useful functions: to train a horse to be a pleasant ride and in so doing to give satisfaction to the creative urges of the trainer. True creative art can never be competitive in the normal sense of the word, since it cannot be restricted by rigid rules. But dressage, which has certain purely functional aspects, is another matter and those aspects are sufficiently strong to render competition useful in providing a stimulus and a meeting-ground on which progress and technique can be tested and discussed.

The descriptive word 'art-sport' has been carefully chosen because the overriding consideration of good dressage – as of such activities as gymnastics, skating, diving, etc. – is that the form of what is done is at least as important as its content. Every action or movement of both horse and rider must have beauty of style as its aim, to be given final expression in the full development of the horse's power. It is never enough to struggle through a series of set movements without regard to their form. This does not apply, for example, to athletics, in which no account whatever is taken of the style – or lack of it – with which a high-jumper clears the bar. In dressage, style is always of prime concern.

It follows that a dressage competition becomes, in effect, a competition in style. Its object is not to discover which combination of horse and rider can perform the longest sequence of flying changes, the smallest circle, even the highest leap in a *capriole*. The prize will go to the combination that performs a series of clearly defined movements in the purest style. A competition also provides a convenient frame for the display of equine talent that will give pleasure to interested spectators without particular regard to the allocation of marks by judges. It also provides a stimulus for bringing horses and riders together to test their comparative progress throughout the long years of their training.

Standards in competitive dressage vary enormously from country to country, and even more from continent to continent. So, too, do the numbers of those taking part. Europe, itself the womb of dressage, remains to this day the only continent able to produce more than a mere trickle of talent. Even there, some countries have virtually no competition riders of international status, and in many, including such great countries as France, there may be only one or two able to perform creditably at the top level. In others, such as Germany, Switzerland, Sweden and the USSR, where for many generations there has been very active interest, there is today strong support from both spectators and riders. In Germany in particular, dressage can be said to have the status of a national sport. German riders capable of showing horses in Grand Prix dressage competitions, which are, after all, synonymous with the Olympic standard, can be numbered in their hundreds, and they will be watched with intense and informed interest by thousands.

It is not easy to rationalise the unusual priority given to dressage by Germany, but it would appear to be due to a combination of certain geographical and historical factors. Fine horsemanship has, until comparatively recent times, been an accomplishment of courtiers, and Germany had a multitude of courts. The climate of central Europe does little to encourage cross-country riding in the long winter months, so the Germans had nothing to lose by concentrating on the perfection of movements suited to the *manège*. They kept a powerful standing army throughout the nineteenth century, and this provided both the means and the encouragement to develop equitational prowess in the form best suited to them. Later, in the present century, when Germany had absorbed the industrial revolution, the wealth of industrial magnates and others ensured that the art was not deprived of patrons by the disappearance of the old establishment. And finally, perhaps as a logical sequence to these other factors, they have deliberately geared their horse-breeding activities to the requirements of pure riding, as opposed to racing, trotting or hunting, which have been given priority elsewhere. So it has come about, whether for these or other reasons, that Germany has established a very clear and authoritative lead in the art-sport of dressage to the extent that that country would be capable today, if the rules permitted, of entering at least three teams in an Olympic Games, all with a chance of a bronze medal or better, whereas no other country in the world could produce more than one.

The strength of any country's output in dressage probably rests on the three cornerstones of public interest, financial backing and professional trainers. And although the trainers would not exist in any numbers alone, they are probably the most important and fundamental of the three factors. Wealth in some form cannot be overlooked because it alone can ensure a really adequate supply of horses of high quality and potential. Just as a human athlete of true international calibre is a rare thing, so a true international dressage horse is equally rare and hard to come by, and those riders who always appear to have one or two at their disposal have almost certainly tried and discarded a great many more. Only people with substantial, though not necessarily great, wealth behind them will be able to do this, although such backing may come in various guises. It may represent a long-established cavalry or national school of equitation, or it may be the private fortune of the rider or his sponsor. In all cases it is an inestimable advantage.

As for the professional trainers, their accumulated knowledge and practised eye is virtually essential, even to the most experienced amateur rider, in avoiding or in overcoming problems and in perfecting the style of both partners in the enterprise. In training horses to this standard there is always the danger that a mistake will creep in, and a bad habit become fixed, before a rider working alone becomes aware of it. Thus a horse can be spoilt for life, or at best his educational and competitive progress will receive a serious setback.

The first impression made on any newcomer to the main competitive arenas in Europe may well be a startled awareness of the very high standards of accuracy and of attack that are necessary to make any real impression. He will find the Germans there in force, with all the power of a nation in arms. The Russians usually appear once a year, usually at the annual championship meeting, on the whole showing individual style in which the horses work accurately but move rather as if on rollers, missing something of the fire that stirs the blood. They will not be in great numbers, but their best will be heading for the top prizes. They provide a shining example of what can be achieved by relatively few individuals operating under conditions that many wealthier nations would consider difficult.

The Swiss, both in the 1930s and again for some years after the last war, had a very strong team based at their cavalry school at Berne, mounted as a rule on Swedish horses. But it seems they can no longer pay the prices currently obtained by the Swedes, whose breeding establishments are rapidly becoming famous for producing suitable and successful dressage horses. The Swiss have thus had to fall back on horses of lesser quality and at the present time have little to show. The Swedes themselves are not at present as strong as they were in the fifties under the leadership of the brilliant

Mrs Jennie Loriston-Clarke competing at Stoneleigh.

Major St Cyr, but they have good horses and plenty of young talent working in the traditions of their old cavalry school.

Until recently, France had for many years produced no riders of any note, despite the continued existence of the great school at Saumur with the Cadre Noir. Like the Swiss, they do not breed many quality horses themselves, and they undoubtedly suffered more than is generally appreciated from the ravages of the long wars in Indo-China and Algeria. But now, within the last two years, Saumur has thrown up one outstanding dressage rider in Adjudant le Rolland, probably the most polished horseman Europe has seen since the days of his famous predecessor in the thirties, Commandant Lessage. It remains to be seen whether his country will be able to supply le Rolland with adequate horses to follow his present mount, and whether his example will spark off a new era of competitive interest in the country which, two hundred and fifty years ago, gave birth to the true dressage renaissance.

Great Britain, which only began to show the first signs of interest in this form of riding immediately prior to the last war, has since then produced, and is continuing to produce, a small number of accomplished dressage riders. The British are certainly not at any disadvantage as far as the availability of suitable horses is concerned, even if the English Thoroughbred is not always the easiest to train. Perhaps for this reason, there is some tendency for British riders to look to Germany or Sweden for new mounts, encouraged also by the knowledge that it is sometimes possible – which it is not in Britain – to find partly trained animals in those countries. Their chief problem to date has been the lack of any experienced trainer domiciled in Britain, added to a persistent disinclination on the part of the riding public generally to take a serious interest in dressage.

What is the aim of dressage, and therefore of dressage competitions? Even the specialised dressage tests, as well as those forming part of combined training events, are basically designed to demonstrate the extent of obedience, suppleness and lightness of the horse. The Prix St Georges test includes the half-pass, half-pirouette, counter-canter and flying change of leg. None of these movements is really specialised in itself, but they are all difficult to perform outstandingly well. The more advanced tests, the Grand Prix and Olympic tests, do include movements of *haute école* standard, such as *passage, piaffe,* changes of leg at canter at every stride, and so on. Nevertheless, even at this advanced level the primary object is the performance of comparatively simple movements superbly well, and success will depend very largely on achieving this.

It is the precision, the aim at perfection, that makes dressage possibly the most fascinating of all specialist branches of equitation.

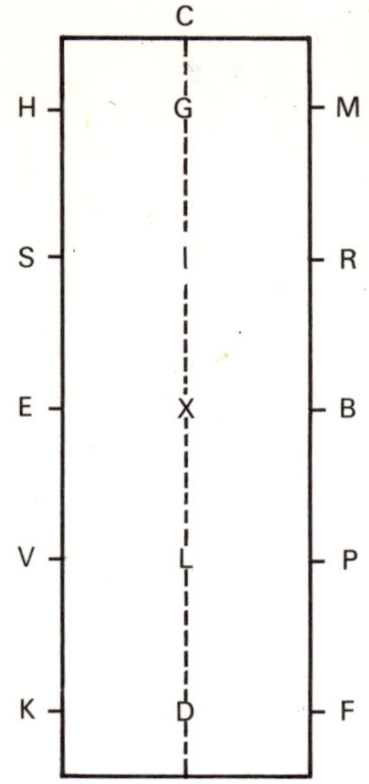

Plan of 60m x 20m arena

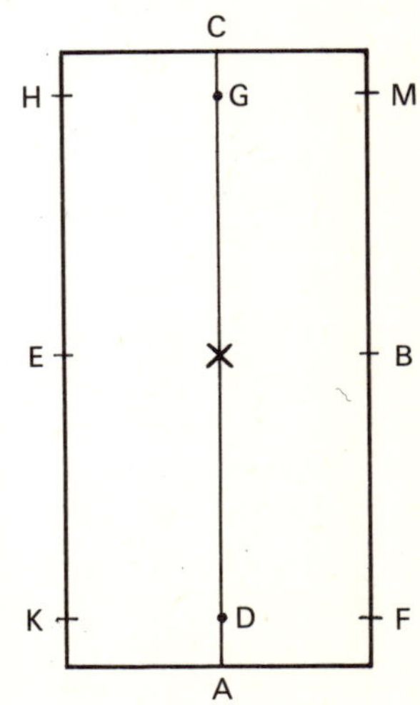

Plan of 40m x 20m arena

THE OLYMPIC GAMES

The original Olympic Games started officially in 776 BC, but the horse did not appear until the 25th Olympiad in 680 BC and he was then confined exclusively to the noble sport of chariot racing. The Games were abolished by Theodosius in AD 393 and were not revived until 1896 by Baron Coubertin; they did not then include equestrian events.

The staging of mounted competitions was only achieved, in the face of monumental apathy and active opposition to the project, by the intervention of Count Clarence von Rosen, Master of the Horse to the King of Sweden. He felt very strongly that the inclusion of ridden events in the Olympic Games would have a beneficial effect upon standards of horsemanship all over the world. His efforts have been responsible for a great upsurge of interest in equestrianism in many countries, and for the founding of the International Equestrian Federation, the governing body of all horse sports other than racing, in 1921.

At the meeting of the International Olympic Committee in Athens in 1906, Count von Rosen proposed that equestrian events should be introduced into the Olympic Games. The majority of the committee initially rejected the suggestion on the grounds of expense, but Baron Coubertin was sympathetic and asked von Rosen to prepare detailed proposals and regulations for the projected equestrian events, and to present them at the congress at The Hague in the following year.

The Swedes immediately formed a committee who devised a programme of equestrian events to include dressage, a pentathlon and a *Jeu de Rose* (a type of mounted game of 'he'), and they were sanctioned for inclusion in the next Olympic Games, to be held in London in 1908. Unfortunately the newly-built White City Stadium was not big enough to hold them and Lord Lonsdale therefore volunteered to be responsible for them at the International Horse Show, which he had started at Olympia in the previous year. But as 88 entries were received from eight nations the project became too daunting, and as Lord Lonsdale felt, probably correctly, that his show could not incorporate

The gold medallist at Rome in 1960: the Australian, Lawrence Morgan, on Salad Days.

them in its programme without detriment to both, the inaugural equestrian Olympics were postponed until the Games were held at Stockholm in 1912.

In 1909, at the Berlin congress, the programme was extended to include riding competitions–'prize riding' (now Grand Prix dressage), 'prize jumping' and 'the military' (now the three-day event). The latter was divided into five phases. In order of performance they were: long-distance ride (33 miles); cross-country (3 miles, incorporated in the long-distance ride); 2-mile steeplechase; show-jumping; dressage. The dressage came last because the Committee considered that the maximum obedience could be obtained after three days of maximum effort. Now, of course, the dressage constitutes the intitial exercise in any horse trial.

Any serving officer on a military horse (other than instructional horses) was eligible. There were twenty-seven starters, of whom fifteen actually finished, and seven nations were represented. Fittingly, having been responsible for the inauguration of the entire exercise, Sweden won both the team and the individual gold medal, the latter going to Lieutenant Nordlander on Lady Artist, an English Thoroughbred. In fact, although the British team, with Belgium and Denmark, failed to finish, fifteen of the horses came from the British Isles.

The first four teams were very close in their markings–judged under a complicated code which a Senior Wrangler might have found perplexing. Sweden's score was 139·06, Germany was second with 138·48, the United States third with 137·33 and France fourth with 136·77.

Three jumping competitions were held in conjunction with the Olympic Games in Paris in 1900, but the first team event was in Stockholm and was also won by the host nation, with France second and Germany third.

The war enforced an eight-year interval and it was not until 1920 that the next equestrian Olympics were staged, in Antwerp, when the Swedes again prevailed in both the three-day event and the show-jumping. This time Britain failed to muster a team, and of the eight competing nations only the United States came from outside the continent of Europe.

The Swedes retained their superiority in the show-jumping in Paris four years later, but the three-day event prize went to Holland, who won the team title again in Amsterdam in 1928. During this time the three-day event conditions and markings were still in the experimental stage and were varied from one Olympiad to the next. The dressage test was replaced by a second endurance test in 1920, and there was also included as a separate event a vaulting competition open to privates and NCOs. Only three nations–Belgium, France and Sweden–availed themselves of the opportunity thus presented and the vaulting competition died a natural death before the Games were held again. Seventeen nations competed–a record number.

Spain and Poland each had the measure of the Swedes in 1928, winning the gold and silver show-jumping medals respectively. A record number of twenty teams went to Amsterdam, a measure of the recovery of the European cavalry, and for the first time the three-day event followed the pattern which it has maintained ever since. Dressage now came first, followed by the speed and endurance, and ending with the show-jumping. The whole exercise was still dominated by the continental riders, but the British team finished in sixth place, which in view of the large entry was by no means discreditable.

In Los Angeles in 1932 no team completed the show-jumping course and the only medals to be awarded were individual ones. A Japanese, Lieutenant Baron Takeichi, won the gold. The host country, the United States, represented by Major Harry Chamberlin, took the silver and the bronze went to the son of the man responsible for the entire Olympic equestrian concept, Clarence von Rosen. He must have been a versatile horseman, for he also rode in the three-day event team and won another individual bronze.

Coming, as it did, only three years after the slump, money was short all over the world and only six teams managed the trip to Los Angeles. France, Holland, Japan, Mexico and Sweden were represented, with the United States; and it was the host team, who had finished third at Stockholm twenty years earlier, who managed to prevent the Dutch from achieving a hat-trick of victories in the three-day event.

The Berlin engagement in 1936 was the Olympic Games to end them all, mounted with an ostentation which has never been seen before or since and attracting riders from twenty-one nations. Nineteen competed for the three-day event, which soon became a field of carnage. The fourth fence, the pond, preceded by a small post and rails, was a real bogey. With a sloping bottom and heavy overnight rain, it soon ressembled a battlefield. Forty-six horses got this far, but eighteen fell, ten lost their riders and only eighteen went clear, which makes it the most lethal fence of all time. The American horse Slippery Sam broke a foreleg and had to be destroyed, and there were two other casualties on other parts of the course. In all, out of fifty starters, twenty-three were eliminated and only four teams finished. The British contingent was able to finish third, winning the first equestrian Olympic medal ever for Britain, even though Captain R. G. Fanshaw and Bowie Knife, who had never before been asked to jump into water, parted company and were not reunited until the horse had been pursued by his rider for over 2½ miles! Victory in all the equestrian events went to Germany. Poland was second in the three-day event, and Holland in the show-jumping, with Portugal third.

The Games were reinstated following the war, being held in Britain in 1948. At this time competitive equestrian sport was still very largely in military hands. On this occasion the show-jumping gold medal went to the Mexican team, with Spain winning the silver and a British Army team the bronze. In the three-day event, with only five countries completing the course, the Americans were victorious, with Sweden and Mexico taking the silver and bronze medals respectively. Only 3½ points separated the first two teams.

By 1952, when the Games were held in Helsinki, military riders were in the minority and there was an upsurge of enthusiastic civilians finding their way into the team of every country whose cavalry had given way to mechanisation. Britain, represented by Llewellyn, Colonel Douglas Stewart and Wilf White, won the show-jumping gold medals from Chile and the United States. The British three-day event team was unhappily eliminated when a rider, concussed in an earlier fall, went the wrong side of a flag when he had completed the cross-country course, but even so the Swedes and the Germans were out in front. Only six teams finished out of twenty-one starters, with the Americans capturing the bronze medal.

In 1956 the equestrian Olympic events were held in Stockholm, though the rest of the Games were in Melbourne, and this time it was the British three-day event team which was to triumph, 120 points ahead of Germany, with Canada third and Australia fourth. Sixteen teams came under orders, with eight finishing. The Germans first demonstrated their supremacy in the show-jumping, winning the gold medal from Italy and Britain, who finished third with the first-ever lady rider in her Olympic team.

The cross-country course in Rome took a heavy toll in 1960, when the gold medals, team and individual, plus an individual silver, were won by a most sporting Australian contingent, led by Laurie Morgan on Salad Days. Switzerland and France filled the next two places to take the remaining team medals. In the show-jumping Germany won their second consecutive team gold medal from the United States and Italy.

Marion Mould and Stroller took the silver medal in the show-jumping at Mexico City in 1968.

Richard Meade and Laurieston.

Germany achieved a hat-trick of gold medals at Tokyo in 1964, with France and Italy following on. In the individual event, Pierre d'Oriola of France, who had also triumphed in Helsinki on Ali Baba, took the individual gold on Lutteur B. Italy won the three-day event from the United States and Germany; nine teams finished, Britain being eliminated in company with Japan and Korea.

Mexico City in 1968 brought a rallying-point for British interests when the team won the three-day event after monsoon conditions for the cross-country test. Once again the United States were runners-up, with Australia coming third. Canada won the show-jumping team gold medal from France and Germany, and the individual gold went to the American team captain, Bill Steinkraus, on Snowbound, who joined Hans Gunter Winkler of Germany with Halla (1956) and Captain Raimondo d'Inzeo of Italy with Posillipo as a single holder of an Olympic gold. The silver medal went to Britain, being won by Marion Mould and Stroller, and David Broome, also of Britain, took his second bronze.

The 1972 Games, held in Munich, proved to be another triumph for Britain in the three-day event. Richard Meade on Lauriestown won the individual gold medal and helped his team to carry off the gold too, winning from the United States and Germany, who took the silver and bronze medals respectively. In the show-jumping the team medals went to West Germany, who just beat the United States for the gold, leaving the latter with the silver and the Italian team with the bronze. Graziano Mancinelli of Italy won the individual gold on Ambassador; Ann Moore and Psalm took the silver for Britain and Neal Shapiro of the United States the bronze with his horse Sloopy.

The appalling conditions at Mexico are tackled by German Gazumov of the USSR.

Foxhunting

HUNTING IN BRITAIN

It is not easy accurately to state just when and where foxhunting started in Britain. Although hunting was carried on all over the British Isles certainly since the Norman Conquest, it was almost certainly not until the eighteenth century that packs were used to hunt only the fox.

By the end of the seventeenth century, however, those who hunted–and they were for the most part the great landlords–were beginning to realise that the hunting of the fox provided better sport than that of any other animal. This was because the fox had greater stamina than the hare, had not so strong a scent as the deer, and was more crafty than either, thus presenting in every way a greater challenge to a huntsman and his hounds.

Just three hundred years ago the great Duke of Buckingham, who hunted his vast territories in the north of England, is alleged to have exclaimed that he would willingly exchange a flock of his fattest sheep for a similar number of foxes. By the middle of the eighteenth century, thanks largely to the enthusiasm of such great aristocrats as the Duke of Buckingham and other members of the royal family, hunting the fox was the most flourishing sport in England.

A hundred years later it had become far and away the most popular. Just when even the great landlords were finding it impossible to maintain the magnificent establishments set up to hunt the fox over large areas–the Dukes of Berkeley hunted from Bristol to London, the Dukes of Beaufort from Bath to Oxford–so, thanks to the Industrial Revolution, there were many more of the new gentry who could afford to hunt. As a result, subscription packs were started, hunting smaller areas but much more efficiently organised–the subscribers felt that their subscriptions entitled them to demand good sport.

There was, obviously, a certain anomaly here. The duke, or other noble landlord, owned the hounds and probably most of the country, but other people paid the hunting bills–a fact which the landlord almost certainly resented, as he did the interference of a committee which the subscribers felt entitled to appoint. The nineteenth century is full of stories of rows and arguments and clashes between landlord and followers. A situation so fraught with incendiary problems was bound to take a little time to settle down. It could be said that in some parts of the country it has never entirely settled down.

So often called the heyday of hunting, the nineteenth century was in fact the most discreditable period in its history. The fields were huge and paid little attention to the farmer; foxes, in short supply because of poaching, were invariably bagged; with no traffic-ridden roads, no barbed wire, no artificial manures to steady the chase, hounds were forced on at a tremendous pace, the field riding as if in a race, on unclipped horses many of which, their owners boasted, died of exhaustion every season.

The march of progress and the first world war brought foxhunters to their senses, and when hunting was resumed after the war it was much more like what those who, two hundred years earlier, had first discovered the joys of the chase intended it to be. Between the wars there were still plenty of people who maintained large establishments and foxhunting in those two decades was of a very high order: well organised, producing excellent sport, the country still rideable.

It was a miracle–the result, in fact, of a few dedicated and tireless enthusiasts–that hunting survived the second world war. But it did, and during the last twenty-five years it has evolved as the sport that we know today. Lacking perhaps the style and bravura and colour of previous centuries, it is now far more democratic and far more broadly based; and, indeed, giving pleasure to far more people than ever before.

This is due, firstly, to the fact that far more people today can actually afford to hunt mounted than were able to even in the days before the war. Secondly, there is far less evidence of great wealth among the country landlords and so, with few exceptions–exceptions incidentally which are of great benefit to foxhunting–hunting is far less exclusive: far less a sport for the privileged few. Thirdly, more and more farmers come out hunting themselves, whereas before, so impoverished was their industry, not only could they not afford to hunt but they were dependent upon the local hunt for their livelihood, which made it virtually impossible for them to complain to the hunt or even seek their co-operation. Were it not for the hunt they would be bankrupt. Now many hunts are run by farmers.

A last important factor is that there has been, during the last twenty-five years, the formation of hunt supporters' clubs, which has meant that each year more and more people feel involved with their local hunt. These supporters clubs play a vital part in foxhunting. In some countries there is a danger of their playing almost too vital a part because they tend to dictate to the hunt. There have been cases where continuity is much more in evidence in the supporters' club, where wealth is of no account whatever, than in the mastership or hunt committee, where money is still a major feature. In such cases the supporters' clubs become very powerful.

On a misty October morning, the Cowdray foxhounds wait to be put in to draw

Fortunately, however, this potentially awkward situation is remarkably rare, and the supporters' clubs do a splendid job for hunting. Many hunts would be quite unable to exist without them, for they fill the role of ambassadors for foxhunting, popularising the sport among a section of the community that has never before been interested and, even more important, providing a most useful antidote to those opposed to hunting.

Would such a situation ever have been envisaged two hundred or even one hundred years ago? It is, of course, unthinkable: but it may well be that hunting is much better for it. It could even be said that were it not for this development hunting would not have survived.

The fact of the matter is that the foxhunter today has far greater responsibilities than in the past. He has responsibilities to the farmer, to the general public and to the sport itself. In the last century he could enjoy his sport wherever he liked. The whole countryside was his playground. Now even in the most fortunate country there are inevitably a few farmers who do not appreciate the hunt. Unless a proper responsibility towards agriculture is accepted there could well be many more farmers who are opposed to hunting and refuse to allow the hunt over their land.

Never has it been more important that those who go out hunting should fully appreciate the importance of good relations between the hunt and the farmers, and of a proper understanding of the modern farmer's problems. Gates left open, seeds and growing crops ridden over, wire fences broken and so on: these are the things that make a farmer feel it just is not worth his while to allow the hunt to ride over his land. No less important is the actual personal relationship between farmers and the people who hunt. It is really the duty of hunting people to get to know the farmers in their country–not just those who come out hunting, and might therefore be expected to support the hunt, but many others who have nothing to do with the hunt, perhaps have no interest in hunting at all, are even basically opposed to it. These are the farmers

It takes a bold partnership to stay with hounds over this sort of country.

that hunting people should take the trouble to know and make friends with.

There is an obligation, too, to ensure that the general public is in no way incommoded, let alone offended, by the hunt. Too often people out hunting allow themselves to seem thoughtless, even arrogant. They ride on pavements; take up the whole road, making it impossible for cars to pass; they expect ordinary people to open gates or hold their horses and then forget to thank them. Such behaviour can only cause resentment and make people dislike the hunt.

The hunt has always been a colourful part of the countryside, respected, even loved, by local people. For generations hunting people have made it their business to see that the hunt is very much an integrated part of the countryside. With the development of supporters' clubs this aspect should be increased.

Finally there is the responsibility of hunting people towards their own hunt. To keep a hunt going today is by no means easy, and everyone's active support is needed. More people, both mounted and on foot, now get pleasure from hunting than ever before, but with this enjoyment goes responsibility. The future of hunting in Britain depends upon an acceptance of this responsibility.

HUNTING IN THE UNITED STATES

Hunting in the United States and in Canada is extremely popular. There are nearly a hundred packs of hounds recognised by the American Masters of Foxhounds Association, a body which was affiliated to the English MFHA in 1914. There is also a Masters of Beagles Association, a sport which also has its enthusiasts.

The origins of hunting for sport are to be found in the richly historical states of Virginia, Maryland and Pennsylvania. Thomas, sixth Lord Fairfax, who went to live in Virginia in 1747, is believed to have been the first person to own hounds for hunting – as a sport – in the United States. In those days, when gentlemen lived on their estates in considerable splendour in an era when the importation of slaves was considered quite normal, it was natural that the vast woodlands of the Eastern Seaboard, which abounded in deer and grey foxes, should be exploited for sport.

George Washington, who loved to hunt, kept a diary on hunting between 1768 and 1789. It shows that before he was called upon to serve his country and became its first President he had a lot of fun with his hounds, and passages in the diary show that he had a considerable understanding of them. The Marquis de la Fayette gave him some hounds from France. His huntsman was Billy Lee, who went to the war with his master and, having survived it, returned with him to Mount Vernon to re-establish the pack and continue to hunt along the banks of the Potomac river.

Lord Fairfax's hounds, and those kept at Mount Vernon, were of course private packs, kept for their owners' sport. The first organised hunt club was the Gloucester Fox Hunting Club, formed by the Philadelphia gentlemen in 1766. They kept their pack on the banks of the Delaware river, and it was said to contain 'the best English blood'.

Those areas where Washington and General Howe fought their bitter battles, often over 'a fair hunting country', are today hunted by such packs as Blue Ridge, London, Middleburg, Piedmont, Old Dominion, Orange County (Virginia), Elkridge/Harford, Green Spring

Valley (Maryland), Andrew's Bridge, Brandywine, Cheshire, Radnor and Rose Tree (Pennsylvania), to name but a few. The little town of Middleburg, situated some fifty miles from the city of Washington, is synonymous with foxhunting in America as Melton Mowbray is in England. The Red Fox Inn there was established in 1728, so it will be seen that the roots of foxhunting in the United States are nearly as old as they are in England.

Two of the best bred packs of American hounds are the Middleburg and the Essex, which hunt in New Jersey; both packs often take premier awards at the Bryn-Mawr Hound Show held at Malvern in Pennsylvania. The American-bred hound is a special breed, and differs somewhat from the English foxhound. The head is longer and narrower, with ears set on very low; the legs are longer, and do not always have quite the highly arched toes of the modern English hound. It is thought that the American-bred hounds have developed feet perhaps more suitable for the terrain they have to cross. This is dry and hard in some areas, and in others marsh-like – particularly in the Chesapeake Bay area. Conditions such as this hard dry ground and the large woodland areas have given the American hounds extra good noses and a tremendous cry. One would imagine there is a good deal of French blood here.

Harry Worcester Smith (left), founder of the American MFHA, talking to Paul Mellon during a day's hunting in Virginia.

In packs such as the Elkridge/Harford, for instance, drafts of the very best Heythrop blood have been imported to make what is called the American cross-bred. These hounds are selectively bred and are very fine, being famed especially for their speed. There is, of course, a very close liaison between many of the best hound breeders on either side of the Atlantic.

There are two packs of pure-bred English foxhounds in the United States: the Cheshire and the Arapahoe. The Cheshire was formed in 1913 by Mr Plunkett Stewart, and hunts in the Unionville area of Pennsylvania, while the Arapahoe hunts over the ranchlands in the upland plains west of Denver, Colorado. Mr Lawrence Phipps, Jr, is their distinguished Master, and he has imported only the best pure English blood, notably from the Old Surrey and Barstow.

There is a third type of hound bred in the United States, called the Pen-marydel, a name derived from the states of Pennsylvania, Maryland and Delaware as the hounds are bred from strains bred in those three states – although one also hears of an Irish hound called Mountain who figures in the old pedigrees!

Like the hounds, the red foxes had to be imported. It is believed that they were brought in by the British, and landed in the Chesapeake Bay area. There are now red foxes in most of the eastern states, a slightly smaller variety than the big red fox associated with English hunting. The grey fox, indigenous to the United States, is not as bold as his red cousin, often running in circles and frequently ending up in the top branches of a tree! The deer, which were abundant in the past, still exist in numbers large enough to be a nuisance.

Looking at photographs of meets in the twenties and thirties it would be difficult to distinguish between one held near Middleburg and one in Leicestershire – both boasted large, well-mounted and well-dressed fields, nearly all the ladies riding side-saddle. Nowadays, in the United States as elsewhere, fashion and custom have changed, and the ladies will mostly be found riding astride; the big, quality hunters, mostly TB, are still to be seen in the fashionable hunting countries of the eastern states.

Both Maryland and Virginia are great horse-breeding areas, not only for racing Thoroughbreds but also for high-quality hunters. The Upperville Show is famous for its young hunter classes, where standards are always high. Of course, quite a number of horses used for hunting do come via the racetrack as well. The hunts in these areas also have hunt race meetings and point-to-points. Some of these are raced over brush fences, and some over timber; the most famous of the latter is the Maryland Hunt Cup. Some hunts run what are known as 'old-fashioned' point-to-points, where competitors ride as nearly as possible straight from one point to another; these races are always very popular.

The American hunt clubs are by English standards expensive to belong to, and it is quite a privilege to become elected; non-members of a club may only hunt if invited to do so. Etiquette is very strict; as in England, dress is formal, with the usual distinctive hunt buttons and collars. One hunt, the Piedmont, has a button with a legend attached to it. The story is that at full moon a fox with two brushes runs, and is never caught; the Piedmont Hunt button shows a mask with two crossed brushes beneath it.

There are a number of packs in the United States which are lucky enough to have had long masterships. This has enabled them to breed their own particular type of hound. Some of these packs – such as Mr Hubbard's Kent County (eastern Maryland) and Mr Jefford's Andrew's Bridge – are privately owned. Incidentally, both these packs have Pen-marydel hounds; Mr Hubbard's are orthodox in colour, while Mr Jefford's are all black-and-tan, and hunt in the southern part of Pennsylvania.

Another famous private pack is situated in Georgia: Mr Ben Hardaway's, whose hounds come from some of the oldest strains in the country, notably the legendary 'July' strain, called after a particularly famous hound of that name.

In the eastern states the red fox, and sometimes the grey, is hunted, but further west both the fox and the coyote are quarry. The coyote, a wild dog, usually runs in pairs, and is to be found in the hills and canyons in Colorado, Arizona and California. They will run a considerable distance, and it takes a clever horse to clamber up and down the hillsides, in and out of the canyons where the coyote like to go. It also takes a hound of exceptional nose to distinguish any scent in these dried-out areas. Mr Harold Ramser, Master of the West Hills Hunt in southern California, has a pack which carries much College Valley and West Waterford blood. Both these packs are bred from hounds used for hunting the hills and fells of northern England. The West Hills hunt on the big ranches in the Los Angeles area. The Los Altos pack operates in the San Francisco district about 400 miles further north. The areas covered by these packs are far greater than those of any pack in Great Britain; the West Hills hounds and horses are sometimes vanned out the night before hunting if, for instance, the meet takes place in the Palm Springs region.

The hunting season in the United States begins in the autumn and generally continues through to the spring. In some areas the severe winters make it necessary to begin cubhunting early, possibly in July, and hunting may have to close down at Christmas or in early January, depending on how far north one is, until a final few days' sport in late March or early April. In the Boston, New York and northern Pennsylvania areas the really keen hunting people move their horses south when winter closes in, and join the Moore County Hunt in south Carolina, which provides good sport in gently rolling woodlands with a milder climate.

Fences in the United States vary, as they do in other parts of the world. The solid timber fences of Maryland are as famous as the banks of southern Ireland, tough and often high. These rails take (and make) a good jumper. In Virginia fences vary more: some people liken the country to parts of Ireland, with its stone walls and rolling country. There are also panel fences or 'chicken coops', erected by the hunt to enable people to get across country without being too troubled by wire. There are streams to ford, usually wide but not too deep.

There are other natural hazards, too. The warthog, a rat-like animal found in many areas, digs small, perfectly straight, deep holes which

A private pack with the Master in the foothills of the Blue Ridge Mountains.

are very difficult to see and very dangerous – a horse putting a foot into one of these holes can be seriously injured. Out west, of course, there are other things such as rattlesnakes to look out for; most people carry the antidote with them, for without it the result of a rattlesnake's strike can be fatal. There are also wild pigs and mountain lions (the lynx) which terrify many horses coming across them.

One cannot discuss hunting in America without mentioning the famous Long Island pack, the Meadow Brook, which is one of the oldest packs in the country and is still one of the most fashionable.

There are hunting traditions in Canada, too. The Toronto and North York were formed over a hundred years ago, in 1843, together with what is now the Eglinton. The Ottawa Valley pack was started in 1873, and there are two packs in the province of Quebec, the Lake of Two Mountains and the Montreal. The hounds used are, like those of the American packs, crossed with English and Irish imports. Naturally the season in Canada is shortened somewhat by the onset of winter, and usually lasts from July to December, with a few days in the spring if the weather permits.

Much further south, in Mexico, there is also a pack of hounds. They belong to Mr Patrick Tritton, an Irishman who has made his home in Mexico and took out some hounds with him which he hunts himself.

People hunt for different reasons – to see their friends, to jump fences, to watch hounds work (which is the best reason of all to the true hunting person); for whatever reason, and in whatever part of the world, the thrill of hearing hounds in full cry is the same the world over.

Chapter four

The show world

BRITAIN

The showing classes, which in years gone by were the sole reason behind the staging of any horse show, at any rate in the British Isles, were originally put on for the benefit of the breeders and for the improvement of the breed. To a very large extent, this is still the case today, although there are people who buy a show horse for the sheer enjoyment of owning an outstandingly good-looking animal and pitting him against the best that other exhibitors can produce.

Showing classes are undoubtedly somewhat limited in their appeal. The horseman finds them entirely fascinating, and can happily spend a whole day at the ringside making his own assessments, discussing the horses with his friends and applauding or criticising the judging: more frequently, it must be confessed, the latter. Perhaps this is the true fascination of showing, particularly as it is all a question of personal preference and opinion and no one can ever, in the last analysis, be proved either right or wrong.

Basically, however, the charm is in the opportunity to see the best specimens that are available, at the time, of every type of horse. The chance to compare their respective merits and demerits also plays its part. In essence, it all stems from the eternal thrill of horse-dealing, of finding the best animal, without ever having to back one's judgement with hard cash. Perhaps this is why dealers invariably make the best judges–they do have to pay, and may at some time have had to live with their mistakes.

To the unitiated, and to those whose interest in the horse is confined to the spectator sport of show-jumping, the showing classes can be merely tedious. Not for them long hours at the ringside, watching an interminable succession of horses whose only distinctive feature is, to them, a possible variation in colour, being ridden by their owners, then by the judges, and finally being stripped, run out in hand and endlessly discussed.

A knowledge of the horse as a species, and of what constitutes a good or a less good individual, is essential to the enjoyment of showing classes. But, in an age when riding is fast becoming a national sport with the coming of a more affluent and leisured society, and more and more people are acquiring their own horses to keep at livery in a local stable or even at home, it is becoming increasingly necessary for anyone who rides to understand the good and the bad points of the horse he contemplates buying. Ignorance is not bliss with the price of horses rising astronomically every year, for it is all too easy to buy unwisely and all too difficult, once a horse has been acquired, to pass him on.

There is no finer place than the show ring–not even, for the pace there is too fast, the sale ring–to learn about the vital statistics of a horse. The conformation of any working animal is, indeed, more than vital. Good make and shape are essential if an animal is to remain sound in work over the years. For an animal designed to carry weight, and to carry it at speed, this is the most important factor of all. Looks are a luxury, but correct conformation is an insurance against disappointment, or even disaster. Whereas quality is desirable, it is the last consideration when compared to classical limbs and a workmanlike frame. If a horse is well made, he is probably a good and straight mover, able to cover the ground and unlikely to strike into himself, brush or over-reach at his faster paces, not prone to the afflictions which go hand-in-hand with faulty action. But even this does not invariably follow.

Judges vary somewhat in the order of their priorities, but nearly every one starts on the ground and works up. 'No foot, no horse' is too basic to require elaboration. Feet that are long and boxy, like those of a donkey, do not provide that all-important contact between the ground and the cushioning frog, nature's provision for the absorption of concussion. The frog atrophies in disuse and the concussion, no longer taken up by a protective cushion, travels straight into the pedal bone; the result is almost always navicular disease. Flat feet, on the other hand, with the outward appearance of soup plates, indicate dropped soles which lend themselves to bruising on rough surfaces and also cause lameness.

The legs are every bit as important as the feet, and for equally obvious reasons. Convex knees that put strain on the tendons of the forelegs predispose a horse to breaking down, while sickle hocks give rise to symptoms of strain in the hind legs such as curbs, spavins and thoroughpins. Pasterns that are too short and upright, or too long and sloping, are prone to produce unsoundnesses such as sidebone and ringbone – bony deposits which interfere with the free movement of a working part, causing pain and lameness that is permanent.

A horse with good limbs can usually be expected, unless it receives an external injury, to remain reasonably sound throughout its career. Thus the conformation of a horse's body above the legs is more a question of personal preference and is open to debate, though depth of girth is essential for heart-room, strength and stamina. Slope of shoulder, elegance of outlook and conformation behind the saddle are considered in different orders of priority by different judges, who also show their individual preference in what they look for in the ride. Some like a horse that takes a strong hold, others prefer the type that requires only a light contact. Horses go differently, too, for different judges.

The ridden hunter classes are divided by weight, with lightweights to carry up to 12 st. 7 lb., middleweights up to 14 st. and heavy-weights from 14 st. upwards. There are also, at the bigger shows, classes for four-year-olds, for ladies' hunters to be ridden side-saddle (one of the most elegant sights in any show-ring) and for small hunters, not exceeding 15·2 hands. The working hunter classes, in which horses have to jump a small course of fences before being judged for their conformation and ride, have also become extremely popular. An importation from the United States, they not only provide a useful outlet for the horse which is a performer but does not come into the category of the top-class show horse, but also a schooling ground for the likely show jumper or potential three-day event horse.

Breeders have plenty of opportunity to show their stock, not only at the annual show of the Hunters' Improvement Society at Shrewsbury but also at all the county and major agricultural shows, which cater extensively for led hunters – brood mares, with their foals at foot, yearlings, two- and three-year-olds. At the bigger shows the youngstock classes are divided into colts and fillies, and at Shrewsbury there are also classes for young Thoroughbreds in each age group.

One of the most interesting shows in the country, also put on by the Hunters' Improvement Society, is the Thoroughbred Stallion Show at Newmarket which opens the season in March. Here some eighty horses compete for the premiums offered by the Society, and the breeders turn out in force to select a mate for their brood mares. Premium stallions undergo a rigorous veterinary examination and are certified free from hereditary unsoundness, which is not the case with any other stallions in the land. In consideration of their premiums, these horses stand at a very reasonable stud fee to members of the Society, who are thus able to use high-quality sires, often with a very useful racing record.

There are, of course, also classes for ponies in profusion, led by the National Pony Society show, the Ponies of Britain stallion show at Ascot in April and summer show in August at Peterborough. Broadly speaking, they follow the same lines as the hunter classes but in the ridden division far more emphasis is – or should be, and usually is – placed on manners. Children

A winning line-up of working hunter ponies.

are seldom strong enough, or sufficiently accomplished with the exception of the semi-professional children, to control a badly-mannered pony, and if a pony takes charge of its rider or bucks him off it is not suitable to be ridden by a child at all.

There are many ramifications in the world of pony showing. For example, there are the show pony classes, under saddle, in which the champion can command a sum which is just about double, at some £6,000 to £8,000, the value of a champion hunter, for reasons which no one seems able to determine. There are the working pony classes, a recent innovation run on the lines of the working hunters, for the ponies who do not quite come into the top-class showing category but usually have more bone and substance and are far more suitable for their young riders, not only in the show ring but in the hunting field.

Then there are the mountain and moorland pony classes, which are either mixed according to size, both large and small, and may be led or ridden – in the former case they may also be divided by sex; or, at the bigger shows, separate classifications are included for all nine of the native breeds of the British Isles – Welsh, Dartmoor, Connemara, Highland, New Forest, Dale, Fell, Exmoor and Shetland.

Arab classes are also held for both led and ridden animals, and they have their own show at Roehampton in addition to classes at many of the bigger agricultural shows. Pure Arabs, part-bred Arabs and Anglo-Arabs all have opportunities for exhibition against others of their kind, and there is a lively export trade both to the Continent and to the United States.

The popularity of the cob has sadly dwindled in recent years and the blood cob is nowadays seldom found. Cobs are confined to a single class even at the leading shows, and if this does not receive more support from exhibitors it seems likely that it will eventually be dropped altogether.

Hacks, however, continue in popularity, and if the standard of training and presentation may be said to have declined in the last ten years or so the classes continue to fill and to provide an elegant contrast to the more workmanlike and less highly educated show hunter. It is hard to describe a hack other than as an elegant and well-made horse, lighter in type than the hunter, with proportions as near perfect as possible, a great front giving a good length of rein, a sweet head, long sweeping movement which stops short of exaggerated toe-pointing and, above all things, that elusive, indefinable quality of *presence*, an essential for any great show horse.

Attached to every category of show horse and pony is a strong professional band of showmen and women, and although the amateur exhibitor is often heard to decry the 'pros', and to maintain that the non-professional is never able to beat them, this is both unfair and untrue. A really good horse will always get to the top, but there are few really good horses about and some of the less experienced amateurs are inclined to regard their geese as swans. In showing, as in everything else, there are tricks in the trade, and the art of schooling for the ring and keeping the horse interested, of trimming and plaiting him to the best advantage, and of producing and presenting him to the judges is half the battle of success. Indeed, a professional showman can often make a second-rate horse look like a champion, and get away with faults which an amateur would find impossible to disguise.

Showing has made the English and Irish horse what it is today – the envy of the rest of the world. For, given the right temperament, the well-made horse is generally the best performer. Without the yardstick and the shop window of the showing classes the quality of horses would inevitably decline, but with the prototype on view for breeders to aim at the standard can only improve.

THE UNITED STATES

Horse shows in the United States must cater not only for a wealth of different breeds but also for three distinct styles of riding–the hunt seat, saddle seat and stock seat, as well as halter classes for all breeds. Because the US is such a vast country, shows cater for many breeds and all styles, and each breed show also caters for the different seats, with the exception of the gaited breeds such as the Tennesse Walker and Saddlebred.

In the hunting country of Virginia, Maryland, Pennsylvania and the Piedmont of North Carolina, as well as California, hunt seat classes largely fill the schedules. California is also a stronghold of the stock seat. The deep South concentrates mainly on shows attracting the three-and-five-gaited Saddlebreds and Tennessee Walkers, which are ridden saddle seat. Florida favours stock seat but also caters well for hunt seat riders.

Breaking the styles and breeds into compartments one finds Morgans mostly shown in the New England states, mainly by saddle seat riders though a few classes are open to hunt and stock seat adherents. Thoroughbreds are always shown hunt seat, confining their talents mainly to hunter and jumper classes, which are considerably different from their equivalent classes in Britain. Arabians are shown mainly under saddle seat and stock seat, with very few hunt seat classes. The Western breeds such as the Quarter Horse, American Paint and Appaloosa are shown predominantly under the stock saddle, though most Class 'A' breed shows include events for jumpers and hunters over fences as well as 'English Pleasure', which may either be open to both saddle and hunt seat riders or have a division for each style. Morgans, Saddlebreds and Arabians also demonstrate their elegant bearing and superb action in fine-harness classes.

The average American judge is called on to cope with a great variety of classes. Large shows to some extent delegate their classes by style and breed, an example being the four-day North Carolina State Championships of 1965, in which I showed the same horse in both jumper and Western divisions. Saddle seat and gaited classes, including fine-harness, hackney, harness ponies, roadsters, Morgan and parade were judged by one person; a well-known Arabian trainer officiated in the Western classes; hunters, jumpers and hunt seat equitation came under three joint judges.

In many one-day shows held throughout the country by riding clubs or as part of town celebrations one judge frequently goes through a whole card, assisted only by a ring steward. American judges do earn the fees paid for their services, but they are not required to ride competitors' horses, their evaluation of merit being visual only, which is more favourable as a horse, particularly a show horse, is highly individual, reacting better to his own rider than to a stranger who may or may not be in rhythm with him.

A typical four-day show offers a kaleidoscope of colour, breeds, talents, elegance and thrills. Each day's programme is split into morning, afternoon and evening sessions, the latter offering the cream of crowd appeal classes with open jumping, walking horse, parade horse, and three-and five-gaited stakes. Some shows mix their schedules so that something of everything is exhibited each day, finishing with the championships; others aim to have hunters and jumpers one day, gaited breeds another, and Western at a different time.

Schedules list classes by breed, riding style

Calf-roping at Sacramento, California.

and judging criteria. If the schedule states, for example, that a pleasure class is to be judged on manners, performance, quality, suitability to rider and conformation, that is the preference order for awarding points. As a rule in junior, amateur, and ladies' classes manners carry a premium, but in the gaited stake classes, where the horse will probably be ridden by a professional, the criteria are performance, presence, quality and conformation – which does not mean manners are unimportant, as they reflect on performance. Many schedules state: 'The judge may require unruly animals to be removed from the ring'.

The hunter division approximates most closely to events found in the British show ring, but the similarity ends with the designation 'hunter'. The American hunter division offers a tremendous variety, being sub-divided into pony hunter, junior hunter, first- and second-year green working hunter, open working hunter, conformation hunter and handy hunter, each class title indicating the category in which an animal may be shown or the prime factor concerning judging.

All American show hunters jump, and a horse entered in the green working hunter division will compete in three classes before the championship and reserve championship is awarded. These three are the first- or second-year green working hunter over fences, green working hunter stake, green working hunter under saddle. In this last the horse, having already demonstrated his ability, is not required to jump.

The striking show stance of an American Saddlebred.

In classes for hunters over fences, horses enter the ring alone and jump a set course at an even hunting pace, the judging criteria being evenness of pace, jumping ability, manners and hunting soundness. In working hunter classes conformation does not count other than as an adjunct to ability and ease of movement. A hunter must jump on very light contact, seeming to flow over fences, taking off and landing without perceptible alteration of stride. His head must be carried low, reaching out over his fences.

In classes for hunters under saddle horses must gallop on and from the gallop come to a rapid halt, as they may need to out hunting. Some hunter classes are conducted in the ring, but many are run over an outside course giving a better chance to competitors to move at a true hunting pace. Basically each hunter division follows the same pattern, variations being accounted for by a horse's experience, the rider's age and classes open to amateurs. First-year horses jump 3 ft. 6 in.; second-year horses 3 ft. 9 in. Open classes have fences of 4 ft. to 4 ft. 6 in. The majority of American show hunters are purebreds standing well over 16 hands, but though Thoroughbreds are of a heavier stamp than those usually found in Britain. Of interest is the hunter appointments class, where a percentage of marks is awarded for correct turnout – including the carrying of a filled sandwich case and flask.

Closely allied to hunter events are the green and open jumper classes. Many show-jumpers gain their initial experience over fences in the hunter division, where they learn to go with much more stability than their English counterparts. The American show-jumper has a variety of events to choose from, the emphasis being not so much on speed as the ability to jump a really clear round. There are several 'tables' under which horses jump. In some, touching obstacles with any part of the horse before the stifle is penalised by one fault and behind the stifle by half a fault, so greater accuracy is required.

The novice is termed a 'green jumper'. Once upgraded he is known as an 'open jumper', with classes being correspondingly more demanding. Green jumpers may also jump in open classes.

An event popular with all three 'seats' is the pleasure horse class, which is judged on performance, manners, suitability to rider and conformation in that order, unless otherwise stated. All pleasure horses except the Tennessee Walker show at walk, trot (jog for Westerns) and canter, great emphasis being placed on ease of handling. Hunt seat horses must move in a relaxed manner and on very light contact, any animal requiring a suggestion of strength for control being considered ill-mannered. Correct turnout is the same as it is for hunter classes, with horses' manes and tails being plaited. Saddle seat entrants show considerably more impulsion, particularly at the trot, with the horses more collected and more on the bit. The correct tack is a straight-panelled saddle with big skirts to protect clothing, and a double bridle in narrow leather. Horses are shown with loose manes and tails, though the forelock and section behind the bridle is braided with coloured ribbon. Western pleasure horses shown under the stock saddle must work on a loose rein, their gaits being easy, low to the ground and supremely comfortable. High head carriage is penalised, as is excessive use of spurs. Whips, permitted in hunt and saddle seat, are absent in Western pleasure classes.

The pleasure class is probably the only class which is included in every breed association's list of classes, as it is the least specialised. For many horses it offers an introduction to the show ring.

The Saddlebred and Tennessee Walking Horse show as riding horses, which makes them ideally suited to the many pleasure classes schedules include. However, although only shown on the flat and not now being considered working horses, the gaited breeds show in many other classes than pleasure classes.

Walking Horses have three gaits – running walk, flat-footed walk and canter. The difference between show and pleasure types is obvious even to an inexperienced eye. Pleasure Walkers do not have the tremendous action or the presence of their more showy brothers. In apportioning merit the show walker, with a fast, extravagant, running walk, has the edge on his rivals, a ful 40 per cent of points being awarded to this one gait; the balance of points is equally divided between the flat-footed walk, canter and conformation. This division offers classes under saddle for two-year-olds upwards, and the monetary awards in the stake classes are high, but the cost of keeping a show Walker is correspondingly expensive, the average show Walker being professionally trained and shown.

Saddlebreds, shown as both riding and fine-harness horses, have two distinct types: the three-gaited horse, with hogged mane and tail, shown at walk, trot and canter; and the five-gaited horse, with full mane and tail, shown at walk, trot, canter, slow gait and rack. Judging criteria for the three- and five-gaited horses are performance, presence, quality and conformation, the whole evolving in a show of controlled energy and brilliant animation whether under saddle or in harness. As with Walkers, Saddlebreds, other than pleasure mounts, are usually produced by a professional.

Saddle seat equitation riders are mounted on three-gaited horses, such animals best showing riders' horsemanship. In both hunt and saddle seats the mounts used are frequently referred to as equitation horses, as they have the utmost reliability blended with correct paces, and although it is the riders under scrutiny that indefinable something attracts the judge's eye.

Thirdly comes stock seat or Western riding. Under this heading is the greatest variety of competitions. Events open to Western riders include pleasure and equitation, judged on a similar basis but with breed and seat requirements different. In more specialised categories there are cutting contests, where the horse cuts a given animal out of the bunch and prevents its return until the rider signals by a hand laid on the horse's neck. Reining classes show the manoeuvrability of the cow horse working at speed. Stockhorse events call for a mount that is nimble, surefooted, able to work a rope and to remain ground-tied when the rider is dismounted. The trail horse negotiates a number of obstacles likely to be found on the trail. Here the judge looks for the horse that is workmanlike and co-operative as well as showing good and easy paces.

No list of Western classes would be complete without a mention of the 'games' events. These are comparable to gymkhana classes but more specialised. Horses are often kept for just one type of event, as competition is needle-sharp and the rewards worthwhile. All Western games are run against the clock. Number one in popularity is barrel racing, a contest over a cloverleaf course. Although entered almost exclusively by girls in larger shows it is no gentle sport, needing courageous riding and split-second timing, the winning ride being separated from the also-rans by tenths of a second. Next to barrel racing comes pole bending–not the type seen in English gymkhanas, but the electrifying sort that makes you wonder if the horse has run at all. A pole-bender changes leads like lightning, then bends nearly double before starting on the homeward run. A crowd-pleaser is Western pickup, evolved from the old idea of rescuing a stranded man. One rider picks another up between given points without slacking pace, necessitating tremendous agility by the man being rescued.

These three events appear on most schedules, but the list of possible games or speed classes is almost endless. The breed most popular for speed events is the Quarter Horse, which possesses the compact body and quick acceleration needed for rapid twists and turns.

Apart from the three distinct riding seats there are many other classes appearing in American show schedules. Colour bursts into the arena with the Parade horses. The stock saddle used bears little resemblance to a working saddle, being heavily embellished with silver in varied designs, each owner furnishing his horse with the most elaborate equipment money can buy. Indeed, in many cases the equipment far outvalues the mount. The horses used are predominantly Saddlebreds of the heavier sort, standing at around 16 hands, essential in view of the tremendous weight of the silvered equipment. The breed's presence is used to full advantage, as Parade horses show marked animation and elevation in the parade gait, a balanced trot not exceeding 5 miles per hour. In judging, 75 per cent is awarded to performance, manners and conformation, and 25 per cent to appointments. Flashy palominos and chestnuts are popular colours with exhibitors, their natural attributes adorned with coloured ribbons in mane and tail, the hooves sparkling with gold and silver paint. Not to be outdone by his horse, the rider wears gleaming jewel-coloured fringed shirt and matching trousers, crowned with a fancy stetson, while his patterned boots jingle with ornate silver spurs.

Harness horses and ponies frequent the show ring but again, as with other phases of American horsemanship, there is a wide variety, from the fine-harness horses shown to a four-wheel vehicle, and performing at an animated walk and trot where the emphasis is on brilliant action rather than speed; through the familiar hackneys; and on to the roadsters. The horses are of Standardbred breeding and show at the jog trot –road gait–which means a fair travelling pace, and at speed where the drivers, dressed in stable colours as if on a racetrack, really turn them loose.

Of all American show horses the roadsters, especially when they turn on that blinding burst of power and speed, are the most fascinating, being run a close second by the agile, clever-thinking cutting horses, who surely contradict the often-propounded theory that horses are none too intelligent.

For sheer wealth of beauty in horseflesh coupled with versatility we must surely look to the Arabian shows, where the desert horse is shown under all three riding styles and, in addition, as a side-saddle mount, the ladies wearing elegant and colourful period costumes. Another event is the Arabian costume class, where horses and riders enter the ring at full gallop fitted out either in Bedouin style or, for the ladies, as a version out of the *Arabian Nights*. One other competition popular with Arab exhibitors is the versatility class, where horses show under Western and English tack, the riders making an unofficial race of the change-over. Horses must show true Western and English gaits to be among the prizewinners.

In a country which possesses more breeds than any other, many of which either originated in Britain or drew on British blood for foundation stock, it is hardly surprising to find all these breeds catered for in such variety. Shows do reflect the regional breeds and styles of riding, but the overall American show scene presents a scope unparallelled in any other competitive horse-loving nation.

AUSTRALIA

Australia has a great variety of activities involving the horse; many of them are adaptations or imitations of spectacles, sports and show events from Britain and the United States, but there are also a number of uniquely Australian sports and competitions.

One of the most colourful and exciting spectacles is provided by rodeo, an important part of the Australian horse scene and a great crowd-pleaser. Rodeos in the various states are arranged in circuits, each fixture lasting for one or more days, with top professional performers travelling from town to town in their big, heavy-laden, dust-covered cars to try to take the attractive prize money away from the local 'ringers' or cowboys.

The most popular events as far as the spectators are concerned, carrying the most prize money and with the finals being billed as the star features of the show, are the buckjumping competitions, divided into bareback and saddled divisions, in which the cowboys try to stay astride head-plunging, back-arching wild horses. Next most popular is the bull riding, particularly since the introduction of the fierce, thick-horned, high-humped Brahman cattle of India. When the big grey Brahmans flip their riders free, they often turn and attack the fallen men, and this is when the rodeo clowns spring into action. Dressed like circus clowns they are in fact brave and skilful bullfighters, and with capes and with their own bodies protected by nimble footwork, they decoy the dangerous bulls away from the thrown contestants – a number of whom may be in no condition to get up and run.

Bulldogging is another rodeo event which is fast and spectacular: the contestant gallops alongside a fleeing bullock, then throws himself from the saddle to grip the steer's horns and wrestle it to the ground. Calf roping, recently introduced from the United States, is fast gaining popularity; it requires a high degree of skill on the part of the rider and of training on the part of the horse.

A uniquely Australian competition, campdrafting, often precedes these rodeo contests. Here, a rider separates a large bullock from a group of cattle, then drives it at the gallop around a large course marked out with upright poles, often using his mount to shoulder the beast over in the direction in which he wants it to travel. The rider, on his speedy, long-striding Waler, must stay right on top of the hefty bullock all the time in order to keep it on course, and if the steer should suddenly decide to change direction and cut under the horse's neck, both man and horse can be in for a nasty spill.

During the last few years American-style cutting contests have been introduced, following the rising popularity of the Quarter Horse in Australia, and this sport is catching on quickly.

Polo is played in many of the country districts, where enthusiasts often use the infield of the local racecourse as a polo field. In certain areas, where the game has been played for generations, the standard is very high indeed, and a number of Australian polo teams have won fame abroad, the best known on the international scene today probably being the Skenes.

Much more popular, though, is 'poor man's polo', polocrosse, which is played by women as well as men. This Australian invention supplies all the fun, action and thrills of polo for the players at only a fraction of its cost, since each player only needs one mount throughout the game. Polocrosse is rather like a horseback version of lacrosse: the ball is scooped up in a small net at the end of a long stick and is then carried or thrown. This game is played in a much more restricted area than is polo, and the ponies do not have to gallop either so far or so hard – hence the fact that a player can manage with only one mount.

A bending race at a gymkhana near Canberra.

There are a number of one- and three-day events held in various parts of Australia, with the greatest interest being shown in the southern states. The standard here is very high, and Australian three-day event teams have done extremely well in international competitions, winning the gold medal at the Rome Olympics in 1960 and the bronze in Mexico City in 1968.

Dressage in its own right – as opposed to its forming just one part of eventing – is increasing in popularity, and the skill of the accomplished dressage rider and the high degree of training of his mount are much admired. However, since Australians are generally geared more towards active sports on horseback, the higher echelons of dressage activity will probably continue to remain the province of only a few riders.

Endurance riding has recently gained a high place in the Australian equestrian calendar. The big event is the Quilty Cup, held in September each year over a rugged up-hill-and-down-dale course of a hundred miles in the Blue Mountains near Sydney. This demanding sport has been catching on fast, and there are now a number of one-hundred-mile and fifty-mile competitive rides held in various parts of the continent.

Surprisingly enough, foxhunting is a traditional equestrian sport in Victoria, South Australia and Tasmania, where the first settlers introduced foxes from England in order to provide sport for themselves. Near Sydney, the capital of New South Wales, draghunting is also enjoyed, the hounds being followed across the rolling countryside by inhabitants of the big city out for some weekend sport.

Racing is a major Australian passion, and carries on all the year round. The biggest event of the racing year is the two-mile Melbourne Cup. Australian racecourses are noted for their excellent facilities for the public, as the sport is considered top-class general entertainment. The city racetracks are situated among lawns and flower gardens, with large modern stands giving patrons of all enclosures the best possible view of everything that is taking place, from the parade of the glistening Thoroughbreds in the paddock before the start of each race to their final furious battle down the straight.

Most training of racehorses takes place at the racetracks in the early morning, with the horses working out on concentric tracks inside the racecourse proper. Most of the time his horses are working, the trainer will have his eye on the stopwatch. As a result Australian jockeys, who ride almost every morning, develop a very keen sense of timing, which is probably an important factor in their very considerable success abroad, notably in Britain and France.

A very colourful facet of racing in Australia is provided by the 'picnic' race meetings, held in the outback of this vast country, where amateur riders and their grass-fed mounts compete against each other for small prizes on primitive, dusty bushland racetracks. The sport is everything at these unique meetings. Women jockeys have been taking part in special 'ladies' bracelet' races at these fixtures for some years, and in the more remote areas special races are arranged for the dusky-skinned Aboriginal stockmen – who form a large part of the labour force on the cattle stations, and who come yelling down the straight as if their lives depended on it! Picnic race meetings often last two, three or more days, and as well as the fierce amateur competition there is much eating, dancing and drinking for the bush-dwellers during this important annual social event.

Trotting and pacing are both very popular activities, and in the big cities the trotting tracks are lavish installations where meetings are usually held at night under floodlighting. Racegoers can watch the action through the plate-glass windows of restaurants at the track, and going 'to the trots' is a popular family outing.

Hard-fought trotting and pacing races also take place at many agricultural shows, and the close-run contests round the quarter-mile showground tracks provide many thrilling spectacles.

The Australian show ring provides a wealth of other events. Two of the most spectacular are high-jumping and the water jump. In the high jump, riders come racing at a six-barred obstacle with huge wings on either side, and send their mounts soaring into the air to clear it and to land in a variety of startling attitudes in the deep sand on the other side. The world record for the high jump, although unofficial because the contest did not strictly conform to FEI regulations, is held by the Australian horse Gold Meade. Ridden by the intrepid Jack Martin, Gold Meade cleared 8 ft. 6 in. at Cairns, in the far north-east of Australia, in 1946. However, high-jumping has been excluded from the programmes of a number of shows in recent years in response to the much-voiced complaint that it often involves cruelty during training.

The water jump is another exciting event. One at a time the horses come at full gallop the length of the arena in an attempt to clear a wide but very shallow water jump. And it is indeed an impressive sight to see a tall, deep-chested, long-striding water-jumper come hurtling down the showground, suddenly to lift into the air above the glinting surface of the water and land without raising a splash on the far side.

Show-jumping is featured at all Australian shows and, as in other countries, is the biggest crowd-puller. Courses are built to international standards and competition is very keen.

Many types of hack event are featured at Australian shows. There are events for 10 st., 12 st. and 14 st. hacks, ladies' and gentlemen's hacks, educated hacks, open classes and those for pairs and teams. However, the Australian show ring hack is very different from its classic English counterpart. The Australian variety is usually a tall Thoroughbred, perhaps 16 to 16·2 hands, and often a retired racehorse. The contestants are first seen by the judge walking, trotting and cantering in a large circle, then they are called in, lined up and give their individual shows. These generally involve a figure eight at the canter, with great importance being placed on a smooth change of lead.

The Galloway hack events are a special feature. The Galloway category is uniquely Australian, based only upon an animal's height: in Australia, ponies are under 14 hands, and Galloways those from 14 to 15 hands. Finally, there are a number of pony hack classes; the ponies are not always ridden by children, though, and sometimes a silver-haired old bushman will be seen putting a beautifully-prepared pony through its paces.

Every Australian show features a considerable number of best rider competitions, starting with those for very small children (though they rarely appear on leading reins), and going up the age scale to include adults. Some shows even have riding classes for parent and child.

Carrying considerable prestige on the Australian show circuit are the turnout classes, in which women riders are judged on their clothing, personal grooming, tack and the presentation of their mounts. There are classes for both formal and informal turnouts, and since it is generally the competitor with the newest and best outfit and tack who wins the class, the contest is strangely 'snob' for a country which generally prides itself on its egalitarianism!

In-hand classes in Australian show rings are rich in variety, as the Australian horse world is now the home of many breeds. There are classes for Thoroughbreds, Arabians, Quarter Horses, Appaloosas, American Saddlebreds, Palominos, various breeds of British native ponies – with the Welsh Section A and the Shetland being particularly popular – and a growing interest in the Connemara, Australian pony and Australian Stockhorse.

All the bigger shows have classes for harness horses and ponies, too. Depending on the locality, these can range from the traditional British hackney to general-purpose driving animals. There are not a lot of harness enthusiasts in Australia, but those who are interested in it take great pride in their turnouts.

The start of a race at a 'picnic' meeting.

A competitor in the spectacular Mareeba Rodeo in Queensland.

One of the most recent innovations has been the introduction of colourful parade and costume classes, modelled upon this event in the United States. They are a great crowd-pleaser, since high-stepping Palominos carrying Spanish caballeros, and fiery Arabians with desert sheikhs astride, add an exotic and entertaining note to the show ring.

Classes for stockhorses are in evidence at most shows. The stockhorse can be of any breed or mixture of breeds, and points are awarded for conformation, temperament and general presentation. The horses must give an individual display, usually making a 'dry run' somewhat akin to an American reining pattern.

At all the 'Royal' shows, which are held annually in the capital cities of the various states, there are classes for police horses. The mounted police detachments of the state governments are very popular, and often give displays such as musical or pattern rides at the big shows. The troopers and their mounts also compete against each other in various events, ranging from those judged purely on presentation to others which resemble fairly advanced dressage tests.

The novelty events are a big feature of most Australian shows. They consist mainly of pole bending and flag and barrel racing. Bending is like its British gymkhana equivalent, and to win you need a fast-starting, agile and responsive animal who has been carefully trained. In flag racing, small flags are placed on a line of poles. The contestants must race out, snatch a flag – one at a time – wheel around the upright, and come speeding back to the starting line to drop the flag into a receptacle before galloping out to collect the next flag. Barrel racing can either resemble bending, or alternatively the competitors race against the clock in a figure of eight round just two barrels.

Across the Tasman Sea in New Zealand, many of these Australian activities can also be found. The New Zealanders also make a feature of working hunter classes at their shows, and dressage is very popular, too, in these lush green islands.

Polocrosse at Sydney Royal Easter Show.

Chapter five

Endurance riding

BRITAIN AND THE UNITED STATES

A phase of horsemanship established in America and now also becoming popular in Great Britain is the sport of endurance riding, possibly the only equestrian event where the initial competition is against the rider's own preparation and training rather than other riders. Finishing the course is of prime importance; awards and merit placings only secondary. This attitude relieves endurance riding of the tensions inherent in other competitions, leaving one's energies to be channelled into completing the course with the minimum of fatigue to one's mount. On this point American and British riders agree, though a closer look reveals many differences between them.

Britain, being fairly new to endurance riding, has fewer events to offer. It also lacks vast areas of rugged terrain, and the same freedom to ride over open country as is found in the United States, nor are there intense climatic conditions to contend with. British rides are considerably less taxing in consequence. On the whole British riders do not expect as much from their horses, although there are some to whom endurance riding, with its heavy demands on horse and rider, is becoming a way of life.

Britain's major ride is the seventy-five-mile Golden Horseshoe, held each September at one of our many racecourses. To date host courses have been Salisbury, Goodwood, Cheltenham and Beverley, each having the stabling and other facilities required as well as being sited in hilly country. To enter the Golden Horseshoe riders and horses must qualify at one of the shorter rides held throughout the spring and summer months.

Qualifiers are one-day rides over a minimum course of forty miles, with a maximum of sixty miles. Horses must be at least five years old and a minimum of 14 hands in height. There are no weight divisions, but riders must be at least seventeen years old. BHS rulings require all contestants to wear a hard hat; spurs are not permitted and only riders of stallions may carry a whip. Entrants must pass all veterinary checks and travel at a minimum speed of six miles per hour, no advantage being gained by speeding, though many treat the ride as a race. Successful entrants are awarded rosettes and certificates. The certificates, signed by a BHS steward, must be produced before the entrant is accepted for the Golden Horseshoe.

During the spring and summer hundreds of horses enter qualifiers, many riders enjoying the shorter ride without intending to tackle the stiffer 'Golden' course. This is held over two days, fifty miles being covered the first day and twenty-five miles the second. In 1972 seventy-nine horses entered for Goodwood; sixty-eight of them made it to the start and thirty-eight finished.

The award system at the Final covers a wide range. The ultimate is a 'gold', awarded to horses gaining 200 veterinary marks and averaging a speed of 9 miles an hour; silvers go to horses with 9 miles per hour and 140 marks, and also to those doing 7½ miles an hour with full marks; bronzes to those doing 7½ miles per hour with 140 marks, and to those doing 6 miles per hour and gaining full marks. Rosettes are awarded to competitors finishing the course at 6 miles per hour and scoring 140 marks. In 1972, for example, four golds were awarded, 20 silvers, 13 bronzes and one rosette.

All types and breeds of horse enter the Golden Horseshoe. To take 1972 again, there were horses of Arabian, Thoroughbred, Welsh cob, Hanoverian, Trakhenen and Clydesdale blood, as well as crosses involving three of the larger British native ponies–Connemara, Highland and New Forest. Many entries were just good all-round horses with no particular breeding.

The breed showing its stamina and ability to withstand stress to heart, wind and limbs to a marked degree is the Arabian and part-bred Arab. In all our 'Goldens' Arabians have accounted for most of the gold awards and a large share of the silvers, three of the four 1972 golds being awarded to pure-breds.

In addition to qualifiers and the 'Golden', many riding clubs hold unofficial long-distance rides, each year showing a steady increase in events and entries. The British system of conducting endurance rides falls between those operating in America, where 100-mile three-day competitive trail rides put the accent on condition and the marathon 100 miles in one day events popular in the western states on speed plus condition.

America's most famous annual endurance ride is the Tevis Cup 100 miles in one day ride. Held in late July from Tahoe City to Auburn, California, it covers ruggedly beautiful terrain traversing the Sierra Nevada and following the old western states trail through forest and mountain, across rivers and into canyons.

The Wells Fargo Company, famed in many a Western film, owed much between 1872 to 1892 to its President, Lloyd Tevis, who developed the company with its express riders, stages and fleet horses. Closely allied with Tevis was James B. Haggin, President of the Anaconda Copper Company and a breeder of fine racing stock successful in California during the days of the gold rush. From this historical background the present ride takes its name and its premier award, the Tevis Cup, the award for the best-conditioned horse, the James B. Haggin Cup; and the route over the Sierra Nevada used by the gold miners of California.

Rules governing the Tevis Cup are that horses must be at least five years old and carry a minimum of 150 lb. Each entrant must be accompanied by a veterinary certificate of soundness. Horses assemble on the Friday prior to the ride for a thorough physical inspection by a team of veterinary surgeons, and on Saturday morning fit entries trot out in the pre-dawn dark at 5 a.m.

There are three compulsory halts of an hour each during the ride, where horses are vetted. Evaluation criteria, in addition to cases of lameness, are pulse and respiration, any horse not recovering satisfactorily within the hour being withdrawn. There are also three additional checks of fifteen minutes each.

Riders finishing within the twenty-four hour limit are awarded a silver and gold belt buckle. The Tevis Cup winner is the first horse across the finishing line at Auburn, and many outstanding efforts have been recorded of less than thirteen hours' riding time. The James B. Haggin Cup is awarded to the horse among the first ten finishers considered to be in the best condition after being vetted the morning after the ride. That is a real test, for with cessation of movement stiffening often comes and only a really fit horse is still limber after a night's rest.

Though the Tevis Cup holds pride of place among American endurance rides there are many others beckoning enthusiasts for this sport. Some are 100 miles in one day rides, others are of fifty miles, these having time limits of ten, eleven, or twelve hours according to the individual ride. Almost all these western states rides are over terrain that I doubt the average British rider would dream of asking his horse to cover.

Eastwards across the United States the type of endurance ride changes to the 100-mile three-day competitive trail ride. These are not as strenuous as the 100 miles in one day variety, but I have ridden on several in Florida's deep sand and North Carolina's Smoky Mountains, and found them a real test of horses and horsemanship. They are far more taxing than Britain's Golden Horseshoe, mainly because of more stringent veterinary inspections and tougher rules–no limbering your horse up prior to starting was one rule causing many downfalls. Horses have to be Fit, with a capital F, to come out of their boxes and snap straight into a ground-eating trot while the vets, who are also the judges, award fitness scores.

The three-day 100-miler season opens with the Florida Ride held in the Ocala National Forest. The operational base is the Sewell Ranch at Umatilla, where horses begin to arrive several days before the ride. All entries must be accounted for by noon on the day before the ride for the preliminary veterinary inspection. There are three divisions: heavyweights carrying 180 lb. and over; lightweights 155 lb. to 179 lb. (riders not making the weight carry lead); and juniors, who ride at catch weight but must be at least eleven years old.

Florida's ride is run in March at orange

blossom time, when the groves are heady with perfume and still bear the previous crop of oranges, grapefruit and kumquats. The marked route through groves and forest covers forty miles on each of the first two days with a seven-hour minimum time, with thirty minutes early allowance, and eight hours maximum. The final lap is of twenty miles in three hours, with a fifteen-minute early allowance and four hours maximum. Horses going too fast or too slowly are disqualified. Any horse exceeding the minimum time is allotted time penalties. Apart from the time factor which, with its generous leeway, is easy to keep to on a fit horse, judging is 100 per cent on natural, superb condition. All entries start with a 100 per cent fitness score. Should any horse have blemishes, peculiarities in way of going, be a fussy eater, etc., the rider declares this beforehand, as everything is taken into account in judging, even to a horse refusing food because he is overtired. A good endurance horse copes with fatigue and eats well. No liniments or salves are permitted, and water must be given only at ordinary temperatures.

Vets constantly check competitors during the ride, sometimes stopping horses at critical points–a favourite being the top of a mile-long climb up a pipeline a foot deep in holding going where horses' heartbeats were recorded. Other checks are visual, judges looking for signs of fatigue; horses travelling eagerly; scouring animals; heavy sweating, etc. Their observations, both favourable and unfavourable, go into each horse's folder and are used in the final assessment.

Horses cross the finish line at a trot, with a vet watching for soundness. Once dismounted, riders cool their mounts out, anything up to three hours being allowed for this according to the temperature. Florida's heat, even in March, can be wickedly high. From a freezing start on one ride the temperature soared to well over 100°F. by mid-afternoon. If night temperatures reach freezing fruit farmers are in trouble, needing oil burners to protect their crops.

It is advisable to keep your horse moving during the cooling-out process, gradually loosening the girth and relieving pressure on his back. Let him unwind slowly, easing his muscles into inactivity. Once in his box he may not leave it until called for next morning's start, except for the nightly veterinary examination when the vet goes over each horse minutely, feeling for hot backs, fever swellings, tender legs, clear eyes, healthy gums, over-reach and forging marks, and complete soundness as the horse is trotted out.

The next day is a repetition of the previous one, but with attendant stiffness for the not-so-fit horses (not to mention riders). On this lap several riders will withdraw as the going takes its toll, but the hardest part is the last twenty miles. If a horse leaves his box freely on the final morning he will most probably make it to the finish. Even with the last mile clocked off the ride is not over. Final judging takes place the morning after, when placings are decided.

Apart from ride placings there are frequently awards for the best Arab, Quarter Horse, Appaloosa, Florida Bred, and best unregistered horse. There are also awards for ladies', men's and junior horsemanship and sportsmanship.

On every ride I have been fortunate to participate in the bulk of awards have gone to Arabians. On the Florida 100 four out of six placings went to Arabs in both the lightweight and the heavyweight divisions, and all Arabs in the junior were placed.

After Florida come the Virginia, New Jersey,

Competitors in the Golden Horseshoe Ride, held in the West Riding of Yorkshire.

The supreme test of fitness: this award-winning Arab, owned and trained by the author, after completing the 100-mile Florida ride.

Vermont and North Carolina rides. The last of these is very similar in form to the Florida ride, the only major difference being stony terrain, necessitating borium on shoes, and mountains involving steady climbs of several miles' duration. Horses competing in at least three rides are eligible for the championship, which is awarded on a points basis.

The value of Arabian blood cannot be overlooked. In all American rides, as well as in Britain's Golden Horseshoe, this breed is outstanding for durability, recovery and stamina. The ultimate aim of any endurance rider is to produce a horse fit enough to do the ride, turn around and go again–in short to endure–and the Arab is the most rewarding horse to work with in achieving this aim.

AUSTRALIA

Competitive endurance riding is a horse sport of recent origin in Australia. Modelled upon the great endurance contests of the United States, such as the famous Western States 100 Miles One Day Ride, popularly known as the Tevis Cup, endurance riding in Australia has caught on quickly, and a number of fifty-mile and 100-mile competitions are now held every year.

The biggest event in the endurance riding calendar is the Quilty 100-Mile Endurance Ride, which takes place over a rugged and very testing up-hill-and-down-dale bushland course in the Blue Mountains inland from Sydney, New South Wales. 'The Quilty' was first staged in 1966, when it was won by the blaze-faced Arabian stallion Shalawi, ridden bareback by his owner, Gabriel Stecher. Shalawi's performance made such an impact that there is now a fifty-mile endurance ride named after him, The Shalawi Memorial, which takes place in the Yarra Valley near Melbourne, Victoria.

In most endurance rides, competitors are judged under two headings: the time taken to complete the ride; and the condition of the animals after the journey. Thus, one winner is always the horse that takes the least time for the trip, but the winner of the award for the horse which finishes in best condition might well be another animal. A really outstanding horse, however, is the one that regularly collects both trophies.

The regulations for a typical endurance ride would demand that all horses taking part must be at least five years old, be shod and carry a minimum weight of around 150 lb. All horses would have to be presented for a pre-ride veterinary examination between certain hours on the day before the ride was due to start, and then and throughout the ride the opinion of the veterinary surgeons would be final in all matters concerning the condition of the horses.

Other fundamental rules would be that the same horse and the same rider must pass all control points and remain on the marked trail in order to qualify for any awards, and that the abuse of horses and the use of stimulants are strictly prohibited.

Contestants leave the starting line one by one, spaced a few minutes apart, and on a 100-mile ride the veterinary checkpoints would ideally be distanced twenty-five miles apart. All riders must stop at each checkpoint for a certain minimum period of time to have their mounts' respiration, pulse rates and temperatures checked by the vets, assisted by helpers who record their findings. Any horse which does not meet certain minimum standards with regard to rate of breathing, pulse and degree of temperature is not permitted to continue. Whenever possible, checks are held at places with water, and with plenty of space for parking cars belonging to spectators, and to friends and helpers of the contestants.

The most important task, and often the most difficult one, in staging an endurance ride successfully is that of the proper marking of the trail to be followed in the unfenced, uncultivated, uninhabited bushland. No matter how well the veterinary checks are staged, and how attractive the trophies may be, the ride will not be much of a success if a goodly number of the competitors get lost along the way! The route to be followed must therefore be well marked with plenty of indicators of a colour that readily catches the eye.

Once all the contestants have left the starting line, two riders follow up in case someone has an accident or is left behind for any reason.

In Australia the most popular mounts for endurance rides are Arabians and horses carrying some proportion of Arabian blood, and the country's own Walers, compact, deep-chested, long-striders with a lot of Thoroughbred blood. It is in these contests, incidentally, that you see the Arabian in his true pristine beauty, not as an over-fleshed, too-rounded show ring bauble, but as a lean, hard, all-quality riding horse with the long-distance heart of a lion and with fine, steely tendons and rock-hard hoofs.

Anyway, whether Arabian, part-bred Arabian, Waler or whatever, the Australian endurance rider looks for an animal that is tough, compact, absolutely sound, with plenty of stamina, lung capacity and heart room, and which is comfortable to ride. An animal is generally considered to be at its best for this work when it is between eight and ten years of age.

Conditioning and training for an endurance contest of 100 miles would normally take from four-and-a-half to five months of consistently steady work. Before the conditioning process begins, special attention is paid to worming and to teeth, and a blood test may be made to determine if the animal is suffering from some degree of anaemia. During the conditioning period some riders have their mounts' blood checked from time to time to make sure that feeding is adequate and that the feed is being utilised properly by the animals' bodies, and also to make sure that the increasingly hard work is not having an injurious effect on their horses.

Good and correct feeding is, of course, fundamental right from the start of conditioning, with sufficient salt always available for horses that will sweat plentifully during the coming months, and possibly a feed additive to boost stamina. Grain feed must be absolutely first class, and hay should be hard, sweet and also of top quality.

Attention to legs and feet is of the utmost importance, too, and in competition the endurance horses wear light shoes that are well nailed on. Some riders prefer to train their mounts in heavier shoes, then swap to light shoes for the actual contest, feeling that the sudden lessening of weight will give an importtant physical and, possibly more importantly, psychological boost to the animals' performances. Many riders fit pads to prevent bruises from stones.

The fitting of saddlery is of course of very great concern for a horse being ridden competitively over such long distances. Preferences with regard to seating vary greatly, from that of Gabriel Stecher, who 'cantered in' the first Quilty Ride bareback, to those who go to the trouble of importing saddles like the fairly rare McClellan military saddle from the United States. The aim is always to make the horse feel just as comfortable as possible. And, naturally, during training the riders keep a very close watch to make sure that there is no rubbing or chafing or undue pressure anywhere the saddle touches the horse.

The essentials for successful conditioning and training in endurance riding are a well thought out, practical plan of conditioning, and all the will power in the world. The endurance enthusiast must ride with almost clockwork regularity, and the luxury excuses of inclement weather, minor illnesses of the rider, boredom, etc., cannot be permitted to alter the training programme. In training for endurance riding, stability and persistence are fundamental.

Typically, a rider would start to condition his or her horse with short rides about four times a week, mainly walking, with some trotting. If the horse is new to the job, right from the start the rider would begin to school it towards the far-reaching, featherlight trot of great ease and almost mechanical regularity and efficiency that is the most important pace of the successful endurance horse. It is at this gait, in which the horse extends himself to cover as much ground as possible as easily as possible in each stride, that most of the distance in the ride will be covered.

As the weeks pass, the rider would ride, say, three days a week, with half an hour on the lunge on off days. At this stage he would be slowly increasing the distance covered in the rides. By the time two months of conditioning had passed, the rider would be covering twenty to twenty-five miles in each of the three days out, each of which would be alternated with a rest day.

By the fourth month of training, horse and rider would be going out five days a week, covering fifteen miles on each one of three of the days and some twenty miles a day on each of the other two. Training sessions would probably consist of trotting three-eighths of the distance, walking half of it, and cantering the other one-eighth very quietly.

During the fourth month, the rider would probably give his mount two fairly serious training rides of fifty miles each. These rides would, of course, be suitably spaced. And they provide sharp and revealing probes of a horse's progress: at the end of a fifty-miler, the rider should be able to assess fairly accurately from the feel and overall condition of his mount whether it would be physically capable of going on another fifty miles; and, equally important, to judge whether the animal is mentally ready to continue.

All the time this conditioning is being done, the rider will attempt to work the horse over hilly terrain that has much rocky going in order to toughen up the animal's feet. Attention to trimming and shoeing, with great care being paid to correct slope and balance of the foot, is essential during this whole period.

Finally, some riders may spend a week or more riding their mounts in the area where the actual contest will be taking place, and they will continue training there right up until four or five days before the ride is due to begin.

Then, with the long, hard slog of months of training behind them, all the endurance riding enthusiasts will hope that their mounts will measure up well in the testing 100 bushland miles that lie ahead.

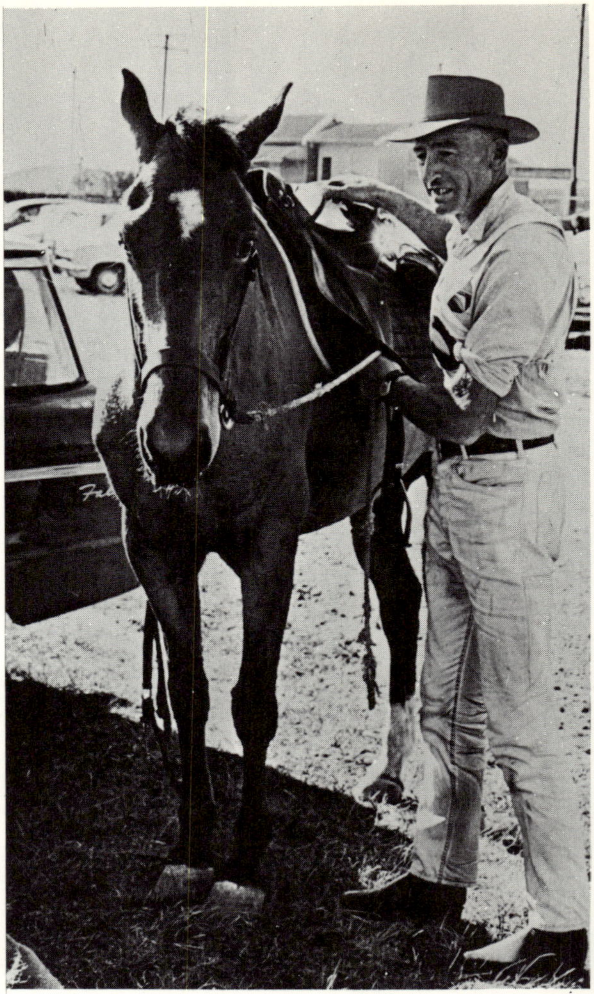

The Tom Quilty 100-mile ride is held in the Blue Mountains of New South Wales. Here a winner unsaddles at the end of the ride.

Chapter six

Mounted games and skill-at-arms

GYMKHANA EVENTS

Mention mounted games to any Pony Club members in Britain and their thoughts will automatically turn to the annual competition organised throughout the United Kingdom culminating in six teams competing in the final of the Prince Philip Cup at the Horse of the Year Show. But the origin of mounted games goes far further back into history than 1957, when this competition was first held.

Equestrian games have formed part of man's riding activities for thousands of years. In the earliest days, games on horseback were used as a form of training for the rigours of war, and cavalry regiments were some of the first to invent mounted games. It was essential, when facing the enemy, that a soldier's mount should be exceptionally quick to turn and manoeuvre in order to avoid the attacks of his opponent. As one watches the agile little gymkhana ponies galloping in and out of a line of bending poles and turning on the proverbial sixpence, it is easy to step back in time and see mediaeval knights jousting for honours in the lists. Indeed, some of the sound effects from today's competitors would have done justice to many an ancient battlefield.

In South America, where ponies are used in everyday work for rounding up vast herds of cattle, some of the oldest mounted games are still played. The agility of the pampas ponies has resulted in a demand for these tough and handy animals by the best polo players all over

Tskhenburti, a national game in Georgia in the USSR.

Points are lost if water is spilt from the buckets carried by these Australian pony club competitors.

the world. Although there is a genuine Pony Club polo tournament organised each year, there are other forms of this game which definitely come under the heading of mounted games. Cushion polo, which should only be attempted by fairly competent riders, is one variation, and involves the passing of a cushion-type object from one rider to another while the opposing team tries to wrest away the 'cushion'. As in polo, riding-off is permitted, but a firm umpire is absolutely essential to check any dangerous or rough play. Paddock polo can also be played and this requires a larger, softer ball than that normally used in adult polo. Ex-polo ponies make excellent gymkhana mounts.

Mounted games today consist of the well-loved gymkhana events which have always formed part of the smaller show's schedule. They provide an opportunity for the everyday pony and rider to compete in events expressly designed for the maximum of fun and enjoyment without requiring a top-class or particularly well-bred pony. No matter the shape or appearance, with a little practice at home even the hairiest pony can become proficient at bending and potato races. Gymkhana events are judged purely on performance and although a pony with a little more breeding will probably have an advantage on pure speed over its less aristocratic brothers and sisters, this is not the whole secret of success in such events. The pony must be extraordinarily handy and obedient to hand and leg, willing to stop and go instantly it receives the aids and, in many cases, the less well-bred pony will be calmer and quieter when confronted with the strange objects that appear as part and parcel of some of the events. Approaching a stuffed sack lying on the ground can strike a pony rigid with terror, and it is only by endless hours of practice behind the scenes that one's pony can be persuaded to approach this hideous object. It is in cases like this that the less highly-strung pony will score.

Training at home to improve the pony's response to the aids can be a great help in competitions. A fast start, for example, is a tremendous advantage over one's rivals. Equally, a pony that is trained to stop dead at a given signal will save precious seconds in certain events. It is only with practice that a pony will become used to such things as balloons flapping round, and it is also as well to get the pony used to the sound of a balloon exploding, as this is quite likely to happen during the course of a race. The sound of a potato dropping into a bucket can conjure up unbelievable terrors if the pony is not used to it. As in all branches of equestrian sport, practice makes perfect, and for mounted games there is the added advantage that no expensive equipment is required; just with a few bending poles, the odd bucket and a sack anyone can school their pony to their heart's content.

It was with the ordinary Pony Club pony in mind that Prince Philip presented his trophy for a national mounted games championship. This gives Pony Club members the chance to compete at one of the major shows. The thought of riding at the Horse of the Year Show in the autumn encourages prospective members of branch teams to practise for months in advance and work their way through the preliminary rounds before achieving a place in the coveted last six teams who go through to Wembley. More important are the aims and objects behind a competition of this type – to encourage team spirit and good sportsmanship. As the races are run mainly on a relay basis the team co-ordination has to be first class. Both riders and ponies must be capable of anticipating their team mates' next move and, judging by the support given to teams in the finals, there is no lack of team spirit among the supporters either. From the beginnings of the Prince Philip Cup in 1957, the championship has proved an extremely popular event; nearly 200 branches now take part. Each team consists of five members under sixteen years of age, who ride ponies not exceeding 14·2 hands. They must be ridden in snaffle bridles and no whips or spurs can be used. Each team has a non-riding captain who is probably the team trainer as well.

This well-trained pony is undisturbed by its rider's antics.

Agility and obedience are important when every second counts.

Most Pony Club branches have always used mounted games of one sort or another as a form of exercise and entertainment after the more serious riding which takes place during a working rally. Ponies and riders alike enjoy and need the relaxation after an hour and a half or two hours of concentrated instruction. A quick game of musical sacks, a bending race or a flag relay race helps to provide variety at the end of the rally and will ensure that ponies and riders finish the day on a high note. Similarly, an informal team competition among the different 'rides' can act in much the same way. An impromptu musical ride can also be organised, which never fails to appeal as all the members try their utmost to maintain the continuity of the ride.

Once it has been decided to enter a team in the championship the team trainer will organise the prospective members for practices. Cold winter weekends seem a far cry from the electrifying atmosphere of the indoor arena at Wembley, but it is during the winter and early spring months that the prospective team must be drilled until the five members are finally chosen for the regional round in the Easter holidays.

The different races are devised by a committee and the games for the year are circulated to all branches well in advance of the competition. The events are expressly designed to test the skill and timing of the competitors as well as including favourites such as team bending, which perhaps more than any other event tests the degree of schooling and obedience of the pony and the skill of the rider, and the potato-picking scramble where vaulting on and off a moving pony is essential if success is to be achieved. A pile of potatoes is placed in the centre of the arena, a bucket is set out for each team and for a given time two members of each team gallop back and forth depositing as many potatoes in their own bucket as they can. The secret of this event is continuous motion. The pony must be trained to describe an oblong using the bucket as a pivot while the rider must be capable of leaping on and off at the right moment.

Several races require a particular skill and athletic ability on the part of the rider. For example, no matter how fast the pony, it will not win a sack race unless the rider is capable of hopping at high speed like a demented kangaroo. But again the pony can help. A well-trained mount will assist greatly if the rider clasps tightly round the pony's neck and is half dragged to the line by a willing partner, though an over-keen pony may well leave its rider trailing behind in an untidy and uncomfortable heap. In the stepping-stone dash, to take another example, the balance of the rider as he hops from one obstacle to the next will be to no avail if the pony refuses to trot alongside. Simple variations of relay races are always included in the schedule, and these test the co-ordination of the team; other races may require the pony to carry two riders at once for short distances. A most popular race involving two riders on one pony also requires a good eye and a true aim. This is the sharpshooters race, which involves two members racing on one pony to an 'aunt sally'. On reaching a marked point in front of the aunt sally, one rider dismounts and attempts to demolish the aunt sally with the supply of balls provided. On completion of this task, he remounts behind the other rider and gallops back to the finish. The simplest form of two on one pony races is the Gretna Green race, when one pony and rider gallop to pick up a second rider and simply gallop back to the finish. A simple race to organise, requiring no equipment except a marked finish.

One entertaining event originated in Russia: pushball. Two teams attempt to push a huge soft ball over their opponents' goal line. This requires a bold pony, one who is quite prepared to lend its weight and is trained to do most of the pushing. Pushball is also played in the Netherlands and is always an amusing spectator event. Once that big ball starts rolling it is extremely difficult to stop it except by lining up the defence in its path.

The emphasis in the Pony Club is mainly on skill at particular events which are easily adapted as team races, but at local shows there is still a wide variety of events to be found. Pace races such as trotting and three-pace races are popular, as is the bucket elimination. Here the competitors jump in single file over an ever-decreasing line of buckets or cans. The slightest touch to a can, refusal or run-out results in elimination and the event continues until either all are eliminated or there is only one can left. In the latter case the long-suffering steward has the unenviable task of finding a winner by sending the remaining competitors over and over the single can until the numbers are reduced to a winner. As it is essential for the judge to stand directly in the path of the oncoming horse, in order to see if there is any deviation from the central path, it is a stewarding job that is not entirely popular as there is a distinct risk of being mown down as the competitors advance!

Apple bobbing or ducking does not seem to feature in gymkhana schedules quite as frequently as in the past. This was often the last event of the day and, provided the weather had been kind, it was often a welcome end as one plunged one's head into the water trough of bobbing apples, hoping to emerge with the apple safely clasped in one's teeth. Obstacle races too, used to be popular, but they require a lot of equipment to be successful. Crawling under tarpaulins, diving through swinging motor tyres, threading needles, devouring currant buns and perhaps jumping an unusual obstacle – all are part of this most comical of races, for the spectators if not for the competitors. Inevitably one competitor would become almost irrevocably wedged in a tyre or the tarpaulin would claim a victim as some poor unfortunate thrashed about in a vain attempt to find a way out into the open. But they were fun, these complicated events, and if not wholly a test of horsemanship, they tested one's ability to remain calm under a variety of trying circumstances.

Varieties of polo have already been mentioned; a chapter on mounted games would not be complete without describing briefly some of the slightly less hectic but equally enjoyable organised games. Mock pig sticking can provide a fast and exciting game as riders attempt to 'stick' a stuffed sack which is dragged along on a long cord by one rider at a fast canter. Mock hunts can help to introduce younger and inexperienced members to the mysteries and thrills of the hunting field and are to be recommended, especially in areas where members are likely to get little or no real hunting. Again this takes a vast amount of organisation but such an event can provide much-needed instruction to any members who are not likely to be able to hunt regularly, who are nevertheless able to learn about the sport.

Criticism is often levelled at gymkhana events in that they may produce rough and careless riding, but a well-trained, well-ridden pony is a joy to watch and will always score over the inadequately trained combination. It goes without saying that such events can improve a rider's balance and agility in the saddle, and there can be little doubt, too, that the majority of ponies enjoy the excitement and fun of this branch of horsemanship.

SKILL-AT-ARMS

Ever since the first soldier mounted a horse and used that horse to convey him into a position of maximum advantage when he came within weapon's length of an adversary, competitions have been devised to test the cavalry soldier's proficiency in the use of his various weapons.

With his lance firmly gripped, the rider aims at the ring, only $2\frac{1}{4}$ in. in diameter.

Mock battles, melées, jousting, tilting, riding at the quintain or Turk's Head and running at rings – all these in one form or another were in use from the beginning of equestrian history, and it naturally followed that the pre-war cavalry soldier was no exception to accepted practice. In the main, tests of proficiency in the use of firearms alone were confined to the range, but various competitions were devised for individual arms such as the sword and the lance.

For the sword alone there were such competitions as dummy-thrusting and slicing the lemon, and for the lance tentpegging in all its

Skilled precision is needed to draw the peg from the ground and convey it over the 'carry-line' on the point of the lance.

varied forms: tentpegging by sections, Indian file, crossover, tentpegs fastened together by ribbon and taken as a half section, burning tent-pegs and two rings and a peg. Tentpegging was also done with the sword, and some say with an Army issue jack knife, but the competition that brought all the soldiers' weapons together to produce the supreme accolade of champion man-at-arms was the sword, lance and revolver competition. Each theatre of British command – the Middle East, Far East, India and the various Home Commands – held a competition to produce its own man-at-arms, but it was generally held by most cavalrymen that the soldier who was champion at the famous Royal Tournament, held in London at either Olympia or Earl's Court, was recognised as the champion of the British Army.

Sword, lance and revolver competitions are held at a number of shows in Britain – agricultural, police, military and ordinary horse shows and are usually conducted in such a way that the competitor who hits the most number of targets, no matter how inefficiently, is the winner. It is at the Royal Tournament that the competition is staged in its purest form. Remembering the original reason for the competition, that it was to be the supreme test of a soldier's skill, every facet of proficiency, horsemanship, marksmanship and weapon-handling is tested to the full. The methods of conducting the competition never vary; the diagram showing the plan of the course gives the measurements used at the Royal Tournament. Usually the accepted distances are 15 yd. between targets and jumps and a 45-second time allowance to complete the course, though the time allowance does depend on the length and width of the course from start to finish and this can vary from show to show depending on the area available for use. Three judges are required, and it is essential for them to have a good knowledge of the correct handling of a cavalryman's weapons.

The contestant comes to the starting line mounted, carrying his sword at the slope and with his revolver loaded with blank ammunition fastened in its holster on his belt. His lance is stuck in the ground or in a special block, point uppermost, some 75 ft. from the first target on the centre line (see diagram). On the signal to start he salutes the judges and sets off at a canter along the first line of targets. This line is the test of his sword-handling. He jumps a 2 ft. 6 in. brush fence, bringing his sword to the 'engage' position as the horse lands over the fence. He is now set to attack a 'cavalry dummy' on his right—a pad, 18 in. long and 12 in. wide with a black disc, 7 in. in diameter, set in the centre. The pad is mounted on a spring-loaded post which gives to the thrust of the sword, and is driven into the ground with the top of the target pad 8 ft. 6 in. above ground level. Now the competitor goes into the actual attack, bringing his sword from the 'engage' to the 'sword in line' position for the thrust into the target. The horse must increase its pace for it is the momentum of the horse and rider behind the sword which drives the point into, and through, the target disc.

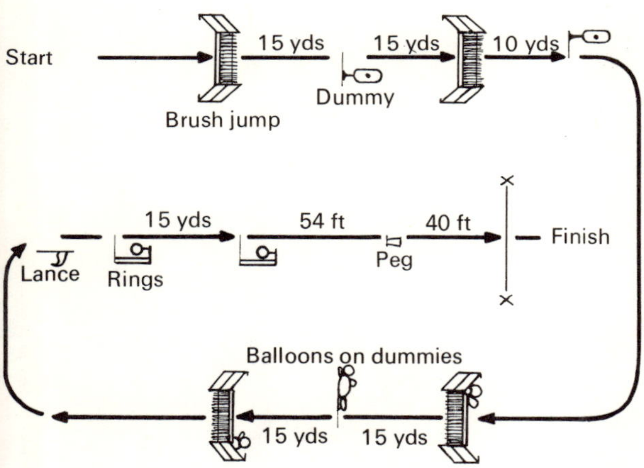

Twelve points are awarded for each of the two sword targets, giving a possible total of twenty-four. Each set of marks is made up as follows: three points for the 'engage', the point of the sword being brought forward quickly without unnecessary flourish, forearm and blade in one straight line pointing at the target, the elbow slightly bent and clear of the body to allow freedom of movement to parry or thrust, and the sword edge to the right. The wrist may be bent laterally but must not be flexed. Three points for the 'momentum': the horse must show a distinct increase of pace from the 'engage' to the moment of impact with the dummy and after the impact must resume the normal gait of canter. Three points for 'body and arm': the rider transfers his weight from his seat to his knees and stirrups, and turning from the hips brings the sword forward into the 'sword in line' position, straightening his arm vigorously so that forearm and sword make one straight line, and pointing at the target with the full extent of body and arm. Three points for the 'sword edge': without shifting his grip the rider turns the back of his sword hand to the left so that the sword edge is vertical on contact with the target and the arm is locked. The whole picture must be one of maximum efficiency and controlled aggression. Should the horse knock down the fence, one point is lost. If the sword point does not touch the black target disc, two points are lost and no points are awarded at all if the dummy is not attacked or is hit from the wrong side.

After attacking the dummy the competitor recovers his sword to the 'slope' position, jumps another brush fence and engages a second cavalry dummy, this time to the left. The system of marking and the points that the judge will be looking for are identical to the first target but this time the sword is left in the target. It often happens that after the attack has been carried through, the spring dummy will fly back and cause the sword to fall out of the target pad. This loses the competitor no points, for providing the black disc has been pierced from the correct side and the sword has been handled in a proper manner a 'kill' is judged to have been made, and the competitor will get his twelve points.

Still maintaining his horse's canter the rider swings right-handed in a half circle and approaches the line of targets on the far side of the course. This line is the test of marksmanship and the handling of the revolver. As he rides away from the last sword target the rider draws and cocks his revolver with the right hand alone. There must be no assistance from the bridle hand at all. He will then fire one round of blank ammunition at each of three balloons. The first is fastened to the top right-hand side of a 2 ft. 6 in. brush fence and is attacked as the horse jumps over the fence. The second is fastened to a block of wood at ground level on the right and is burst as the horse canters past. The third and last of the target balloons is fastened to the top left-hand side of a second brush fence and is attacked similarly to the first balloon. For each balloon burst the competitor gains two points, for a maximum total of six. If a round misfires it cannot be remedied, neither can the target balloon be attacked for a second time. The revolver must, from the time it leaves the holster, be carried over the right shoulder, pointing upwards except when engaging a target. Points for the style and safe handling of the weapon are awarded by the chief judge. After the third balloon has been attacked the rider replaces the revolver in its holster, fastening the flap. This is important, as the dropping of any weapon closes that competitor's score at the point where that weapon fell.

Again the rider swings his horse on a right-handed half circle and comes up to the centre and final line of targets, this time for the lance. With the horse still at canter he seizes his lance at the point of balance from its position in the ground and in one movement brings it up to the position of 'engage'; as he does so the horse must increase its pace to a gallop. The targets to be taken on the lance are two rings of $2\frac{1}{4}$ in. internal diameter suspended from gallows 45 ft. apart, with the bottom of the rings 6 ft. 6 in. from the ground. 54 ft. on from the second ring, a tentpeg is driven into the ground so that 5 in. are above ground level. The rider attempts to take both rings on the point of his lance, drop the lance hand down until level with the instep of his boot to take the peg from the ground and carry all over the 'carry line' some 40 ft. further on when the lance is recovered and trailed. There is a maximum of twenty points made up as follows: for each ring carried on the point of the lance, five points. If a ring is struck with the point of the lance, even if it does not fall to the ground, then two points are awarded. If the peg is taken and borne on the point of the lance over the carry line, five points. If the peg is drawn from the ground but not taken over the carry line, three points; if the peg is struck and remains in the ground, two points. A further five points are awarded for the style in which this weapon is handled – two points for the position of the rider's hand at impact, two points for the pace of the horse and one point for the manner in which the rider recovered the lance after the peg had been taken.

The Chief Judge then has a further twelve points to award for 'performance': six for horsemanship and pace of the horse throughout the run and six for neatness and ability in the rider's handling of his weapons, from the time the revolver is drawn to the taking of the peg. This gives a possible maximum for the whole competition of sixty-two marks. He also deducts the penalty points: for each second over the time allowed, one point; for each fence knocked down, one point; for the first and second refusal or other disobedience before a fence, three and six points respectively. A fall of either horse or rider, a third disobedience, or an unrectified error over the course all mean elimination.

The first of three balloons: two points to gain.

Chapter seven

Bullfighting

Bullfighting was introduced into Spain by the Moors in the eleventh century, and some of the most highly skilled and dedicated horsemen in the world are still the gentlemen horseback bullfighters of Spain, the *rejoneadores*. Riding valuable, highly-trained horses, the *rejoneador* courses and kills a fast, powerful and supremely dangerous fighting bull in a performance which combines the impeccable execution of *haute école* with the feat of galloping close past the bull's reaching horns.

The high degree of horsemanship of the *rejoneador* is seen from the moment he rides into the bull-ring on his high-stepping Andalusion stallion in the opening parade into the arena. Then he rides arrogantly ahead of the silk-suited men on foot, his black Cordoban hat slanted forward, his right hand resting gallantly on the upper part of his thigh, and his mount performing a slow and lofty *passage*. The *rejoneador* will always be dressed in the stylish garb of the Andalusian bull rancher: Cordoban hat, short, trim jacket, and leather chaps over slim trousers.

Once the colourful parade has crossed the yellow sand to halt and salute the bull-ring president high in his box above the arena, the *rejoneador* becomes the focus of all eyes. As the other bullfighters slide behind the red barrier at the edge of the ring, the horseman will set out to circle the *plaza* at the canter, his black hat raised high in salute to the audience, and his long-maned stallion full-passing round the whole circumference of the arena so that he and his rider are always facing the crowd.

Then the *rejoneador* will whirl his stallion gracefully on its broad haunches, and will begin an impressive exhibition of *haute école*, linking the Spanish walk with the Spanish trot, then the *passage* with the *piaffe*, and finally, perhaps, sending his mount soaring high into the air in the breathtaking climax of the *capriole*! This exhibition over, and with more serious business soon to follow, the mounted *matador* will back his horse right across the arena and out by the gate at the other side without touching the reins, his hands high in the air in salute, the horse stepping elegantly backwards in one long-reaching, supple stride after another.

Soon the *rejoneador* will ride back into the ring on a mount with the lean, slick lines of a highly-trained athlete. This horse will often show a lot of Thoroughbred blood, since speed is needed during the first stage of the spectacle, when the bull rushes fast and full of force into the ring.

The horseman will canter across the ring to where the other bullfighters are grouped behind the red fence in a splash of clashing colours. His manservant will hand him the first *rejon*, a long wooden dowel covered with brightly-coloured paper and ending in a steel blade. When the blade is driven into the charging bull's withers it breaks off and a small flag, usually in the red and yellow colours of Spain, will unfurl at the end of the stick. This can be used to decoy the bull's attacks as the *rejoneador* rides away.

Armed with the first *rejon*, the gentleman bullfighter will usually adopt one of two ploys: either he will take up a position at the centre of the ring, facing the red door through which the bull will soon come rushing into the ring or, more spectacularly, he will ride his mount right up to the bull-pen door and turn it to face the centre of the arena. The *rejoneador* will sit looking back, holding his mount poised on the bit, the super-alert bunch of muscle between his leather-clad legs ready to spurt forward just as soon as the bull-pen door swings wide and the bull can be glimpsed surging towards the sunlight.

As the bull explodes high-headed into the ring, horse and rider will sprint just ahead of it, galloping around the arena several times with the bull reaching vainly for the speeding horse's streaming tail. With several laps of the ring completed, the bull's impetus will slacken somewhat and the *rejoneador* will lead him towards ring centre, then suddenly spurt right away to the other side of the *plaza*.

Then, with the noisy crowd for once hushed, the horseman will turn and challenge the now-stationary bull, calling loudly to it as he raises the *rejon* high in the air, the blade pointing downwards. Suddenly, when the bull's fierce gaze is riveted on man and horse, the *rejoneador* will urge his mount forward into a fast run straight towards the *toro bravo*. The brave bull will also leap into action, speeding right for the on-rushing horse and man!

As the bull drops his huge head to gore, the man will curve his mount close in past the *toro* so that the horns almost graze the horse's flanks and thighs, then drive the *rejon* blade into the bull's black withers. As the horse speeds past the up-hooking bull, the little flag at the end of the *rejon* stick will flare loose, and leaning back in the sheepskin-covered saddle, the *rejoneador* will use it to control the bull's charges in much the same way as a foot bull-fighter will employ cape or *muleta*. Then he will suddenly whirl his mount away from the bull again, which slows to a trot and then stops, and will ride across to the *barrera* fence for another *rejon*.

Three *rejones* are placed in this fashion, before the horseman rides out of the ring to change horses. The animal he uses for the second stage, the placing of the pairs of shorter *banderilla* sticks, will usually be shorter-backed and more heavily muscled than the previous horse. He will be an animal that starts fast, stops quickly and can spin on a *peseta*. Such a mount is needed now, for the bull will have slowed somewhat and will have started to go on the defensive, and consequently the horseman will have to get much closer to him to incite a charge, then be able to spurt away quickly.

This stage can be the most spectacular part of the whole performance. Picture the black bull at ring centre, his white horns curved up on either side of his broad brow, staring ferociously at the man astride the heavy-hammed Andalusian-Arab near the red fence. The man calls to the bull, and sends his mount into a high-stepping Spanish walk towards him as he challenges with the paper-frilled *banderillas*. The band, high in the stands, crashes out a rousing *paso doble*.

Suddenly the horse breaks into a fast canter straight towards the bull. The bull hurls himself forward too and, much more slowly this time, the man skilfully tempering the speed of the bull's attack, the horse brushes safely by the horns as the *rejoneador* places the *banderillas* right on top of the withers.

After the placing of two more pairs of normal-sized *banderilla* sticks, one pair of which may be planted with the man holding a stick in each hand, the reins on the horse's neck and the horse being guided by leg pressure alone, the man may elect to bring his mount in extra close to the bull and, leaning far out of the high-pommelled, high-cantled saddle, place a very short pair of the barbs. As he drives them home, the thick horns go slashing up past his chest!

With the placing of the *banderillas* completed, the *rejoneador* will change mounts once again for the killing of his adversary with the long, slim *rejon de muerte*.

To dispatch the bull honourably in this ancient manner the horseman will have to bring his mount right in close to the bull, and he must therefore be able to rely implicitly on his horse. Momentary hesitation or disobedience on the part of his mount at this crucial stage would mean almost certain disaster for horse, man or both.

The training of a *rejoneador's* mount is a lengthy and costly business, and it is not until much of the training has been completed that the horse's suitability for the demanding task can be confirmed. In the first stage of preparation the horse is educated to a very high standard of 'Andalusian dressage', the end-product of which is a cross between an *haute école* horse and a highly-trained cow pony. As well as being able to whirl on its haunches, start, stop and weave in fluid, almost instant, transitions, the horse will be able to perform a number of *haute école* airs to perfection.

The animal is next introduced to the bulls. It will be ridden out quietly into the huge pastures where the bulls are grazing, and later be used to take part in herding them from one part of the ranch to another. Once this familiarity with fighting cattle has been established, work will start in training the horse to react to the charges of an attacking bull. However, the charging *toro bravo* is at first simulated by the *carretón*. This is an apparatus consisting of a set of bull's horns mounted above a bicycle wheel, with long handles extending at the back. A man pushing the *carretón* can imitate perfectly a bull's attacks, thus allowing the rider to train the horse in the basic movements of the bull-ring performance.

Finally, after many months of patient training, the horse's first real test takes place when it is ridden against a live fighting animal, in this case a lean, agile fighting cow in the small, whitewashed arena of a bull ranch. The speedy cow's sharp horns will be covered with leather sheaths to ensure that the horse cannot be hurt, but even so, the actual presence of a live, loud-snorting and highly-aggressive adversary will show clearly whether the horse has the necessary 'heart' for the job. It is at this stage that many horses are discarded, since it is absolutely essential for the *rejoneador* to have a mount on which he can rely utterly in the heat of combat and, if possible, one that is not only unafraid of the big black bulls but which actually goes to meet them with zest!

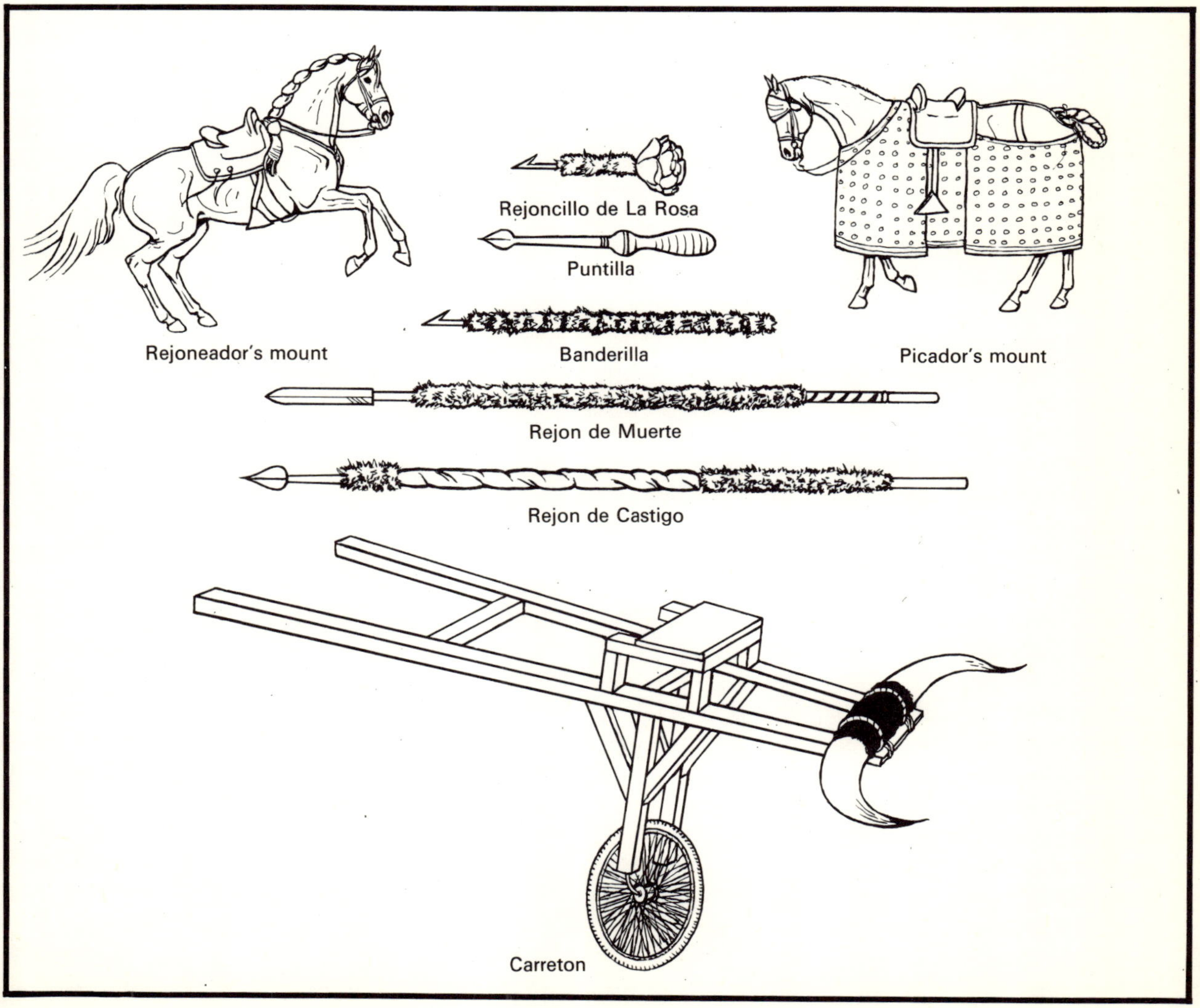

Chapter eight

Polo

Polo is the fastest team game in the world. And when you consider that it involves hitting a ball of a mere 3 inches in diameter with a long and strangely-balanced bamboo mallet from a galloping pony, with an opponent always trying to thwart that procedure so that the play is constantly twisting and turning, it is also, perhaps, the most difficult game of all to master. It is certainly one of the finest spectator sports.

Where did it originate, this tough and complex game, which is played between two teams of four on a grass ground, 300 yd. by 200 yd. in area, with goalposts set 24 ft. apart? Probably in ancient Persia where it was known as *changar* – a mallet. The word polo is derived from the Tibetan *pulu*, meaning a ball, and developed its present character in India where the British Raj learned it from the native princes. Indeed it was introduced into the western world by the English, a match staged between cavalry officers at Hounslow in 1868 being the first sign of the game in Europe. The strokes, tactics and style of play were further developed by John Watson, who became known as the father of English polo. Up to the first world war the grounds were all in the London area – Hurlingham, Ranelagh and Roehampton – and it soon became a highly fashionable game, an integral part of the London season.

Polo was first played in the United States in 1883, Meadow Brook being the American headquarters, and the initial match for the Anglo-American Westchester trophy was contested in 1886. Owing to the war, and thereafter to the overwhelming superiority of teams fielded by the United States, this famous competition was discontinued in 1939. But with the contemporary upsurge of British polo, it has been revived in the form of the Coronation cup, which has since 1971 been competed for annually at Windsor Park or Cowdray Park.

Perhaps because of the continentals' preoccupation with pure equitation, polo was never played across the Channel on as large a scale as in England or the United States; but in South America the Argentines, with their wealth, their natural facility for ball games and their handy little ranch ponies (*criollos*), soon came through as undisputed leaders of the polo world. By 1970 there were 3,000 players in Argentina, 1,000 in the United States and 400 in Britain.

In 1899, the English imposed a pony height limit of 14·2 hands, and owing to their indigenous breeds and particularly strong supply of ponies of the small type, they provided the chief market for mounts in the polo world. But in 1916, when polo had fallen into abeyance in Europe, the Americans abolished this rule, other countries conformed, and the average height went up to from 15 hands to 15·3 hands. In 1909, the American handicapping system was also adopted internationally. It rates players from –2 to +10 goals, but is no indication of goal-scoring capacity: it is simply a yardstick to indicate a player's grading in the polo world. The term 'high-goal polo' means that the aggregate handicap of each team entered in a particular championship is around nineteen and upwards. In the case of medium-goal this would be in the region of fifteen to eighteen.

The winning team in polo is that which scores most goals, that is to say the one that contrives, according to the rules, to strike the ball most often between their opponents' goalposts. The duration of a match is generally a little under one hour. And, because polo is played at a more or less continuous gallop, it is necessary to change and rest ponies; the game is therefore split into periods, or 'chukkas', of seven-and-a-half minutes' duration. The big tournaments are now divided into five or six chukkas and the smaller contests into four.

The players are numbered one and two (forwards), three, and Back. The duties of number one are: in defence, to ride off the opposing Back and prevent him from having an uninterrupted hit at the ball; and, in attack, to give him the slip and await a pass, or to ride him away from his goal mouth and so leave it open for the number two to score. The number two is usually the stronger of the two forwards and should be the driving force of an attack and the principal goal-scorer. Number three, the pivot of the team, is usually their best player, so this is the most suitable place for the team captain to play. He will often initiate attacks, and his first objects will be to send the ball up to the forwards and to intercept attacks. As principal defender, the Back must be thoroughly reliable. He must be a safe hitter, especially with backhanders. His aim will be to pass the ball to his number two; he will mark the opposing number one. The opposing number two and three also mark each other. But the positions are not rigid and the essence of good polo is in flexibility of teamwork, in changing positions as the game dictates.

At the start of the game, the two teams line up in the middle of the ground, each team being on its own side of the halfway line, numbers one, two, three and Back facing their respective opponents. An umpire (there will always be two in a match) then bowls the ball underhand between the ranks, and the game begins. A goal is scored when the ball passes between the posts and over the goal-line. Teams change ends after each goal. Whenever the ball goes over the side line an umpire throws in a fresh one between the lined-up teams. When it is knocked over the back line, it is then formally hit in by the defenders. If, on the other hand, the ball is knocked over the back line by one of the defenders, the attackers are given a free hit at goal, 60 yd. in from the point where it crossed the line.

A tiresome feature of polo from the spectator's point of view, even in matches of the highest class, can be the frequent temporary cessation of play by the umpires owing to fouls. There is no offside rule, players may impede their opponents by knocking and hooking their sticks, lean into them and 'ride them off'. In fact they will endeavour to do so at every opportunity. But owing to the inevitable hazards of a fast-moving stick-and-ball mounted duel, certain rules have to be stringently enforced: the penalties for riding across another player's 'right of way', misusing one's stick, bumping and zigzagging, are given instantly and severely, in the form of 60-yd. and 40-yd. free hits at one's goal, or free shots from the place of the foul. The 'right of way' exists during every moment of the game and is possessed, by and large, by that player who is riding most closely in the direction in which the ball was last hit. It is an offence for another player to cross this line.

A bell is sounded at the start and end of each chukka, though play is continued until the ball is hit behind or over the side lines or, in the opinion of the umpires, lies in a neutral position. If a match ends in a draw, additional chukkas are played until one more goal is scored.

The successful polo player invariably possesses a natural ability to co-ordinate limb and eye, and a natural athleticism. He will have a flair for ball games; he would be good at cricket, squash or tennis. Next, since all players are obliged to wield the stick with their right hand, one who by instinct holds, say, a tennis racquet or fencing weapon in the left hand will be at a marked disadvantage with a polo stick. (The author, himself thus handicapped, writes with certain feeling on this subject: it took him twice as long as most players to master the strokes and hit the ball! But you need only watch that agile performer, the American Joe Casey, to know how it can be done.)

It is much harder to teach a non-ball-game player to hit a ball from horseback than it is to teach any ball-game player to ride well enough to play polo. The beginner need not be a high-class horseman, but he must be able to concentrate on hitting without having to think about his riding, must also be a sufficiently competent horseman to get the best out of his ponies. The same principles apply as in most other spheres of equitation. A good seat is the top requirement. This is particularly important in polo, since many shots have to be taken leaning right out of the saddle, at the gallop. The best way to achieve both firmness and balance is to ride regularly without stirrups, and also to do plenty of schooling exercises to keep supple at the waist and shoulder: the polo player must be capable of moving the upper and lower parts of his body more or less independently.

The polo pony must be able to move off sharply from the walk to the gallop, to stop abruptly, rein back, turn very tightly and suddenly on either the hocks or the forehand, pirouette and swerve in a manner that is probably quite unnatural to him. So the polo horseman must fully understand the aids, and develop good hands as soon as possible; on the one hand he must have very good control and, on the other, generally avoid spoiling the mouth, temperament and performance of nicely-made ponies.

The beginner will probably learn the strokes from a dummy horse in a polo pit, a small wire-enclosed court with a sloping perimeter floor that returns the balls, and a flat centre on which is placed a wooden horse with a saddle. There are four basic polo strokes: the offside forehand, which is the equivalent of the racquets game's forehand and is the most generally used, the offside backhander, the nearside forehander and the nearside backhander. The first aim in hitting should be to develop a stylish swing, so that the stick head may properly gather momentum and the shaft guide it to the ball. The head will then expend most of its energy in driving the ball, and the remainder of it in the course of the follow-through. But the pit is inclined to give a feeling of false accomplishment, so as much practice as possible should be done from a quiet pony – as soon as possible at the canter and gallop.

The polo pony must possess speed, stamina, great agility and good response. He will need a similar quality of conformation as that expected of the hunter and event horse; he should have a nice straight humerus, a deep stifle and plenty of depth through the heart and lungs; he ought to be well ribbed-up to afford a

A fine back-handed stroke during the final of the Smith's Lawn Cup.

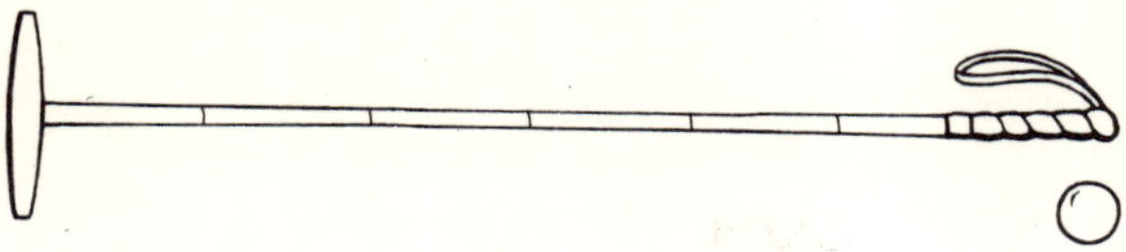

A polo pony with all the essential characteristics of conformation for success.

rounded loin; he should have a round well-shaped foot set on a long flexible pastern; and he will show an alert, kindly eye. He must have enough weight and stature to carry his rider well and to be affective in riding off, so he should not be much under 15 hands; at the same time he should remain a pony, with a pony's smooth and low-galloping action, so that the base from which the stick-stroke is made stays reasonably constant. He should, therefore, be well under 16 hands.

Where are the pony paragons to be found? Among the British owners, Lord Cowdray breeds his own English Thoroughbreds, of which he possesses a string of thirty, and some of the Americans breed mainly from American blood. But the great majority of ponies in play today come from Argentina where, with a vast amount of space available and limitless cheap labour, the *criollo*-English Thoroughbred cross is easily mass-produced and shaped. These ponies, made and polo-schooled, can be bought for £600 to £1,000.

There are four chukkas to a practice match and no pony is up to more than two chukkas a day. So the beginner will ride at least two ponies in each game. But he may not have to buy them. Some clubs hire ponies for £3 to £4 a chukka, which is quite a reasonable price when you reflect that, anyway in England, each costs £300 to £400 a year to keep in forage, farriery, wages, transport and equipment. But it is surprising how the keen and promising beginner is encouraged and actively helped, and how little he may pay for his polo. Many who have advanced on the handicap list have been sponsored for teams to the extent that their sport has come more or less free.

The most useful influence in offsetting the cost of club ponies and other expenses is spectator attendance. Considering what an eminently exciting game this is to watch, it is surprising how small a following it has, though the audiences are growing every year.

A game of polo in progress in India in 1650.

A member of an Argentinian international team fights off his opponent from Chile.

Chapter nine

Driving

Driving a horse appears to have been a sport as well as a means of conveyance from the early centuries, the Greeks and Roman charioteers being perhaps the earliest known exponents of driving more than one horse as fast as possible. In England, the setting up of wagers of every description was a major sideline among sportsmen during the seventeenth and eighteenth centuries, and it was not long before horses, both ridden as well as driven, were being exploited by their owners into earning considerable sums of money by this means.

Newmarket, which has always been the centre of horse-racing, was often the venue of these activities, and in the latter part of the seventeenth century a Captain John Gibbs won £500 by driving a four-in-hand to a light chaise up and down the steepest part of the formidable Devil's Ditch on the Heath. For this feat it is reported that he had the chaise built on to a jointed 'perch' (the main connecting rod between the front and back axles) so that the vehicle moved more easily over the undulating ground, the horses being very loosely harnessed, and without a pole between the two wheel horses.

Almost a hundred years later, in 1750, another chaise match, which was of sufficient interest to warrant its being illustrated by the artist James Seymour, also took place on Newmarket Heath. This was for the much higher stake of 1,000 guineas, and was between the Earls of March and Eglintowne against Messrs Theobald Taafe and Andrew Sproule. Four horses postillion driven should draw a four-wheeled carriage containing one passenger for a distance of nineteen miles in under an hour. In fact this journey was performed in 53 minutes and 27 seconds, and since the horses were Thoroughbreds who virtually ran away for the first four miles, it was not perhaps so surprising that they achieved this record. This match was the first of many triumphs for the two noblemen: Lord March, who became Marquess of Queensberry, and was later known as 'Old Q,' had a strange spindly-looking vehicle specially built as lightly as possible to his own specifications, and it was his idea also to use well-known racehorses ridden by professional jockeys.

By the late eighteenth century driving had become a very fashionable pastime which was sponsored by the then Prince of Wales – later George IV – and many young men about town started to bribe the professional coachmen into teaching them to drive a team by allowing them to 'have a handful'. This was of course strictly forbidden for safety reasons, but it took place nevertheless, and so a great many first hand, and at times hair-raising, stories of the coaching days have thus been handed down. This fashion for driving also led to the formation of several driving clubs, the last of which was the Coaching Club, which recently celebrated its centenary year. The British Driving Society is another modern club, formed in 1955 and now with some fourteen hundred members. It was inaugurated

A pair of Cleveland Bays, owned by H.M. the Queen and driven by H.R.H. the Duke of Edinburgh.

in response to the recent renewal of interest in driving, and provides for members who drive singles, pairs and tandems.

While the names of several people have been handed down as wagering for feats of coachmanship or speed, others have been recorded merely as performing the unusual. The Prince Regent himself on one occasion drove a random (three horses harnessed in front of each other, tandem-fashion) from his home in Carlton House, London, to the Pavilion at Brighton. Another sportsman to become famous for driving feats–although more for his daring than for skill–was Squire John Mytton from Shropshire, rather aptly nicknamed 'Mad'; the artist Henry Alken painted at least two of his exploits.

A perfectly turned-out pony and gig drive out into the country.

Despite the fact that the appearance of the railways virtually put an end to coaching as a means of travel, the urge to drive four horses remained, and a revival of the sport took place with many old stage coaches being run on their original routes. The eighth Duke of Beaufort was one of the instigators–in 1866, in partnership with a Captain Haworth, he devised a scheme of putting a coach on the road to Brighton, with subscribers paying for the privilege of driving it on different days. This system was soon copied by other sportsmen, and more and more subscription coaches took to the road, although for limited seasons and in summer only, until in 1908 one of many events to be recorded in road history took place. This was the arrival from America of Mr Alfred Vanderbilt with two road coaches, the Viking and the Venture, and teams of horses. Mr Vanderbilt ran his coaches between London and Brighton during the summer until 1914, when, with the outbreak of war, it was ended abruptly by finding that at one of the stages his horses had suddenly been commandeered by the Army!

With the arrival of peace, enthusiasts became busy again, and coaches were soon running to Brighton, Hampton Court, Oxford and Tewkesbury. This in turn ended in 1939, and although with petrol rationing during the war a few people took to driving horses, no one visualised that it could, or would, ever become popular again under modern road conditions. But old traditions die hard, and it was not long before one or two teams of horses were being put together, and yet another era of driving had begun.

Horses had, however, by now become treasured possessions, so while it was unlikely that old-time achievements of driving excessive distances and speeds would be attempted, yet it was felt that something more than sedate trotting was needed in order to maintain driving as a sport. At some shows, therefore, driving competitions, both competitive as well as against the clock, have been staged. These involve the negotiating of small obstacles such as driving through narrow markers; stepping over raised poles; and backing into and out of gateways, etc.

On the Continent, in the meantime, all types of harness, including teams of up to twelve stallions in hand, had been displayed, and eventing–involving negotiating rough ground and steep hills as well as driving through water–have been achieved. This activity is now

Emerging from the water during a marathon drive in Poland.

officially recognised as a sport by the FEI, and rules for both pairs and four-in-hands have been drawn up. These competitions are similar to those held for riding horses, and involve presentation, dressage and cross country work at varying speeds, as well as an obstacle course, and are held over several days.

Apart from these competitions held on the Continent, in which the Hungarians, Germans and Austrians are the principal participants, driving as a sport is actively pursued in a number of other countries as well. In the United States a lavishly-produced quarterly journal containing illustrated articles of driving interest is published by the Carriage Association. This society also arranges courses of lectures and meets of carriages in historic settings, and despite the fact that distances are so great its many members manage to attend horse shows and organised functions, as well as driving entirely for pleasure in their own areas.

In Scandinavia, too, there are many enthusiastic driving sportsmen, and the Danish Driving Society is a thriving concern. Similarly, in Belgium, Holland and Switzerland, Spain, Portugal and Italy, keen drivers are also to be found, although France, which produces so many elegant and attractive items on harness and carriage themes, appears to be sadly lacking in drivers, although harness racing is, as in the other countries, an active pastime.

Although it is true to say that most horses can be driven as well as ridden, certain breeds show a greater aptitude for harness, and a number of these have become popular all over the world. In Holland, both Friesian and Gelderland horses are extensively used in harness, and the latter, with their outstanding looks and showy action, have become very popular in England as coach horses. Similarly, the German breeds of Oldenburg and of Holsteins, which are broken to harness on their stud farms, are also exported to other countries, and a team of Oldenburgs is at present in the Royal Mews in London. In Austria, Lippizaners and Haflinger ponies are the main harness breeds, and these too are finding their way into England. Elsewhere, Arabian horses, as well as both Anglo-Arabs and part-bred Arabs, are bred and used extensively for sporting driving in many parts of the world. Another universal harness breed is the hackney, which, though primarily intended as a roadster, is now largely used as a show animal.

In almost every part of the world driving enthusiasts are now to be found, and these range from the elderly to the very young. Competitions such as those staged abroad are now also held in Britain, and with H.R.H. the Duke of Edinburgh as an active participant, they may well prove that while racing has always been considered the sport of kings, driving is, perhaps, the sport of princes.

Harness racing

Harness racing is one of the oldest horse sports. Chariot racing certainly existed in Roman times, and probably in the days of the Greeks, Egyptians and Assyrians. It now takes place all over the world, particularly in Europe, North America and Canada, Australasia and New Zealand.

Modern harness racing is divided into trotting and pacing. In both forms, a single horse pulls a very light sulky and is guided by one driver. Trotters move diagonally, as horses generally do, while a pacer's legs move laterally – i.e. the two nearside legs move forward together, followed by the two offside legs. Some people mistakenly regard pacing as an unnatural gait. It has been accepted for many centuries: Chaucer wrote about 'a proper amblynge little nag', meaning a pacer, and royal courts frequently mounted their ladies on pacers, for the movement is more comfortable than that of the normal trot.

The United States is undoubtedly the leading country in harness racing, and nearly all international records are held by American-bred horses. The world pacing record is held by Steady Star, an American horse, with a time of 1 minute 52 seconds for the mile. This gives an average speed of just over 32 m.p.h., which compares favourably with the Derby record, held by Mahmoud, of an average speed of just over 35 m.p.h. The world trotting record currently stands at 1 minute 55·2 seconds over a mile.

The predominance of American-bred harness racehorses is now being challenged by pacers from Australia and New Zealand and by trotters in France. (Harness racing, though popular on the Continent, is mostly limited to trotters; though in Britain pacing races are held.) The first dollar millionaire stakes winner in pacing history was the horse Cardigan Bay, bred in New Zealand, while a French trotter, Une de Mai, is now well past the $1½ million mark. $2 million in prize money looks within her compass, which would make her not only the biggest money winner in harness racing history but in world equine history, for no Thoroughbred racehorse has as yet reached this figure.

The harness racing breed is known as the Standardbred, a name that originated because the aim was to mate a horse and dam capable of producing offspring that could trot or pace over a mile within a standard of time. The founder of the breed was an English Thoroughbred, Messenger, imported into the United States. The famous sire, Hambletonian, established through four of his sons the main sire lines of virtually all the American trotters and pacers of today. Standardbred stud books are now

A fine pacer on the racetrack during training.

American pacers during a race.

The finishing post: trotters on the Vincennes track in France.

maintained in all the countries where harness racing is popular.

The breeding of harness racehorses divides itself naturally into trotting and pacing strains. It is evident from watching even the youngest foals that pacing is a natural gait, and it is in fact very difficult to persuade a natural pacer to trot conventionally. The hobbles generally worn by pacers are not a restrictive device, but merely act to accentuate an already natural action. Trotters wear no form of hobble; they are much more likely to break their trot. In a race this does not involve disqualification, but the driver must pull back in order to re-establish the trot, and in doing so will lose valuable ground.

Pacers closely bunched at the Australian Horsham Show.

The conformation of a good harness racehorse is different from that of a Thoroughbred. The criteria, however, are also different, perhaps the most important being that the horse does not have to carry the weight of a rider on its back. The set of the head and neck, the absence of a good riding shoulder and the sloping rump are all factors which perhaps contribute to the performance of the Standardbred at its particular sport, though they would be regarded with disfavour in other spheres. In spite of the iron-hard going of limestone racing tracks, neither trotters nor pacers seem to suffer unduly from leg trouble, though great attention must always be paid to their action, so that they do not damage themselves, and to the way they are shod. Free and level action, which can be assisted or marred by shoeing, is the most essential attribute of a successful harness racehorse.

Harness racing is today one of the growth sports. Any competent horseman can, with a little familiarisation, drive a racehorse, for whether the reins are two or ten feet long many of the same principles apply. To sit behind a good-quality, well-behaved horse and be pulled along the track at nearly 30 m.p.h. is exciting in itself; in a race it is that much more so, and this excitement communicates itself to the spectators.

PART NINE

The horse at work

Chapter one

The role of the working horse

Horses will themselves not distinguish between what man in his wisdom has designated 'work' and what he considers to be pleasure. As far as the horse is concerned, what it is asked to do by its rider, driver or overseer will constitute work whether the activity to be performed is a dressage movement or dragging logs through a forest. For the purpose of this chapter, however, one can perhaps define the working horse as one that helps man to perform his own work more easily or efficiently, or in other ways helps him to earn a living.

The advent of the motor age has obviously eliminated one of the important roles the horse fulfilled in the past: that of everyday transport. Not so long ago, all land transport was carried out with the help of the horse, and even after the establishment of the railways horses were still used to collect and disperse goods carried over long distances by rail. Many of those who now own a car would in the past have had a horse of some kind instead, ranging from a stableful of carefully bred, matching carriage horses attended by grooms in livery to one small pony and cart. Tradesmen's deliveries in towns were made with the help of the horse, and until quite recently ponies still did the early morning milk round in large cities.

Nowadays, however, at any rate in the advanced industrial nations of the world, horses are much less frequently used for these purposes than they used to be. There are exceptions. The sight of a beautifully turned-out team of dray horses delivering barrels from the brewery is still a familiar one, and is indeed now a successful piece of publicity on the part of the brewer. In many countries, both in Europe and elsewhere, where heavy horses have either traditionally been bred or are imported, they are still used in preference to mechanical equipment in rough or mountainous terrain and in remote areas where machinery would be prohibitively expensive to buy, difficult to maintain and of limited use.

Ponies, donkeys and mules are also still used for pack work, the latter being particularly frequently seen in the Mediterranean countries.

In rural environments all over the world horses still form an important and integral part of life as a whole. In Hungary, for example, the *czikos* (cowboys) tend their herds on the plains during the summer from horseback. The *gardiens* of the Camargue district in France herd the small, fast, fierce black bulls from horseback. In Australia and North America stock horses still form an essential part of the agricultural and herding pattern, shepherds and drovers tending sheep, horses and cattle on the vast expanses of outback or prairie. The same is true in the South American continent, where criollos are used to tend the valuable beef herds to be found, especially in the Argentine.

It is the small farmer and peasant who perhaps most of all still relies on his horse. What piece of mechanical equipment can be pressed into help over transport, ploughing, threshing, fetching, carrying and all the other activities of daily life, demands relatively little care in return (a good deal less than it deserves, sometimes), lasts for upwards of twenty years and is a good friend into the bargain? The versatility of the horse has as yet proved irreplaceable.

There are some activities where the use of the horse has practically died out. The most obvious example of this is in the coal-mining industry. About ten years ago thousands of ponies were to be found working underground, and though the miners took very great care of their ponies it was a dangerous, unnatural life for them—as, indeed, it is for the men.

On the other hand, there are many spheres where the change in man's routine and way of life has meant an increased use of the horse where one would have expected its disappearance. In all the major tourist-attracting countries of the world, horses and gaily-painted carriages are available to take visitors on con-

Pit ponies, once vital to the coal mining industry.

ducted tours of the city. The drivers compete with each other for trade by having the smartest turnout, which ensures that the horses are properly looked after.

One fast-growing tourist and holidaying activity is pony-trekking. The Austrian Tyrol, Spain, Portugal, Hungary, Great Britain and the United States, to mention but a few, are all countries where trekking centres are now well established. As many of the riders are people riding for the first time, it is important that the animals used should be quiet and reliable, and it is one area of equine activity where regular inspections of establishments and the horses they use should be carried out to make sure that the horses are being properly treated.

Another activity where controls are needed is that of riding schools. The tremendous upsurge of interest in horses and riding as a sport since the second world war has come very largely from a section of the population that previously had little contact with them. The demand has been so great that there are now about 20,000 riding school animals in Britain alone, and the activity is rapidly growing elsewhere, particularly in the United States, Canada and New Zealand. Many riding school animals are beautifully cared for, but because in the short run the school will make more money the harder each animal is worked, there is a temptation to over-use them, and to give them inadequate stabling and attention, particularly when the school is situated in a town, where space is almost bound to be limited.

One relatively new activity which has developed out of the riding school is riding for the disabled. This idea originated in Scandinavia and has spread rapidly, giving valuable exercise and an enormous amount of enjoyment to many people who are deprived of it in other ways.

There are horses to be found in all sorts of other aspects of human endeavour. The film world is a good example. All Westerns, all historical films, and many others, use horses both to depict life as it used to be and for stunt work. Some barbaric practices used to be carried out in order to achieve some of the most spectacular scenes one witnesses on the screen. Nowadays, however, horses are very carefully trained and a good stunt horse is greatly valued.

Teams of farm horses harvesting wheat before the days of the combine harvester.

It is probably true to say that the value – and not just in monetary terms – of horses has perhaps never been better appreciated than it is today. This augurs well for the horse for, no matter what work it is engaged in, if it is valued there is a good chance that it will be properly cared for. Perhaps the increasing use of the horse in leisure occupations has been a good influence on man's treatment of it.

Chapter two

The circus

If you talk about circuses in Britain nowadays you will hear someone say: 'Oh, but surely the circus here is a thing of the past. After all, the Bertram Mills show is finished; Billy Smart's show has gone off the road . . .'. It is true that the circus in Britain, together with other forms of live entertainment, suffered a severe blow from the advent of television, but to say that the circus is dying is very misleading. Both Mills and Smart were 'jossers' – that is, they were not born in the business; both succeeded brilliantly in creating huge glittering shows. It was Mills who stocked his stables with horses which for quality had never been equalled in Britain, and never surpassed anywhere in the world. In the opinion of many circus people, both Mills and Smart became too big; the expense involved in running such enormous shows was astronomical, and these big circuses finally became completely uneconomic. Smart's are, however, making a comeback with a smaller show. In the meantime, the stock of elephants and trained horses carried by the original Smart circus have been sold to the Ringling circus in the United States.

In America, the circus is thriving as never before, in spite of television. The Ringling-Barnum circus is the biggest in the world. At the time of writing, Ringling's are operating two complete circuses, known as the Red Unit and the Blue Unit, and both shows are playing to capacity houses. The horses purchased from Smart are still in Ringling's winter training quarters. It is said that the company soon intends to put a third huge show on the road, in which the Smart horses will be featured. Both Ringling-Barnum circuses carry performing horses of extremely high quality. Some of their liberty groups are formed from pure-bred American saddle-horses; others are groups imported from European circuses, consisting mainly of Arab and Lipizzaner stock. The Ringling-Barnum circus advertises itself as 'the greatest show on earth'. Circus people refer to it briefly as 'the big one'. However, size is not everything. European liberty trainers

Rosinbacks, broad, placid and with a smooth, low stride, are essential for equestrian circus acts such as this, performed by the Enrico Caroli troupe.

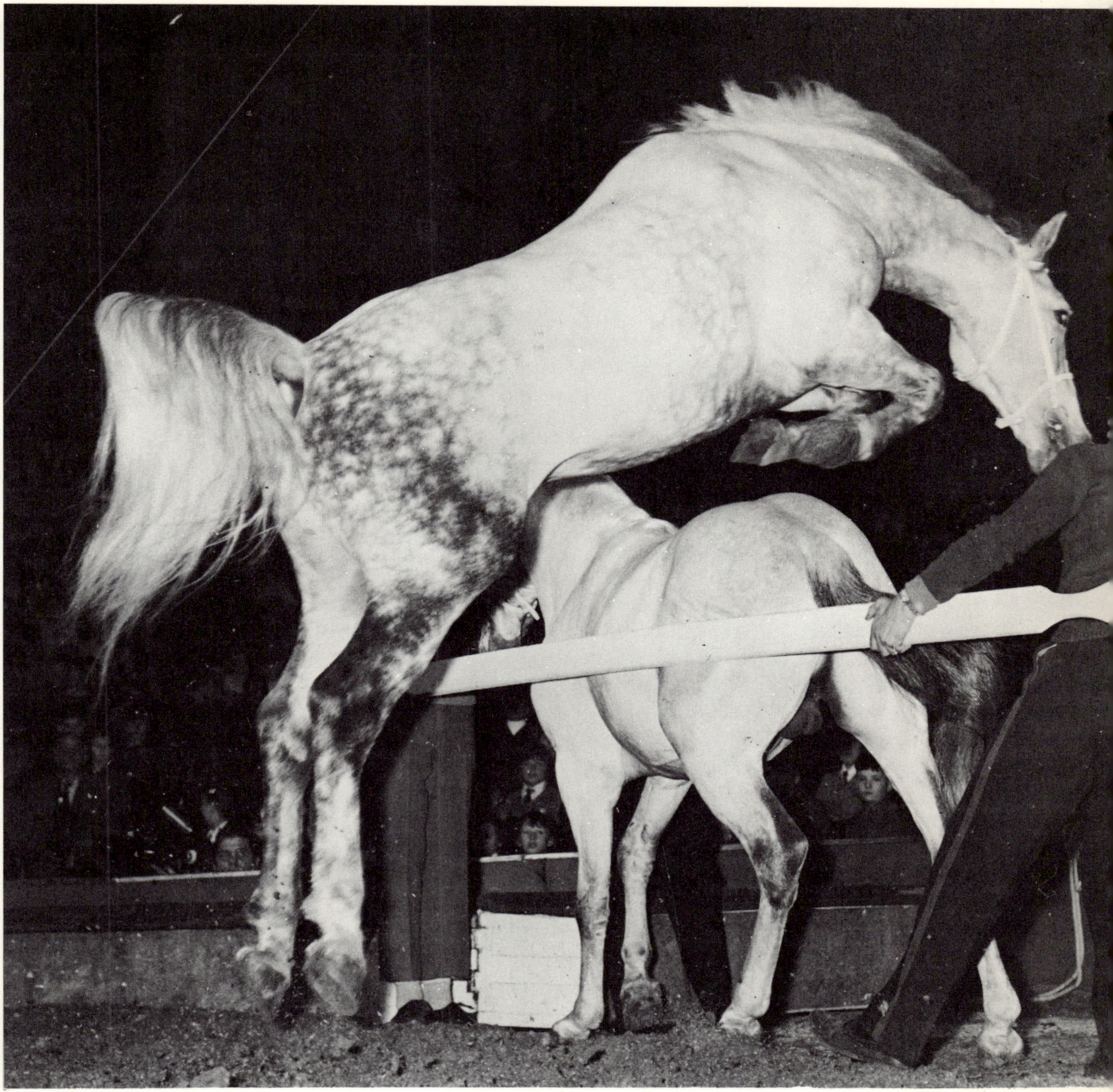

A liberty horse makes a startling exit from the ring.

who have worked in the Ringling shows and with other American three-ring circuses often find such shows a trainer's nightmare. Invariably three acts of a similar kind take place simultaneously. This means that the 'star' team of liberty horses works in the centre ring while two other groups perform in the flanking arenas. The routines of these acts are rarely identical, and it is impossible for the enthusiast to follow the animals' movements without missing something that is going on in an adjoining ring.

The famous Schumann family parades.

It is also initially very difficult to get animals working satisfactorily in a three-ring circus. The presence of other horses working nearby tends to distract them from their own routines, and once the liberty group in the centre ring has finished working the ringmaster's whistle sends all the groups hurrying back to the stables, whether or not their own routines are completed. These problems also occur with bareback riders working simultaneously in three rings. All the acts are geared to the pace of the star number in the centre ring, and this leads to frustration, jealousy and occasionally open warefare between the temperamental artistes of the riding teams. Yet, although the big American circuses are criticised by European artistes, there is much to be said in their favour. They employ a large number of performers and animals; they cater for a public which in general prefers quantity to quality; and they act as a lucrative outlet to the top trainers and horsemen of the overcrowded European circuses.

Mills, Smart and Chipperfield are household names in Britain, but other circuses deserve attention too. For many years, for example, circuses run and owned by members of the Fossett family have entertained British audiences. The Fossetts are of Irish descent, and are real horse people. Fossetts have supplied the sawdust world with some of its finest riders – the late Bobby Fossett was famed for his 'jockey' routine which he worked at tremendous speed. One of his more hair-raising feats was a jump-up to a standing position on his horse's back with his feet fastened into baskets and his eyes blindfolded.

Audiences tend to regard such acts as the high trapeze, the tight-wire and the wild-animal numbers as the danger highlights in a programme. The bareback riding act is in fact one of the most risky features in a circus as far as possible injury to the artiste is concerned. Some of the smaller tenting shows frequently do not pay enough attention to the ring surface and, particularly for one- or two-day stands, the ring is just given a load of peat and sawdust and roughly raked; long grass in the ring area is cut and left lying. Concealed potholes and uneven surfaces can play havoc with a bareback riding number where pace and timing are all-important; worse still, a bad ring can result in severe injury to the rider if the horse misses its footing.

There are several circuses on the Continent, particularly famed for their superb collections of trained horses, which are well worth a visit. The Circus Knie in Switzerland, for example, is always a rewarding show for the horse enthusiast. Each season Circus Knie has a new treat to offer those who are interested in horses. The show recently featured a big liberty act with twelve lovely Lippizaner horses in a number called 'Carrousel Baroque'. The animals were harnessed in trappings which were accurate replicas of horse clothing depicted in the paintings of the Spanish artist Velasquez. Nothing is left to chance in Circus Knie, and designers were specially commissioned to study the works of Velasquez and to reproduce exactly the colours and designs of the old Spanish harness.

The extensive stable tents of the Circus Knie house the largest collection of horses owned by a private concern in Switzerland, and are also well worth visiting. Not only can the horses currently performing be admired, but also groups of animals in training for future seasons. Many breeds are to be found here, all lovingly cared for and in superb condition. Among others, there are Shetland and Welsh ponies, Knabstruppers, Lippizaners, Polish-bred and Russian-bred Arabs, heavy black Frieslands with coats polished to resemble gleaming ebony, dainty golden Akhal-Teke stallions from Russia, native Swiss mountain ponies.

In France, an outstanding circus is the Cirque Jean Richard. Jean Richard is a famous

French comedian who fell in love with the circus, bought his own show and graduated to being a lion tamer. A man of many talents, Monsieur Richard is not always with the show – he has many commitments, with television appearances and other business. Though France is not a country famed for its circus horse acts, there can be little doubt that those of Alexis Gruss of the Cirque Jean Richard put him at the head of France's horse trainers. In recent years this trainer in the classic tradition has produced some truly splendid liberty routines. He uses only pure-bred stock – Arabs, Frieslands, Lippizaners – and his liberty and *haute-école* productions are always highly polished and a delight to watch.

Another name to watch out for in France is that of the Cirque Rancy. This show is probably the oldest-established circus in France. The show's present directors are Sabine Rancy and her husband Dany Renz; Madame Rancy's ancestor, Antonio Franconi, founded a circus in France in 1737, and she claims that in doing so he created not only the first French circus but also the first circus in the world. Historians disagree, pointing out that the Englishman Philip Astley established the first circus in London, thus giving Franconi his inspiration.

Like most French circus big tops, that of Rancy is set up with four king-poles in line, unlike the British and German shows which usually favour a four-pole square canvas. The disadvantage of the French type is that from a position at the extremity of the big top it is quite difficult to follow what is going on in the ring. With Rancy you can always be assured of some horse numbers well worth close attention, for example the group of twelve truly beautiful Arab stallions recently starring in a number described in the printed programme as 'une horde de chevaux sauvages' – a slight exaggeration, to call a team of highly-trained circus horses a herd of wild horses, but the animals are an impressive sight, working entirely without harness and at a very fast pace. The finale to the act is a *cabrade* in which the horses rear on their hind legs; some of the stallions were inclined to fight during this part of the number, perhaps because a hind-leg stand is the natural combat position for a stallion, and several of the horses wore muzzles for safety. Sabine and Dany, by working in the ring themselves as well as carrying out the onerous duties involved in directing a big circus, follow a long tradition of the circus world on the Continent.

Spanish horses from Andalusia are favourite mounts for the *haute école* in the circus world. Like the Lippizaner horses of Austria, to whom they are related, Andalusians seem to be born to this kind of work. Andalusians are of course mostly used in Spain and elsewhere by the *rejoneadores* of mounted bullfighters. Amazingly supple, and as agile as polo ponies, they are ideally suited to circus work. It seems strange that in Spain itself it is very rare to find Spanish horses employed as circus performers – in fact it is rare to find horses of any kind in a Spanish circus. In a country of fine horses and horsemen circuses tend to ignore equestrian numbers and concentrate on acrobatic acts and wild-beast numbers. Recently, however, the two Schickler sisters did appear at the Circo Price in Madrid with their high-school horses – and the applause almost stopped the show.

There are superb circus horses in Germany, too. Circus Krone carries the biggest and best stud of horses in Germany; Circus Barum-Safari boasts a team of trained Norwegian stallions, Don horses from Russia, Hungarian Gidran stallions and Russian Karabakh horses. In Italy you will find top-class horse acts with the Circo Orfei, and the Circo Nacional de Mexico which, in spite of its name, is also to be found in Italy, features magnificent groups of liberty horses and a fiery display of high-school riding by the Mexican Emiliano Lopez and his partners.

The circus ring provides a perfect setting for the horse. The natural beauty of the animal is enhanced by the brilliant lighting which picks out his gay trappings, emphasising the disciplined movements and controlled power and energy. Beautiful, well-bred and finely-trained horses are among the greatest attractions any circus can offer. It is nonsense to say, as I have heard some circus people say, that the public has grown tired of horses in the ring. A really polished circus-horse act will always be a popular item.

The police

A preoccupation with tradition and an affinity with horseflesh are two British characteristics that could partly explain why there are still 197 police horses at work in the larger provincial towns and cities, eight of them on duty in the even less likely environs of the City of London, and around 200 collaborating in the keeping of law and order in the Metropolitan area. There would also appear to be a natural niche for them in countries such as Jamaica, where mounted police are used for the normal duties of their branch, but where the emphasis is on enhancing the spectacle on ceremonial and public occasions. The same applies where the terrain is obviously suited to horses–the more remote areas of South Africa, for instance, and in Lesotho where the entire police force is mounted and sturdy country-breds carry the long arm of the law on arduous treks into the mountains. And although the Jordanian police are now almost entirely mechanised, it still makes sense to retain a few horses for making contact with nomadic Bedouin following the desert grazing trails.

But otherwise what are police horses doing in the space age, where machines of increasing complexity are the gods and an ever-growing volume of mechanised traffic chokes the highways and urban streets? And the reason why, in many of the seemingly most unlikely places in the world, the mounted policeman still has a worthwhile, as yet irreplaceable job of work to do, lies partly with the working partnership possible between a rider and his horse, partly with the healthy respect horses still command from the majority of people.

Even on traffic control the mounted officer has an advantage over his motorised or foot-slogging counterpart. He is not only clearly visible himself but can also see far better himself, with the added advantage of being able to squeeze his horse between vehicles or go up onto the sidewalk to reach and sort out the trouble. And there are police horses coping with traffic problems in places as far apart as Stockholm and Sydney, Birmingham and Boston. In Tokyo, where pedestrians and traffic have almost reached saturation point, horses are still considered indispensible for controlling street crossings, particularly those used by schoolchildren, just as they are in Melbourne, Australia, at the St Kilda Road crossing for blind persons. And although the International Association of Chiefs of Police in the USA can be a little coy about admitting to the fact, New York still possesses the largest mounted police force in the world–and among those long congested avenues and intersecting streets the horses undertake traffic control as well as the many other duties they so admirably perform.

Like the original two English 'pursuit horses', ridden by 'proper persuers', that preceded them by some fifty years, the animals of the regular London Bow Street Patrol in 1805–the prototype for all mounted police forces–were chiefly concerned with patrolling the main roads within twenty miles of Bow Street. Through the years the force proved so successful that the depredations of the footpads and 'gentlemen of the road' ceased, the turnpikes were rendered safe and the horses became almost exclusively urbanised–unlike now, when they are once more proving their worth by materially helping to keep parks and tracts of countryside close to towns safe for the use of law-abiding citizens, in addition to street patrol.

In Australia the mounted force began as a thirteen-strong patrol, raised by Governor Brisbane in 1826 to perform duties previously undertaken by the New South Wales militia. These Trooper Police often took their lives literally in their hands to guard the early settlers from the notorious bushrangers and cattle thieves. As highways were constructed their duties were extended to protect both travellers and merchandise, and to convey the gold-trains safely from the diggings. For many years the Trooper Policeman was often the only contact a squatter 'out back' might have with the outside world, and this, usually welcome, limb of the law took on most things from local law administration to the issue of birth and wedding certificates.

Although under modern conditions the Australian Mounted Police force has been reduced, most states retain their own units, although with few exceptions the horses are on duty in urban areas. They have also changed in type, being much lighter and better bred than the heavyweights necessary when a Trooper Policeman and his horse were vital to each other's survival in the bush. Today the quality horses, many colour-matched, include escort and ceremonial work among their duties to the exclusion of the military, and the New South Wales police horses alone yearly include more than fifty ceremonial parades.

In London the police horses supplement the Household Cavalry and King's Troop on such occasions as royal weddings and state funerals, and the important annual ceremonies such as the State Opening of Parliament. Two police officers and their horses always escort the royal carriage bearing a new ambassador to and from Buckingham Palace when he goes to present his credentials to the Queen, and each day they are on duty outside the Palace at the changing of the guard. The Lord Mayor's Show is the special province of the City of London's mounted police–augmented on that occasion by members of the Metropolitan force–and nine grey police horses provide part of the escort for drives in state of foreign royalty and heads of state. They make a brave sight, if dwarfed by the forty-eight mounted policemen who accompany the Dutch Queen when she drives in state through the streets of the Hague.

Once a year it is the proud duty of the Metropolitan Mounted Police to supply many of the horses ridden by members of the royal family and household, and by senior army officers, at the ceremony of Trooping the Colour. For many years this also included the horse ridden by the Queen at this her official birthday parade. Two of these animals, first Winston and then Imperial, had this special honour on many occasions and became well-known public 'figures', but sometimes lameness or other

cause makes a substitute necessary. There are usually two or three horses suitable for the task, since the police keep an eye out for candidates of the right temperament and bearing who are taught to carry a side-saddle and receive a little extra schooling nearer the day.

The Royal Canadian 'Mounties' are perhaps the most famous police force in the world. Their roots lie with the small, semi-military force of North-West Mounted Police, less than 300 all told, which was established in 1873. Their horses' beat was an ungoverned territory of some 300,000 square miles, their duties to keep the peace between the 30,000 Indians who roamed there and the increasing flow of traders and settlers encroaching on the Indian homeland and threatening the buffalo herds that were their source of life and livelihood. The environment alone called for great endurance and courage from both men and horses of this small force; there was constant danger, and daily situations demanding their fullest capabilities and resourcefulness, but they gradually brought peace, and a wise and just treaty with the Indians, in a vast territory at first 'without law, order or security for life or property'. And in time the scarlet-coated policeman and his horse became a familiar and welcome figure on the plains.

In 1920 the jurisdiction of the North-West Mounted Police was extended to the whole of

The Canadian Mounties, perhaps the most famous mounted police force.

Canada, their title changed to the Royal Canadian Mounted Police. Although rapidly mechanising, the Mounties still used horses on general duties between the two world wars, and retained them for some years more for recruit training. Nowadays they still breed their own black horses at Fort Walsh, which was from 1878 until 1882 the headquarters of the North-West Mounted Police, but the animals are not operational. They are kept to uphold the Mounties' high standard on some ceremonial parades, for prestige and to fill the ranks of the spectacular and highly skilled musical ride, staged all over Canada and in many other countries to help promote good relations between those who uphold the law and those who can live in security of life and property because of it.

Public relations is a very important aspect of police work, and always furthered by such events as the Metropolitan Mounted Police's enjoyable annual show, or by displays like the breathtaking performances of trick riding popular with Australian units. But the most exacting duty is that of crowd control.

When a highly inflammable Iranian crowd tries to storm the entrances of the huge stadium in Tehran during the Shah's magnificent birthday celebrations, a posse of the Tehran Mounted Police riding fiery Persian stallions can disperse them in a matter of minutes. The old-time formation, the 'flying wedge' perfected by the New York mounted branch, is still occasionally used by them to break up a potentially dangerous mob. In England the mounted branch of the Lancashire Constabulary has the responsibility of controlling the press of people attending race-meetings at Haydock Park and Aintree, and police horses in Melbourne are used to prevent people crossing the track at Flemington racecourse or the motor racing track at Sandown Park, Calder. Where big crowds are expected, whether for a political demonstration or a sporting event, in many countries the horses and their riders are an indispensable part of crowd control.

In London especially, and despite the marbles and missiles used against them during violent demonstrations, police horses are now used more frequently than before. This is due partly to their proved effectiveness, partly because of the shortage of manpower for foot-patrol duties, but also because the tendency is now to employ the horses as a normal part of policing rather than as a reserve. And for the horses to be employed on this kind of work, not to mention on all their other exacting duties, they have to be exceptionally calm, obedient and well behaved.

A few horses do receive specialised training – a Dutch unit is on summer patrol at the Hague, the horses trained to stand while their rider throws a lifeline to a swimmer in distress, or to plunge into the sea beside their rider to fetch out a drowning child; the Tehran police stallions remain motionless when a rifle is fired from the saddle – but otherwise the basic training is much the same as that given to animals of the Metropolitan Mounted Branch at their training quarters at Imber Court.

Police horses are usually a good stamp of animal with the essential quality of a quiet, co-operative temperament. Trainers experienced in the art, and using much patience, firmness and kindness (allied to liberal rewards of oats), first bring them to a high standard of riding horse good manners. Then comes more specialised training such as swinging the quarters to push, without a thought of kicking, 'leaning' against rowdies trying to jostle them off their feet, and stepping over 'bodies' in their path. They must learn to ignore noises ranging from gunfire to brass bands and football fans' 'rattles', and cope equably with such hazards as waving umbrellas, fires and fire-engines, and the intangible but upsetting mass emotions of a crowd. At the end of their training each horse is given an experienced rider with a temperament that suits its own. And soon the police horse becomes so dependable and co-operative that its rider can, theoretically, forget it and get on with the job in hand.

A few years ago an official in Stuttgart, after watching police horses at work on a tricky situation of crowd control, included in his report the observation that 'men have more respect for the creature than for their own kind', and that 'the police of a large city should never be afraid of the cost and subsequent expense of police horses.'

Chapter four

The cavalry

French *Gardes Forestières* patrolling at Fontainebleau.

The more one studies the evolution of the horse and the deeper one delves into the history of equitation, the more one recognises them as being part and parcel of the development of the mounted branches of armed forces. Without the incentive of military power the equine beast might possibly never have developed as a racehorse, show-jumper, hunter, eventer or polo pony. Even in the very beginning the influence is there, for before the wild Asiatic horse was tamed by tribesmen at the end of the Stone Age its first use was to be driven in herds against the enemy. The ancient civilisations of Assyria, Babylon, Egypt and Libya put the horse in harness primarily to produce a shock force of mounted troops and chariots – the forerunners of modern armour. They also broke their horses in order to hunt. Hunting took place not only to procure food or to make sport: it was practised then – and all down the ages – as training for war.

The Greeks learned the art of horsemanship from the Egyptians, and it was the Greek general Xenophon who wrote the first definitive

A charge by the King's Troop, Royal Horse Artillery.

A detachment of the Indian cavalry of the Mexican army in action in 1924.

study of equitation, his knowledge being derived from his experience as a cavalry leader. The soldiers of Rome, applying the Greek lessons throughout their empire-building years and gaining much from the horse-minded Gauls, also improved the scope and advantage of mounted troops.

But it was Charlemagne, Christian revivalist, eighth- and ninth-century emperor of the West, professional commander-in-chief and architect of the new order and civilisation that emerged from the chaos of the Dark Ages, who has the greatest claim as the 'father of European cavalry soldiering'. In the first place, Charlemagne welded Europe by the sword. His knights were prototypes for the Carolingian age of chivalry. Besides making superb men-at-arms, they were required to be skilled in the art of hunting and other mounted sports; they also had to be men of honour, possessing a great sense of personal duty to their emperor. And morale, pride and authority demanded that they be superbly mounted and turned out. Under Charlemagne the cavalryman became the *beau ideal*, and the warhorse the chief symbol of the empire's supremacy and greatness.

Early in the tenth century the Saxon prince, Henry 'the Fowler', in order to improve his defences against the massive Hungarian invasions when he became emperor of Germany, followed suit. He raised a force of cavalry from his aristocracy, a mounted order with a strict code of honour to which it was the greatest privilege to belong. And this concept of mounted units whose wartime hallmarks were dash, valour, integrity and efficiency, and whose peacetime ideal was chivalry, elegance, fine horsemanship and the fraternity of caste, spread right through the Continent. They began the cavalry spirit which still exists today.

It was knights such as these who, under the leadership of William of Normandy, invaded Britain in 1066 and subdued King Harold's massive Anglo-Saxon army. All through the Norman, Plantagenet and Tudor periods improvements in the English world of the horse stemmed almost entirely from the soldiers' interest. The Conqueror's son, William Rufus, produced a better knight's charger by importing Spanish Jennets—crosses of the Iberian native horse with Barbs and Arabs. Arab blood was further infused during the Crusades. To develop a more formidable battle-charger, King John imported huge numbers of horses from Andalusia. Edward III formed a body of mounted archers, the original dragoons and the first really effective light horsemen.

Throughout this period men of property and standing were obliged to keep a horse ready for war and to take their place in the ranks of the cavalry, while military mounted sports appear to be the only events at which equitation skills were practised. Nevertheless, in the 1580s the Spaniards, whose cavaliers had driven out the Moors and who were determined upon Catholic supremacy in Europe, fielded mounted forces much superior to our own; and Queen Elizabeth I was so disgusted when she reviewed her army at the time of the Armada threat that she bade her generals set to and breed chargers good enough to face those of Philip II.

In the seventeenth and eighteenth centuries, with the development of artillery and battle mobility and the wane of the set-piece siege, war on the land became a more flexible affair. That military genius, King Gustavus Adolphus of Sweden, who led the Protestant armies against the Jesuit Austrians and Spanish in the Thirty Years War (1618–1648), trained his speedy regiments to charge home with the sword. This produced the brisk shock action which was to prevail in cavalry tactics for the next two hundred years. Faster, handier horses with more stamina were now bred for the military. With this movement, as adjuncts to their national cavalries, grew the great French, German, Italian and Spanish riding schools. The English school also developed at this time, under the guidance of the Duke of Newcastle, courtier and general to Charles I and tutor to Charles II. Helped by Prince Rupert, the Royalist cavalry commander, and later by Charles II and others, he was largely instrumental in laying the foundation of the modern blood horse.

The French mounted royal bodyguard, the *Maison du Roy*, which had been founded in the thirteenth century, was the most resplendent regiment Charles II had ever seen. This prompted him, when in exile in France in the

The Royal Scots Greys
during a training exercise.

1650s, to form the first troops of Life Guards, the original Household Cavalry and the first mounted troops of Britain's standing army. The Royal Regiment of Horse Guards, Blue, followed at the Restoration. Then came the horse grenadiers, dragoon guards and light dragoons, and the various tactical roles emerged among the cavalry of the line. At the end of the eighteenth century the Royal Horse Artillery, the 'Galloping Gunners', were raised to support the cavalry; the regiments of light dragoons evolved into hussars, and some of them–after the battle of Waterloo–into lancers. Through inter-regimental rivalry and through vieing to cut a better dash than their Continental counterparts, the horseflesh as well as the excellence of the units achieved very high standards. The cavalry became known as the *arme blanche* and, as in Charlemagne's day, the prestige of the horse was made paramount by the army; and vice versa.

Ever since the birth of modern foxhunting at the end of the eighteenth century, the cavalry have brought their high regimental standards to the hunting field and have provided countless Masters of hounds; and ever since the first race meetings, retired cavalrymen have become officials in the racing world and in other ways have patronised the turf. At the end of the nineteenth century it was the British army that introduced polo from India to the West, and for at least the next half century fielded the nucleus of the English polo-playing fraternity. It was Continental and British cavalry officers, too, who initiated the sport of show-jumping and who gave most of the original lead for eventing.

Keeping their titles of lancers, hussars and dragoons throughout these last forty years or so of mechanisation, the tank and armoured car regiments have maintained their cavalry reputation for dash and discipline, personal initiative and elegance for which the horse first made them famous. The horse still provides cavalry officers with their favourite sports. And the old glamour has not entirely disappeared. At the head of the British army still ride the mounted squadrons of the Life Guards and the Blues and Royals and the King's Troop of the Royal Horse Artillery, all of whose soldiers have a fighting as well as a ceremonial role. They now provide some of the best of London's pageantry and remind the present generations a little of what the Army has done and is still doing for the world of equitation. In very many other countries all over the world, from Denmark to Ethiopia, the horses of the cavalry still have an important role, even in the latter part of the twentieth century.

PART TEN

The horse and the arts

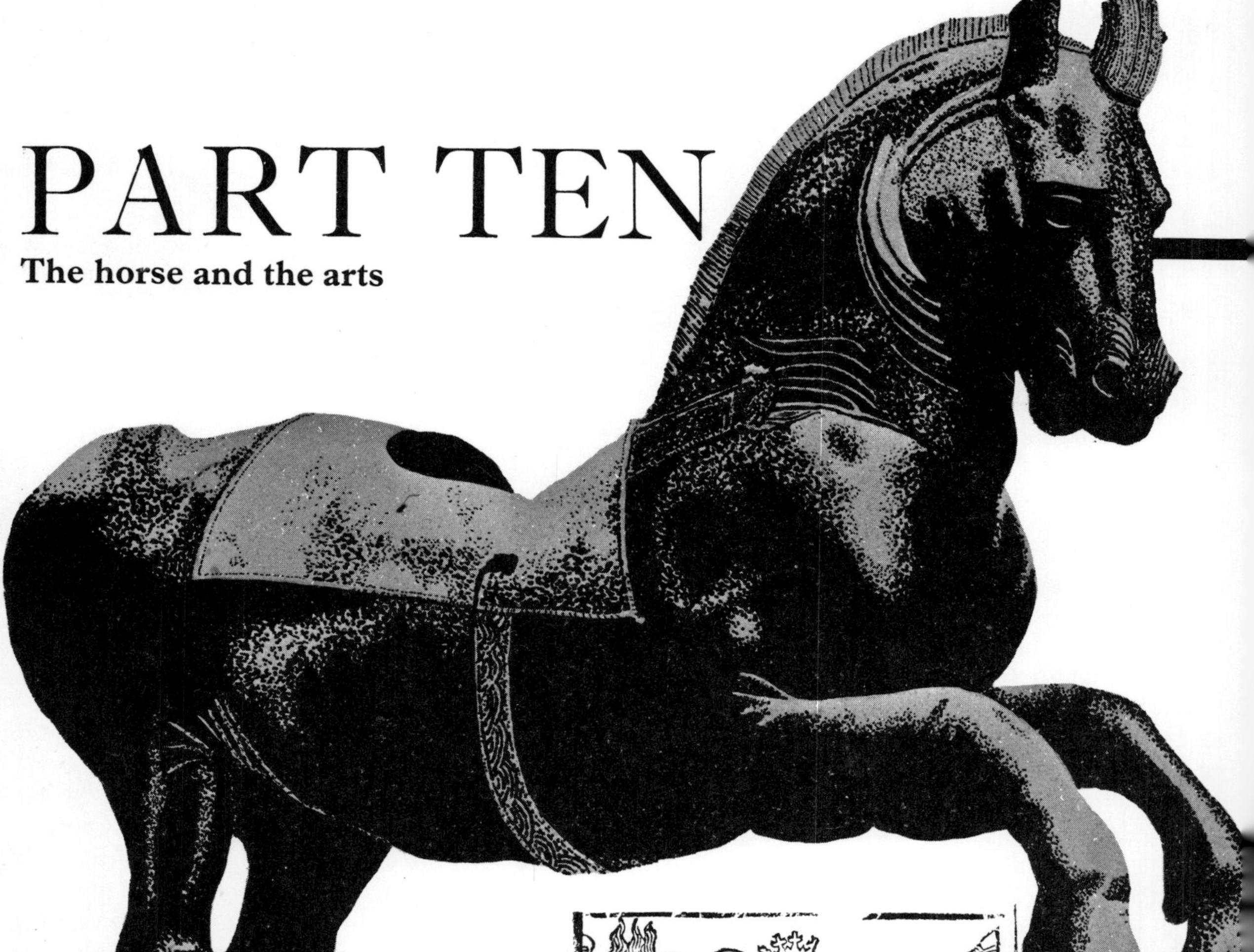

Chapter one

The horse in English literature

The horse has been an inspitation in all forms of art, not least in literature. It is interesting, however, that in English literature the style of the writing depends very much on the status of the horse at the time of writing. Chaucer, for instance, had little of a romantic vein in his descriptions of horses, though naturally they feature prominently in his famous *Canterbury Tales*. The horse is as 'lean as a rake' (I wonder how many twentieth-century horsemen realise that they are quoting a fourteenth-century poet when they use that phrase); or a 'stot'; even an 'ambler' is scarcely complimentary.

But in Chaucer's time the horse was purely functional, used as a beast of burden and as such taken for granted. By the time of Shakespeare, however, things have changed, The beautiful Arab had already been imported into Britain, while all over the Continent the nobility were establishing at their courts or palaces magnificent riding centres where talented riding masters instructed *haute école.* The horse had suddenly become socially important.

Thus Shakespeare in his first major work, *Venus and Adonis*, describes the stallion in the following exquisite stanzas:

Sometimes he trots, as if he told the steps
With gentle majesty and modest pride:
Anon he rears upright, curvets and leaps
As who should say, 'Lo, thus my strength is tried'.

Round hoofed, short jointed, fetlocks shag and long,
Broad breast, full eye, small head and nostril wide,
High crest, short ears, straight legs and passing strong,
Thin mane, thick tail, broad buttock, tender hide.

Reading this it is tempting to believe that he himself had seen the famous Lippizaners at the Spanish Riding School in Vienna, for it had been founded just about the time of his own birth. The description could well be that of one of the Lippizaners.

In different vein, however, Shakespeare shows his brilliant powers of description. In *Henry V* there is the following remarkable passage depicting the horses of the exhausted English army:

. . : and their poor jades
Lob down their heads, dropping the hides and lips,
The gum down roping from their pale dead eyes,
And in their pale dull mouths the Gimmel bit
Lies foul with chewed grass, still and motionless.

Has one not seen just this picture of dejection among some of the wearier specimens of cab horses in some of the seamier cities of Europe?

But Shakespeare was a psychologist, too, as shown in the brilliant little cameo of a scene between Richard II and the groom, when the latter clumsily describes how Bolingbroke rode Richard's favourite horse at his coronation.

Richard would like to hear that the horse refused to go for his new usurping master, but no: he went 'so proudly as if he disdained the ground'. This is the last straw for Richard, who rails on the poor horse:

That jade hath eat bread from my royal hand;
This hand hath made him proud with clapping him,
Would he not stumble? Would he not fall down,
Since pride must have a fall, and break the neck
Of that proud man that did usurp his back?

But he relents:

Forgiveness, horse! Why do I rail on thee
Since thou, created to be awed by man,
Was born to bear. I was not made a horse,
And yet I bear a burden like an ass.

How many children have not felt just this bitterness when a beloved pony has passed to someone else and gone equally well, even better perhaps, for the new owner?

The eighteenth century for the most part paid scant attention to the horse. It was a peaceful age and the horse was very much taken for granted: after all, they provided the only means of transport.

In the following centuries, however, the pressures were on: the industrial revolution had arrived. The horse was no longer a means of leisurely transport, it was a beast of burden. We find, therefore, that there is a hard, practical streak in the writing about horses: little of romance.

Even the great Robert Smith Surtees, whose Jorrocks stories have so endeared him to horsemen, seldom describes a horse with any sense of affection or admiration. His horses are 'rakes', 'ugly', 'unwilling'. Think of 'Come up, I say, you ugly beast'.

Dickens is little better. His knowledge of the horse, unlike Surtees, was limited: limited to the fact that it was a necessity of life. ''Orses and dogs are some man's fancy'–but not Uriah Heap's.

The wonderful description of Mr Pickwick's drive to Dingley Dell with Mr Snodgrass and Mr Tupman in *The Pickwick Papers* is anything but flattering to the horse, which displayed various peculiarities *highly interesting to a by-stander, but by no means equally amusing to any-one seated behind him. Besides constantly jerking his head up in a very unpleasant and uncomfortable manner, and tugging at the reins to an extent which rendered it a matter of great difficulty for Mr Pickwick to hold them, he had a singular propensity for darting suddenly every now and then to the side of the road, then stopping short, and then rushing forward for some minutes, at a speed which it was wholly impossible to control.*

Nevertheless, Dickens showed the wonderful perception, common to all great writers, in the second chapter of *The Tale of Two Cities* with this thrilling description of the Dover Mail labouring up the hill in the fog and bitter cold:

With drooping heads and tremulous tails they mushed their way through the thick mud, floundering and stumbling between whiles, as if they were falling to pieces at the larger joints. As often as the driver rested them and brought them to a stand, with a wary 'Wo-ho! so-ho then!' the near leader violently shook his head and everything upon it–like an unusually emphatic horse,

An illustration by R. Seymour in a 1837 edition of Dickens' *The Pickwick Papers*.

'The Mail'—as depicted by 'Phiz' in *The Tale of Two Cities* by Dickens.

denying that the coach could be got up the hill. Whenever the leader made this rattle, the passenger started, as a nervous passenger might, and was disturbed in mind.

There are, of course, many delightful passages about horses from Victorian times. There is the wonderful description of the runaway, for instance, in Blackmore's *Lorna Doone*:

Then she took the court-yard gate at a leap, knocking my words between my teeth, and then right over a quickset hedge, as if the sky were a breath to her; and away for the water-meadows, while I lay on her neck like a child at the breast, and wished I had never been born. Straight away, all in front of the wind, and scattering clouds around her, all I knew of the speed we made was the frightful flash of her shoulders, and her mane like trees in a tempest. I felt the earth under us rushing away, and air left far behind us, and my breath came and went, and I prayed to God, and was sorry to be so late of it.

All the long swift while, without power of thought, I clung to her crest and shoulders, and dug my nails into her creases, and my toes into her flank-part, and was proud of holding on so long, though sure of being beaten. Then in her fury at feeling me still, she rushed at another device for it, and leaped the wide water-trough sideways across, to and fro, till no breath was left in me. The hazel boughs took me too hard in the face, and the tall dog-briars got hold of me, and the ache of my back was like crimping a fish; till I longed to give up, and lay thoroughly beaten, and lie there and die in the cresses. But there came a shrill whistle from up the home-hill, where the people hurried to watch us; and the mare stopped as if with a bullet; then set off for home with the speed of a swallow, and going as smoothly and silently. I never dreamed of such delicate motion, fluent, and graceful, and ambient, soft as the breeze flitting over the flowers, but swift as the summer lightning. I sat up again, but my strength was all spent, and no time left to recover it; and at last, as she rose at our gate like a bird, I tumbled off into the mixen.

Or George Borrow's magnificent cob in Lavengro:

The cob was led forth; what a tremendous creature. I had frequently seen him before, and wondered at him; he was barely fifteen hands, but he had the girth of a metropolitan dray-horse, his head was small in comparison with his immense neck, which curved down nobly to his wide back. His chest was broad and fine, and his shoulders models of symmetry and strength; he stood well and powerfully upon his legs, which were somewhat short. In a word, he was a gallant specimen of the genuine Irish cob, a species at one time not uncommon, but at the present day nearly extinct.

Particularly interesting is Charles Kingsley's chapter in *Hereward the Wake* about the dealer trying to sell 'the ugliest as well as the swiftest of mares': for Kingsley, though a parson, not only knew exactly how dealers behaved but also, had a remarkable eye for a horse:

. . . for it was not until the stranger had looked twice at her that he forgot her great chuckle head, greyhound flanks and drooping hind quarters, and began to see the great length of those same quarters, the thighs let down into the hocks, the compact loin, the extraordinary girth through the saddle, the sloping shoulders, the long arms, the flat knees, the large well-set hoofs, and all the other points which showed her strength and speed and justified her fame.

Compare this with Shakespeare's description quoted above.

Even Lewis Carroll was anything but ignorant of riding. His description of the White Knight with Alice is not only hilariously funny, but extremely penetrating:

Whenever the horse stopped (which it did very often), he fell off in front; and whenever it went on again (which it generally did rather suddenly), he fell off behind. Otherwise he kept on pretty well, except that he had a habit of now and then falling off sideways; and as he generally did this on the side on which Alice was walking, she soon found that it was the best plan not to walk quite *close to the horse.*

'I'm afraid you've not had much practice in riding,' she ventured to say, as she was helping him up from his fifth tumble.

The Knight looked very much surprised, and a little offended at the remark. 'What makes you say that?' he asked, as he scrambled back into the saddle, keeping hold of Alice's hair with one hand, to save himself from falling over on the other side.

'Because people don't fall off quite so often, when they've had much practice.'

'I've had plenty of practice,' the Knight said very gravely, 'plenty of practice!'

Alice could think of nothing better to say than 'Indeed?' but she said it as heartily as she could. They went on a little way in silence after this,

Alice and the White Knight: an engraving by Tenniel in Lewis Carroll's *Through the Looking Glass and what Alice found there.*

the Knight with his eyes shut, muttering to himself, and Alice watching anxiously for the next tumble.

'The great art of riding,' the Knight suddenly began in a loud voice, waving his right arm as he spoke, 'is to keep—' Here the sentence ended as suddenly as it had begun, as the Knight fell heavily on top of his head exactly in the path where Alice was walking. She was quite frightened this time, and said in an anxious tone, as she picked him up, 'I hope no bones are broken?'

'None to speak of,' the Knight said, as if he didn't mind breaking two or three of them. 'The great art of riding, as I was saying is – to keep your balance. Like this, you know—'

He let go the bridle, and stretched out both his arms to show Alice what he meant, and this time he fell flat on his back, right under the horse's feet.

'Plenty of practice!' he went on repeating, all the time that Alice was getting him on his feet again. 'Plenty of practice!'

'It's too ridiculous!' cried Alice, getting quite out of patience. 'You ought to have a wooden horse on wheels, that you ought!'

'Does that kind go smoothly?' the Knight asked in a tone of interest, clasping his arms round the horse's neck as he spoke, just in time to save himself from tumbling off again.

'Much more smoothly than a live horse,' Alice said, with a little scream of laughter, in spite of all she could do to prevent it.

'I'll get one,' the Knight said thoughtfully to himself, 'One or two – several.'

By the twentieth century the horse has become a luxury and so, as in the time of Shakespeare, the writing about the horse is flattering and even inspiring: but not for the most part especially distinguished from a literary point of view. This, of course, is because in the twentieth century the poet or the painter probably has little contact with horses. Other means of transport are now available, so those who can paint or write generally know little about horses, and those who can ride cannot necessarily paint or write.

Nevertheless there has, in the last fifty years, been a quite pleasing output of equestrian literature. Masefield himself, Poet Laureate, wrote – a little unexpectedly – *Reynard the Fox* and *Right Royal*. In the latter he thus described the favourite:

His beautiful hips and splendid shoulders
And power of stride moved all beholders
Moved non-betters to try and bet
On that favourite horse not beaten yet.
With glory of power and speed he strode
To a sea of cheering that moved and flowed
And followed and heaped and burst like a storm
From the joy of man in the perfect form.

More the poet than the horseman, perhaps; perhaps more the versifier than the poet.

In his own genre Will Ogilvie, the New Zealander, is quite without equal. His poems on the equestrian scene, so easy to enjoy, have given pleasure to thousands and are much quoted:

The Hunter gallops on the lea,
The Garron treads the ling,
The Hackney touching nose and knee,
Will make the roadway ring.
But apart from ply and pleasure,
With sweat upon their sides,
Where the furrow is to measure
And the earth to turn for treasure,
Serfs of little leisure,
Go the Clydes.

In our own time there is an almost limitless wealth of prose output, but though there is, fortunately for the authors, a huge growing audience, few books are destined for the immortality of Anna Sewell's *Black Beauty* written nearly a hundred years ago. Its perennial popularity should not blind one to its excellence. It is well written, has great feeling, is surprisingly full of practical knowledge (it is the only work of an invalid) and without any doubt it has played a big part in influencing people to treat their horses more humanely.

It is unlikely that the world of today is capable of producing a climate in which another *Black Beauty* could flourish. Present-day authors should be grateful that the popularity of riding ensures a certain sale of their books; present-day readers should be grateful that through the ages the horse has so inspired the greatest writers, and take pleasure in discovering for themselves the many priceless pearls to be found in our literary heritage.

Chapter two

The horse in art

Having been eaten, sacrificed, worshipped and ridden in battle since pre-history, the horse is a recurring motif in primitive art. He appeared in Babylonia in about 2000 BC and was introduced into Egypt several hundred years later, yet by the time he first appears in a convincing anatomical form, in an Assyrian bas-relief of the seventh century BC, he has already been saddled with that humiliating secondary role which so often prevents him from coming properly into focus until the sporting squirearchy of eighteenth-century England began to commission pictures of him alone and in his own right. Too often, until then, he is no more than a fashionable and potent accessory, as glamorous and suggestive as the Hispano-Suiza behind the celebrity in a twenties photograph – a symbol and, most importantly, a creature whose function in art, both physically and psychologically, is largely that of elevating a man (and especially a small man) over the head of his inferiors – a kind of prancing throne. Of course, there are distinguished early exceptions, like the elaborately accoutred but riderless horses of the 'Tang and Wei periods and those stocky Mongolian ponies of the Chou dynasty, but it is important to remember that many of the horses depicted in the classical period had in fact thrown their riders or broken loose from their chariots.

Wild horse taken from a drawing in the cave of Combarelles.

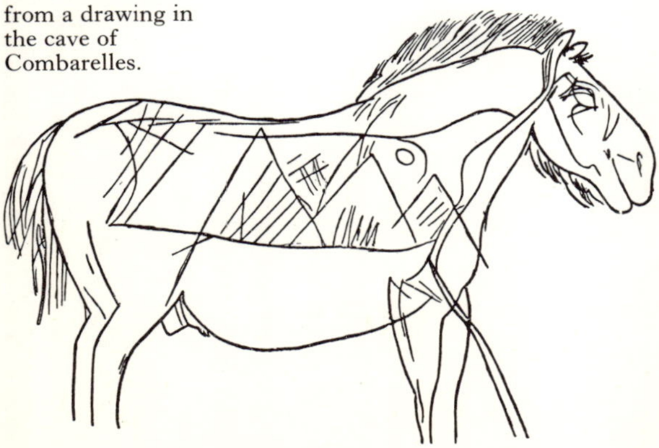

A Tang pottery figure of a woman polo player at full gallop, now in the British Museum, London.

Some of the most vivid, and certainly the largest, of early riderless horses in England are cult figures, like the stylised White Horse of Uffington in Berkshire, cut in chalk by Britons of the first century BC. Simple figures like this, probably suggested in the first place by the bare patches resulting from weathering on a slope, are horse-icons as vital as any in easel painting. It has gone now, but until the eighteenth century a great red horse was cut in the ironstone hillside at Tysoe in Warwickshire. As is the way with horses, it was probably an emblem – in this case of the war god, Tin – but to the horse-mad eighteenth century images like these symbolised only 'horse', and had the effect of covering hills in Wiltshire, Dorset and even Aberdeenshire with gigantic imitations, many of which were begun in the 1770s.

So, from the first century BC and the late eighteenth century, these are horses in their own right; but between come all those riders. From Raphael's mount for St George, with its strange coiling conformation and a little head not unlike the dragon's, up to Jacques Louis David's explosive plunging creature on which Napoleon is shown trying to cross the Alps – Napoleon, incidentally, doing everything wrong: but then he did have a reputation as an incompetent horseman – they make a splendid, if improbable, cavalcade. What so often happens is that, even where an artist has gone to some trouble to make drawings from life in order to help him paint the horse, his careful observations go by the board in favour of the grand design of the finished picture – putting the art before the horse. Quite right, too, of course, but some sitters end up with very strange rides.

Take, for instance, Velasquez's *Infante Balthazar Carlos*, a wonderful picture by any standards, with the child posing as though for the grandest of adult equestrian portraits, but on a tiny pony which strikes exactly the right grandiloquent pose, that of a great charger – in miniature – except for having a strangely squat and deep barrel chest. So pronounced is this it amounts to a disfigurement, but Velasquez planned his picture to be hung very high on a wall and has compensated for the resulting perspective. Or look at that most poignant and majestic of all paintings of a man on horseback – van Dyck's *Charles I*, in the National Gallery in London. The whole arrangement is designed to put a small man up in the air in the grandest possible way – but what a horse! The artist made a study from life in the royal stables with an anonymous rider, and there is a smaller version of the portrait in the Royal Collection;

even then he was tempted to adapt what he saw to conform to the heroid idea of a royal mount with a massive overbent neck and a head disproportionately tiny so as not to dwarf the King's, rather than to imitate too closely what he saw in front of him.

The reverse of the great seals of England always contained an image of the monarch on horseback, following continental models, and there are clear precedents for van Dyck's portrait in Isaac Oliver's Venetian-style picture of about 1612 of the Prince of Wales, derived from Clouet's portraits of the Valois kings; and for Rubens' portrait of 1627 of the Duke of Buckingham. There was no noble or courtly skill to compare with that of riding a great horse in the seventeenth-century sense: 'Doth any earthly thing breed more wonder?' asked Morgan in his *The Perfection of Horsemanship* of 1609. Van Dyck's picture shows Charles as a ruler by divine right, a little god – a pious, virtuous, heavy-lidded, melancholy king, an aesthete and yet a knight and a hero. It is a deliberately theatrical portrait.

Rubens' heraldic prancing steeds owed a good deal to Titian's, and Titian in turn was not the first to take advantage of heavy and elaborate caparisons, for instance in his portrait of the *Emperor Charles V* in the Prado, Madrid (which, according to Bellori, directly inspired van Dyck's picture) to conceal any gaps in his mastery of the complicated subject of equine anatomy.

There were, however, two contemporaries of Titian who in their different ways could never be accused of avoiding any aspect of drawing that had to do with scientific observation: Dürer and Leonardo. They are both supreme examples of artists who had a masterly understanding of equestrian anatomy but did not hesitate to jettison it if they thought it necessary to do so. In his Great Horse etchings of about 1505, Dürer faithfully investigated the horse of his day in much the same spirit as Leonardo was painstakingly drawing, in apparently endless studies now at Windsor, the skeleton, muscles, conformation and action of the Neopolitan type of horse he so much admired and on which he was himself such an accomplished performer. Both of them nevertheless subjected horses to all kinds of distortion to stress what they wished to emphasise – as in *Knight, Death and the Devil* or the Uffizi *Adoration*.

However, in Leonardo's case there is one stupendous exception in the gigantic equestrian statue for the Sforza monument for which he completed a stucco model in the 1480s. Vasari considered it so grandoise in conception that it would probably never be finished – like most of Leonardo's projects – and he was right, though the reason this time was that when the artist told Lodovici Moro, who had commissioned it, that the sculpture would need eighty tons of bronze for casting, his patron said that he could only afford ten. The model, which must surely have been one of the most extraordinary horses in all sculpture, was destroyed by the Gascon bowmen when the French entered Milan, but – even if that one got away – Leonardo's technique of using a heavy, or perhaps an anchored, tail to balance the prancing pose on the hind legs without a front support, was adopted in the seventeenth and eighteenth centuries and very widely used. The enormous Leningrad *Peter the Great* of 1766, by Falconet, is an imposing example.

Leonardo's short-lived horse itself had some distinguished ancestors. It must have been not far in spirit from Donatello's *Gattamelatta* in Padua of forty years earlier, which in turn – despite a gap of over a thousand years – is directly related to the first great statue of its kind, the *Marcus Aurelius* now in the Piazza del Campidoglio in Rome. First put up in AD 173, it is the sole survivor of all the full-scale equestrian bronze sculpture (it is 17 ft. high)

A Victorian engraving of the Seal of Simon de Montfort, of about 1250.

Charles I on Horseback by van Dyck.

Gentleman holding his horse by George Stubbs.

that must have existed in classical Rome. The secret of its success was that it survived Christian hostilities by being mistaken for many years as the Emperor Constantine.

There are, too, marble horses on the great west frieze of the Parthenon, and the superb fragment of a horse of 480–490 BC that was found in the debris after the Persian sack and is now in the Acropolis museum, as well as all those bits of centaur from the Temple of Zeus at Olympia which I suppose count as horses too. But nothing in horse sculpture surpasses those other famous survivors now on St Mark's in Venice. Again, we think of them, forgivably, as complete in their present form, but in fact they originally drew a car driven by Victory, cast in honour of Nero.

Most equestrian sculptures of this scale (they each weigh 1,700 lb.) do not move very far; these, without their car, galloped from Trajan's Arch in Rome to Constantinople, where they were mounted on the tower of the Hippodrome, before coming to Venice. Also, they cantered round Venice for many years before finally lining up in front of the Basilica where they remained until Napoleon took them

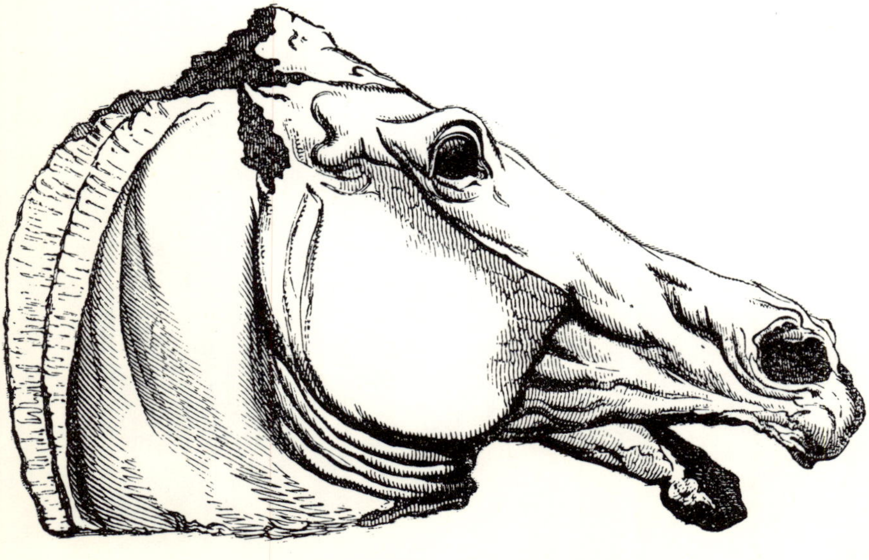

A horse's head from the Elgin marbles, originally part of the Parthenon frieze.

Mares and Foals, perhaps the best-known of all Stubbs' works.

with him to Paris and stabled them for thirteen years in the Place du Carousel. The Austrians took them back after Waterloo, and the only times since then that they have been disturbed was for safety during two world wars – in 1914 they shared the garden of the Palazzo Venezia in Rome with Verrochio's *Colleoni* and the *Gattamelatta*: who could imagine a more exhaustive little equestrian gathering than that?

The only remotely comparable horses in painting are those eighteenth-century mares and foals under a peaceful English summer oak tree by Stubbs, and here at last we are in the realm of pure horse painting. It is thought of as an English phenomenon, though its first successful practitioners were all from the Continent or heavily dependent on continental formulae. Wootton (born in 1686) used a highly artificial pattern for his single-horse portraits, into which he fitted the appropriate colours and markings, strongly exaggerating Arab points in order to please his patrons. Both he and his contemporary, Tillemans, painted Newmarket Heath, but Seymour, who was twenty years younger than Wootton, was more fastidious an observer, if a less talented composer, and seems to have been in greater demand. All three, though, with their stereotyped Thoroughbreds – human-eyed and with spindly legs, their cannon bones elongated out of all proportion – are primitives in relation to Stubbs. Born in 1724 he was, like his predecessors, dependent to some extent on European manners and landscape (his Italian tour took place in 1754–55) but he resembles Leonardo in his painstaking, indeed ruthlessly exhaustive, investigation of equine anatomy.

His first experience was with human anatomy, for he illustrated a book on obstetrics that involved almost Burke and Hare-like adventures in obtaining suitable subjects for dissection. His dissection of horses, taking them down muscle-layer by muscle-layer and drawing each with great precision until he reached the skeleton, began in 1757. It was this three years of remorseless study for *The Anatomy of the Horse* (which did not find a publisher until almost ten years later) that is the great turning point, not only in his own work but in the history of horse painting in general. Stubbs was almost lost to sight in the nineteenth century and only gradually rediscovered in the 1920s; and of course it is only very recently, thanks to the saleroom competition of American collectors like Mr Paul Mellon, that English sporting pictures have been taken at all seriously by museum directors in England. At that time, museum people seemed to have the impression that horse pictures were for the country-house horsy and had nothing much to do with aesthetics. They did not notice that. At the height of his career, however, Stubbs was charging – and getting – a good deal more for horse portraits than was Reynolds for his human ones.

Having mentioned Stubbs it is impossible, without going into detail, to find a way from Ben Marshall through the English horse pictures of the eighteenth and nineteenth centuries – through the Alken family, with all those terrible hunting accidents, and in near caricature the Ferneleys, the Herrings – because they all, in their own way, show the same simple delight in the horse for his own sake, in a good day's sport in the open air either on the race-course or in the hunting field, and in a desire to immortalise specific animals and exploits. One of the most outstanding is Géricault's *Horse Frightened by Lightning* in the National Gallery in London. Géricault died in 1824, aged only 33, after a riding accident.

There is one thing, though, that Stubbs did not get right, not for want of trying, and that was the action of the gallop. For all his anatomical knowledge, he persisted with the conventional rocking-horse action, very often with all four of the animal's hooves off the ground at once as though it had been strung up, left till it was rigid and then painted on its side. Nevertheless, with him and all the artists who followed him until the late nineteenth century, the almost hieroglyphic image – as much as to say, this

Horse Frightened by Lightning by Géricault.

means horse going fast – does in fact convey rapid movement very successfully, especially when, as with Herring or Delacroix, the horse

A portrait by Sir Alfred Munnings, painted in 1919.

is made absurdly long to stress the speeding horizontal axis. It was not until 1887, when the photographer Muybridge concluded his studies at the university of Pennsylvania and published 781 plates in a book called *Animal Locomotion* that the truth of the matter was known and the problem at last solved. What had always before been too quick for the human eye was now set out in clear sequence by thirty-six cameras with tripwires and clockwork shutters set up in line, past which galloped a race-horse called Annie G. The results were suddenly to produce a colossal deluge of horse pictures, very often battle paintings crammed with chargers in all the best Muybridge-accredited positions, as though in an attempt to make up for centuries of horses painted in impossible poses. The battle of Waterloo had not been so popular since 1815, but any battle would do.

The truth was known. But did it really make horse painting any better? I do not think so. The rocking-horse gallop was defended by Rodin and Degas on the grounds that there are times, especially where motion is concerned, when an artistically valid image can transcend optical logic. An isolated moment in the gallop can look like an arrested moment. Two of the artists who have succeeded best at motion since then have both evolved highly personal and rapid caligraphic styles, based – as always – on a thorough knowledge of equine anatomy: Joseph Crawhall, who was an amateur, born in 1861, and Raoul Dufy whose racing pictures, like those of Stubbs, entirely fulfil the requirements of both sorts of horse image discussed in this chapter: the horse as a vivid and frequently recurring ingredient of great art, and the horse in his own right.

Index

Acknowledgements

The editors would like to record their gratitude to all those who have helped this book to see the light of day. We would like to thank in particular the contributors, who spared us time out of their own busy lives; the numerous photographic agencies who assisted in the picture research; and all those who were involved in the design and production of the book. The generous encouragement and help given by practically the entire staff of Ward Lock made possible what at times seemed impossible; special thanks go to Frank Herrmann, Peter Lock and Geoffrey Charters. Apologetic thanks are also due to various families and friends of those involved, whose lives were disrupted by our preoccupation. Jennifer Baker, Gillian Illman and Juliet Walker deserve a mention of their own. We are very grateful to Maureen Charters for compiling the index.

The editors and publishers gladly acknowledge the owners and copyright holders of photographs reproduced in the book. In some cases we have failed to trace the source of photographs, and we hope that both they and any others inadvertently omitted from this list will accept a general acknowledgement. Her Majesty the Queen, the Shah of Iran, the Associazione Nazionale dell'Asino di Martina Franca e del Cavallo delle Murge, Associated Press, the Australian News and Information Bureau, Authenticolor, Geneva, Barnabys Picture Library, A. S. Barnes, K. Bright, the British Museum, London, Buckingham Public Relations, Central Press Photos, Findlay Davidson, Bertram Davis, the Deford Studio, California, the Director of National Studs of France, Sam Farr, Charles Fennell, Fiona Forbes, Fores Ltd, Foto Bureau Sport, Doetinchen, Foto Werner, Freudy Photos, New York, Jean Froissard, Stephen Goodger, Clive Hiles, Major R. A. Hill, John F. Hughes, Sheila Hughes, Ann Hyland, the Iceland Tourist Office in London, Oscar & Peter Johnson, Keystone Press Agency, Leslie Lane, London News Agency, G. A. Marston, Frank H. Meads, Monty, the National Gallery, London, John Nestle, Novosti, PA Reuter Photos, M. V. L. Pearce, Photostore, Biarritz, Pictor Ltd, Planet News, Peter Roberts, A. Ruddle, A. Russell, Jack Scheerboom, *The Scotsman*, G. W. Serth, Sport & General, Len Sirman Press, Geneva, Peter Sweetman, the Tate Gallery, London, Francis Thompson, Topical Press Agency, United Press, Paris, UPI, the US Trotting Association, the Victoria and Albert Museum, London, the Wallace Collection, London, Mrs M. Watney.

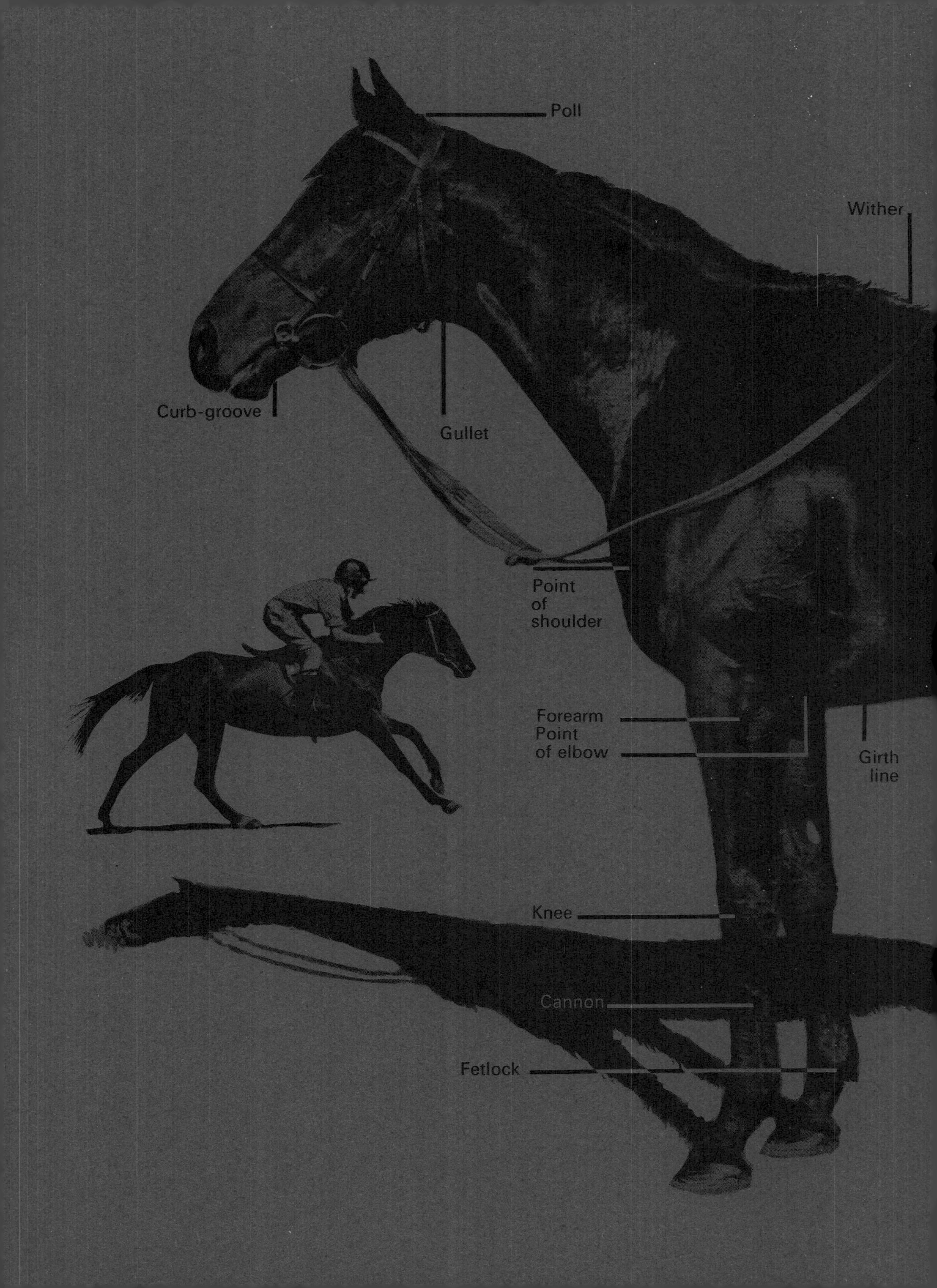

Poll
Wither
Curb-groove
Gullet
Point
of
shoulder
Forearm
Point
of elbow
Girth
line
Knee
Cannon
Fetlock